WITHDRAWN

PHYSICAL BEHAVIOR OF PCBs IN THE GREAT LAKES

PHYSICAL BEHAVIOR OF PCBs IN THE GREAT LAKES

Edited by

DONALD MACKAY
SALLY PATERSON
STEVEN J. EISENREICH
MILAGROS S. SIMMONS

Copyright © 1983 by Ann Arbor Science Publishers
230 Collingwood, P.O. Box 1425, Ann Arbor, Michigan 48106

Library of Congress Catalog Card Number 82-72347
ISBN 0-250-40584-9

Manufactured in the United States of America
All Rights Reserved

Butterworths, Ltd., Borough Green, Sevenoaks
Kent TN15 8PH, England

PREFACE

In December 1981, a group of 35 scientists from the United States and Canada met in Toronto for two days to discuss the physical behavior of PCBs in the Great Lakes. This book is a collection of papers, most of which were presented at that meeting. The issue is of obvious importance; PCBs are among the most notorious environmental contaminants and the Great Lakes are invaluable as a resource for fishing, recreation, transport, and as a source of potable water. But the meeting and this text have a deeper purpose. Perhaps from the understanding of the environmental behavior of PCBs in the Great Lakes will come concepts and information that will be valuable in elucidating and controlling the behavior of other contaminants in other aquatic systems.

The focus of this book is physical rather than biological, with the notable exception of Veith's paper on bioconcentration. This focus does not imply a lack of concern for the biosphere or for the effects or toxicology of PCBs; it merely represents an attempt to tackle a smaller problem of manageable proportions. Besides, the environmental fate of PCBs is largely controlled by physical processes, with biodegradation of lower chlorine congeners being the outstanding exception.

Many lessons have been learned from PCB studies in the Great Lakes. The importance of having reliable physical–chemical property data for individual congeners can not be overestimated. Analytical chemistry, as always, plays a key role. The need for individual consideration of each congener (and not a vague group of PCBs) is now obvious. Processes in the sediments, water column, atmosphere, and at their interfaces must be understood individually and collectively. Mathematical models can synthesize component studies into an overall picture of the contaminant's chemodynamic behavior and determine which processes and compartments are of greatest importance. Failure to reconcile environmental reality with a model (as occurs here with PCB volatilization rate estimation) highlights erroneous measurements or predictive equations. It is ironic that PCB contamination has been the cause of a significant advance in our understanding of the Great Lakes and, it is hoped, will result in an improved ability to prevent future contamination.

Donald Mackay
Sally Paterson
Steven J. Eisenreich
Milagros S. Simmons

ACKNOWLEDGMENTS

We are indebted to the sponsors of the meeting: Ontario Ministry of Environment, Environment Canada, Great Lakes Environmental Research Laboratory (U.S. National Oceanic and Atmospheric Administration), Michigan Sea Grant Program, University of Toronto, University of Michigan and University of Minnesota, and give them our sincere appreciation for their support. We thank the authors and reviewers for their contributions and hope that this book will be reward enough.

    Mackay   Paterson   Eisenreich   Simmons

Donald Mackay is Professor for the Department of Chemical Engineering and Applied Chemistry of the University of Toronto. He is also associated with the university's Institute for Environmental Studies. Dr. Mackay was born and educated in Glasgow, Scotland, where he obtained BSc and PhD degrees in Chemical Engineering. Before his present position, he was a research fellow at the University of Toronto, working on the kinetics of high-temperature reactions in shock waves. He then returned to the United Kingdom to work with Imperial Chemical Industries Ltd. on industrial petrochemical reactions and separations. Dr. Mackay's research interests include the fate and effects of toxic organic compounds in the environment, acquisition and correlation of physical–chemical data for such compounds, and development of models of organic toxicant partitioning, transport and reaction.

Sally Paterson is a Research Associate with the Department of Chemical Engineering, University of Toronto. Her work involves modeling the physical behavior of oil spills on water, and the fate and effects of toxic substances in the environment. She attended the University of Toronto, where she obtained a BA in Mathematics and Physics.

Steven J. Eisenreich is Associate Professor, Environmental Engineering Program, University of Minnesota. He received a PhD in Water Chemistry, an MS in Analytical Chemistry and a BS in Chemistry from the University of Wisconsin. In addition to his teaching assignments, Dr. Eisenreich has also served as consultant to the U.S. Army Corps of Engineers, the National Science Foundation and the National Oceanographic and Atmospheric Administration. Dr. Eisenreich's research areas have included water chemistry, chemical limnology, transport and fate of chemical pollutants in the environment, and atmospheric pollutant–natural water interactions. He has extensive publications in these areas in professional journals and is the editor of *Atmospheric Pollutants In Natural Waters*, published by Ann Arbor Science.

Milagros S. Simmons is Associate Professor in the Environmental Chemistry Program of the Department of Environmental and Industrial Health, School of Public Health at the University of Michigan. She also works as Associate Research Chemist for the University's Great Lakes Research Division, Institute of Science and Technology. Her research interests include analytical chemistry of organic pollutants, and fate and effects studies of persistent chemicals in the Great Lakes.

CONTENTS

1. High Resolution PCB Analysis 1
   S. Safe, M. Mullin, L. Safe, C. Pochini,
   S. McCrindle and M. Romkes

2. Influence of Volatility on the Collection of Airborne PCB and Pesticides with Filter-Solid Adsorbent Samplers 15
   T. F. Bidleman, N. F. Burdick, J. W. Westcott and W. N. Billings

3. Vapor Exchange of PCBs with Lake Michigan: The Atmosphere as a Sink for PCBs 49
   T. J. Murphy, J. C. Pokojowczyk and M. D. Mullin

4. Physical Chemical Properties of Polychlorinated Biphenyls 59
   D. Mackay, W. Y. Shiu, J. Billington and G. L. Huang

5. Reactivity and Environmental Persistence of PCB Isomers 71
   W. B. Neely

6. Reversible and Resistant Component Model of Hexachlorobiphenyl Adsorption-Desorption Resuspension and Dilution 89
   D. M. Di Toro and L. M. Horzempa

7. PCBs in the Lake Superior Atmosphere 1978-1980 115
   S. J. Eisenreich, B. B. Looney and G. J. Hollod

8. Processes Determining the Flux of PCBs across
   Air/Water Interfaces 127
   A. W. Andren

9. Evidence for the Atmospheric Flux of Polychlorinated
   Biphenyls to Lake Superior 141
   S. J. Eisenreich and B. B. Looney

10. Role of Surface Microlayers in the Air-Water
    Exchange of PCBs 157
    C. P. Rice, P. A. Meyers and G. S. Brown

11. PCB Dynamics in Lake Superior Water 181
    S. J. Eisenreich, P. D. Capel and B. B. Looney

12. The Role of the Benthic Boundary in the Cycling of
    PCBs in the Great Lakes 213
    B. J. Eadie, C. P. Rice and W. A. Frez

13. PCB Accumulation in Southern Lake Michigan
    Sediments: Evaluation from Core Analysis 229
    D. E. Armstrong and D. L. Swackhamer

14. PCBs in Sediment and Fluvial Suspended Solids
    in the Great Lakes 245
    R. L. Thomas and R. Frank

15. Estimating Bioconcentration Potential from
    Octanol/Water Partition Coefficients 269
    G. D. Veith and P. Kosian

16. Steady State Modeling of Toxic Chemicals—Theory
    and Application to PCBs in the Great Lakes and
    Saginaw Bay 283
    R. V. Thomann and J. A. Mueller

17. Model Simulation of PCB Dynamics in Lake Michigan .. 311
    P. W. Rodgers

18. Dynamic Mass Balance of PCB and Suspended Solids
    in Saginaw Bay—A Case Study 329
    W. L. Richardson, V. E. Smith and R. Wethington

19. Survey of Polychlorinated Biphenyls in Ambient Air
    across the Province of Ontario 367
        E. Singer, T. Jarv and M. Sage

20. Monitoring of PCBs in Water, Sediments and Biota
    of the Great Lakes—Some Recent Examples 385
        P. B. Kauss, K. Suns and A. F. Johnson

21. Volatilization of PCB from Sediment and Water:
    Experimental and Field Data 411
        T. J. Tofflemire, T. T. Shen and E. H. Buckley

22. Application of a Sediment Dynamics Model for
    Estimation of Vertical Burial Rates of PCBs in
    Southern Lake Michigan 423
        *D. Weininger, D. E. Armstrong and
    D. L. Swackhamer*

INDEX 441

CHAPTER 1

HIGH RESOLUTION PCB ANALYSIS

S. Safe[1][2], M. Mullin[3],

L. Safe[1], C. Pochini[3],

S. McCrindle[2], M. Romkes[2]

(1) Department of Veterinary
    Physiology & Pharmacology.
    Texas A & M University.

(2) Dept. of Chemistry,
    University of Guelph.

(3) U.S. Environmental
    Protection Agency.
    ERL-D at Grosse Ile,
    Grosse Ile, Michigan

HALOGENATED AROMATICS

Polychlorinated biphenyls (PCBs) are members of a group of chemical pollutants generally referred to as the halogenated aromatics. These chemicals include the polychlorinated naphthalenes (PCNs), polychlorinated benzenes (PBs), polychlorinated terphenyls (PCTs) and polybrominated biphenyls (PBBs) which are all primary industrial chemicals. Other members of this group of pollutants include the polychlorinated azo and azoxybenzenes (PCABs and PCAOBs respectively) which are formed as by-products in the preparation of chlorinated aniline-derived chemicals and the polychlorinated dibenzofurans (PCDFs) and dibenzo-p-dioxins (PCDDs) which are formed in the synthesis of chlorinated phenol-derived chemicals and in the combustion of organics (1,2). A major problem associated with the analysis and toxicology of halogenated aromatics is the multiplicity of possible isomers and congeners. For example, the commercial PCBs are highly complex mixtures of individual compounds whose identities have not been unequivocally determined (1,3-5). Moreover, the composition of PCB mixtures which have been identified in environmental samples is different from that of the commercial mixtures (5-8). This can be attributed to numerous factors such as preferential bio-and photodegradation of specific PCBs and the preferential uptake and retention of individual PCBs in different environmental matrices. For example, compared to commercial PCB formulations, atmospheric PCB residues are enriched in the more volatile lower chlorinated congeners (9); the PCBs which persist in fish extracts tend to be the more highly chlorinated congeners; PCB residues in humans are more highly chlorinated and enriched in PCBs which are not readily metabolized (5-7). Moreover, PCB residues in humans contain a lower percentage of the more highly ortho-chloro substituted isomers and congeners compared to the commercial PCB formulations (5). It is also apparent that the toxicity of PCBs, in common with other halogenated aromatics, is dependent on the structure of the individual compounds. Thus the potential environmental and health impact of PCBs must account not only for the integrated quantitation of the PCBs mixtures but also the concentrations of the specific congeners which are potentially toxic.

PCBs: IDENTIFICATION OF THE POTENTIALLY TOXIC CONGENERS

Structure-activity studies within the class of halogenated aromatics suggest that there is a correlation between those compounds which are toxic and those which induce a microsomal cytochrome P-448 dependent monooxygenase enzyme designated aryl hydrocarbon hydroxylase (AHH) (2,12,13,16). Table 2 summarizes all the PCBs which are AHH inducers (2, 16-22). Although the toxicities of these compounds have not been systematically investigated there are numerous reports which confirm the toxic effects elicited by the 3,3',4,4'-tetra-, 3,3',4,4',5-penta-,2,3,3',4,4'-penta-,2,3,4,4',5-penta-, 3,3',4,4',5,5'-hexa- and 2,3,3',4,4',5-hexachlorobiphenyls (2,11-18,23,24). Clearly the environmental concentrations of those PCBs noted in Table 2 should be a major concern for environmental analytical chemists.

PCBs: LOW TO HIGH RESOLUTION GC ANALYSIS

Table 3 summarizes the major approaches to the analysis of PCBs. The two major low and medium resolution techniques utilize packed column separation coupled with detection by an electron capture (EC) detector or mass spectrometry (MS).

The results then are quantitated by peak or pattern matching with the commercial formulations or known mixtures of these formulations (1,25,26). These low resolution methods can be carried out in most analytical laboratories and provide semi-quantitative integrated concentrations of PCBs in environmental samples. The disadvantages of low and medium resolution PCB analysis include; difficulties in quantitation of PCB residues which have undergone extensive degradation, the lack of quantitative data on individual isomers and the lack of quantitative data on the potentially toxic PCBs.

TABLE 1. Halogenated Aromatics: Multiplicity of Isomers and Congeners

Halogenated Aromatic	PCBs	PBBs	PCABs	PCNs	PCDDs	PCDFs
No. of Congeners	209	209	209	135	75	135

TABLE 2. A Summary of PCB Congeners Which Induce AHH Activity.

Number of Chlorine Substituents and Substitution Pattern

Cl_4	Cl_5	Cl_6	Cl_7
3,3',4,4'	3,3',4,4',5'	3,3',4,4',5,5'	2,3,3',4,4',5,5'
3,4,4',5	2,3,3',4,4'	2,3,3',4,4',5	2,2',3,3',4,4',5
	2,3',4,4',5	2,3,3',4,4',5'	
	2,3,4,4',5	2,3,4,4',5,6	
	2',3,4,4',5	2,3,3',4,4',6	
		2,2',3,3',4,4'	
		2,2',3,4,4',5'	

High resolution isomer-specific PCB analysis (4,10,27-30) is now a viable analytical procedure. The method incorporates a high resolution glass capillary column separation technique coupled with conventional EC or MS detection and requires all 209 PCB reference standards for preliminary calibration and quantitation. Although the synthesis of the standards is a prodigous task we now can report considerable progress in this area (30). Table 4 summarizes our current status in which

182 of the 209 congeners have been prepared using the convential Cadogan coupling (31) of a chlorinated aniline with a chlorinated benzene (excess) in the presence of amyl nitrite (Scheme 1). The crude PCB reaction products are purified to give single congeners or defined mixtures which are characterized by their proton magnetic resonance (PMR) and mass spectra (30).

TABLE 3. Characteristics of Low, Medium and High Resolution PCB Analysis

Separation	Detection	Quantitation
Packed Column	EC	[a]Peak(s) or Pattern Matching
Packed Column	MS	[b]Peak(s) or Pattern Matching
Capillary Column	EC and MS	[c]Isomer-Congener Response

[a]low, [b]medium and [c]high

TABLE 4. Synthesis of PCB Congeners - Progress

No. of Cl	1	2	3	4	5	6	7	8	9	10
No. of Congeners	3	12	24	42	46	42	24	12	3	1
No. Synthesized or Available	3	9	8	39	42	41	24	12	3	1

HIGH RESOLUTION ISOMER - SPECIFIC ANALYSIS OF COMMERCIAL AND ENVIRONMENTAL PCB MIXTURES

The proposed numbering scheme (4) denotes specific PCB isomers and congeners. Using the synthetic PCB standards, their relative retention times and molar response factors were determined. Octachloronaphthalene (OCN) was used as the reference standard for determining relative retention times and molar response factors.

Analysis of a number of PCB mixtures was determined using a Varian 3700 gas chromatograph equipped with an EC detector and a 50 M fused silica column coated with SE-54. The column temperature conditions were; initial temperature,

100°C; final temperature, 240°C; temperature program, 1° min^{-1}. The high resolution gas chromatographic separation of Aroclor 1260, Aroclor 1254, Aroclor 1016 and a mixture of Aroclors 1260 plus 1254 plus 1016 (2:3:5 by weight) are displayed in Figures 1,2,3 and 4 respectively. Based on the relative retention times of the synthetic PCB standards the identities of many of the GC peaks has been assigned. Figure 5 illustrates the more complex gas chromatograph of a cleaned-up PCB fraction obtained from a fish extract. A more comprehensive structure assignment of the GC peaks in this sample will require further GC-MS analysis to identify the chemical structures of the highly complex coextractives.

Our results thus confirm the feasibility of high resolution PCB analysis of the commercial and environmental PCB mixtures. We anticipate using this approach to accurately measure the concentrations of the specific PCB isomers in environmental and biological samples and to determine the effects of PCB structures on the ecological dynamics of this complex group of chemicals.

ACKNOWLEDGEMENTS

The financial assistance of the United States Environmental Protection Agency is gratefully acknowledged.

REFERENCES

1. Hutzinger, O., Safe, S. and Zitko, V., "The Chemistry of PCB's" CRC Press, Inc., Ohio (1974).

2. Parkinson, A. and Safe, S., "Aryl Hydrocarbon Hydroxylase Induction and its Relationship to the Toxicity of Halogenated Aryl Hydrocarbons" Toxicol. Environ. Chem. 4 1 (1981).

3. Sissons, D. and Welti, D., "Structural Identification of Polychlorinated Biphenyls in Commercial Mixtures by Gas Liquid Chromatography, Nuclear Magnetic Resonance and Mass Spectrometry" J. Chrom., 60, 15 (1971).

4. Ballschmiter, K. and Zell, M., "Analysis of Polychlorinated Biphenyls (PCB) by Glass Capillary Gas Chromatography. Composition of Technical Aroclor- and Clophen-PCB Mixtures" Fresenius Z. Anal. Chem., 302, 20 (1980).

5. Jensen, S. and Sundstrom, G., "Structures and Levels of Most Chlorobiphenyls in Two Technical PCB Products and in Human Adipose Tissue" Ambio, 3, 70 (1974).

6. Yakushiji, T., Watanabe, I., Kuwabara, K., and Yoshida, S., "Identification of Low Chlorinated Biphenyls in Human milk by Gas Chromatography-Mass Spectrometry" J. Chrom., 154, 203 (1978).

7. Kuroki, H. and Masuda, Y., "Structures and Concentrations of the Main Components of Polychlorinated Biphenyls in Patients with Yusho" Chemosphere 6, 469 (1977).

8. Safe, S., "Metabolism, Uptake, Storage and Bioaccumulation" In: Halogenated Biphenyls, Naphthalenes, Dibenzodioxins and Related Products (ed. R. Kimbrough), Elsevier/North-Holland, p. 77 (1980).

9. Harvey, G.R., Steinhauer, W.G., "Atmospheric Transport of Polychlorobiphenyls to the North Atlantic" Atmos. Environ. 8, 777 (1974).

10. Ballschmiter, K., Zell, M. and Neu, H.J., "Persistence of PCBs in the Ecosphere: Will Some PCB-Components Never Degrade" Chemosphere, 7, 173 (1978).

11. Yoshimura, H., Yoshihara, S., Ozawa, N. and Miki, M., "Possible Correlation Between Induction Modes of Hepatic Enzymes by PCBs and Their Toxicity in Rats" Ann. N.Y. Acad. Sci., 320, 179 (1979).

12. Goldstein, J.A., "The Structure-Activity Relationship of Halogenated Biphenyls as Enzyme Inducers" Ann. N.Y. Acad. Sci., 320, 164 (1979).

13. Poland, A., Greenlee, W.F. and Kende, A.S., "Studies on the Mechanism of Toxicity of the Chlorinated Dibenzo-p-dioxins and Related Compounds" Ann. N.Y. Acad. Sci., 320, 214 (1979).

14. Yamamoto, H., Yoshimura, H., Fujita, M. and Yamamoto, T., "Metabolic and Toxicologic Evaluation of 2,3,4,3',4'-Pentachlorobiphenyl in Rats and Mice" Chem. Pharm. Bull., 24, 2168 (1976).

15. Ax, R.L. and Hansen, L.G., "Effects of Purified PCB Analogs on Chicken Reproduction" Poultry Sci., 54, 895 (1975).

16. Poland, A. and Glover, E., "Chlorinated Biphenyl Induction of Aryl Hydrocarbon Hydroxylase Activity: A study of the Structure Activity Relationship" Mol. Pharmacol., 13, 924 (1977).

17. Goldstein, J.A., Hickman, P., Bergman, H., McKinney, J.D. and Walker, M.P., "Separation of Pure Polychlorinated Biphenyl Isomers into Two Types of Inducers on the Basis of Induction of Cytochrome P-450 or P-448" Chem.-Biol. Interact., 17, 69 (1977).

18. Parkinson, A., Cockerline, R. and Safe, S., "Induction of Both 3-Methylcholanthrene- and Phenobarbitone-Type Microsomal Enzyme Activity by a Single Polychlorinated Biphenyl Isomer" Biochem. Pharmacol., 29, 259 (1980).

19. Parkinson, A., Cockerline, R. and Safe S., "Polychlorinated Biphenyl Isomers and Congeners as Inducers of Both 3-Methylcholanthrene- and Phenobarbitone-Type Microsomal Enzyme Activity" Chem. Biol. Interact. 29, 277 (1980).

20. Parkinson, A., Robertson, L., Safe, L. and Safe, S., "Polychlorinated Biphenyls as Inducers of Hepatic Microsomal Enzymes: Structure-Activity Rules" Chem. Biol. Interact. 30, 271 (1981).

21. Parkinson, A., Robertson, L. and Safe, S., "Hepatic Microsomal Enzyme Induction by 2,2',3,3',4,4' - and 2,2',3',4,4',5-Hexachlorobiphenyl" Life Sciences, 27, 2333 (1980).

22. Parkinson, A., Robertson, L., Safe, L. and Safe S., "Polychlorinated Biphenyls as Inducers of Hepatic Microsomal Enzymes: Effects of Diortho Substitution" Chem. Biol. Interact., 31, 1 (1981).

23. Yoshihara, S., Kawano, K., Yoshimura, H., Kuroki, H. and Masuda, Y., "Toxicological Assessment of Highly Chlorinated Biphenyl Congeners Retained in the Yusho Patients" Chemosphere, 8, 531 (1979).

24. Stonard, M.D. and Greig, J.B., "Different Patterns of Hepatic Microsomal Enzyme Activity Produced by Administration of Pure Hexachlorobiphenyl Isomers and Hexachlorobenzene" Chem.-Biol. Interact., 15, 365 (1976).

25. Cairns, T. and Siegmund, E.G., "PCBs-Regulatory History and Analytical Problems" Anal. Chem. 53, 1183A (1981).

26. Sawyer, L.D., "Quantitation of Polychlorinated Biphenyl Residues by Electron Capture Gas-Liquid Chromatography: Collaborative Study, J. Assoc. Offic. Anal. Chem. 61, 282 (1978).

27. Albro, P.W. and Parker, C.E., "Comparison of the Composition of Aroclor 1242 and Aroclor 1016" J. Chrom. 169, 161 (1979).

28. Mullin, M.D. and Filkins, J.C., "Analysis of Polychlorinated Biphenyls by Glass Capillary and Packed Column Gas Chromatography". Advances in the Identification & Analysis of Organic Pollutants in Water at the Second Chemical Congress of the North American Continent, Las Vegas, Nevada, L.W. Keith, ed., Ann Arbor Press (1981).

29. Albro, P.W., Corbett, J.T. and Schroeder, J.L., "Quantitative Characterization of Polychlorinated Biphenyl Mixtures (Aroclors 1248, 1254 and 1260) by Gas Chromatography Using Capillary Columns", J. Chrom., 205, 103 (1981).

30. Mullin, M., Sawka, G., Safe, L., McCrindle, S. and Safe, S., "Synthesis of Octa- and Nonachlorobiphenyl Isomers and Congeners and Their Quantitation in Commercial Polychlorinated Biphenyls and Identification in Human Breast Milk" J. Anal. Toxicol., 5, 138 (1981).

31. Cadogan, J.I.G., Roy, D.A. and Smith, D.M., "An Alternative to the Sandmeyer Reaction", J. Chem. Soc. (C) 1249 (1966).

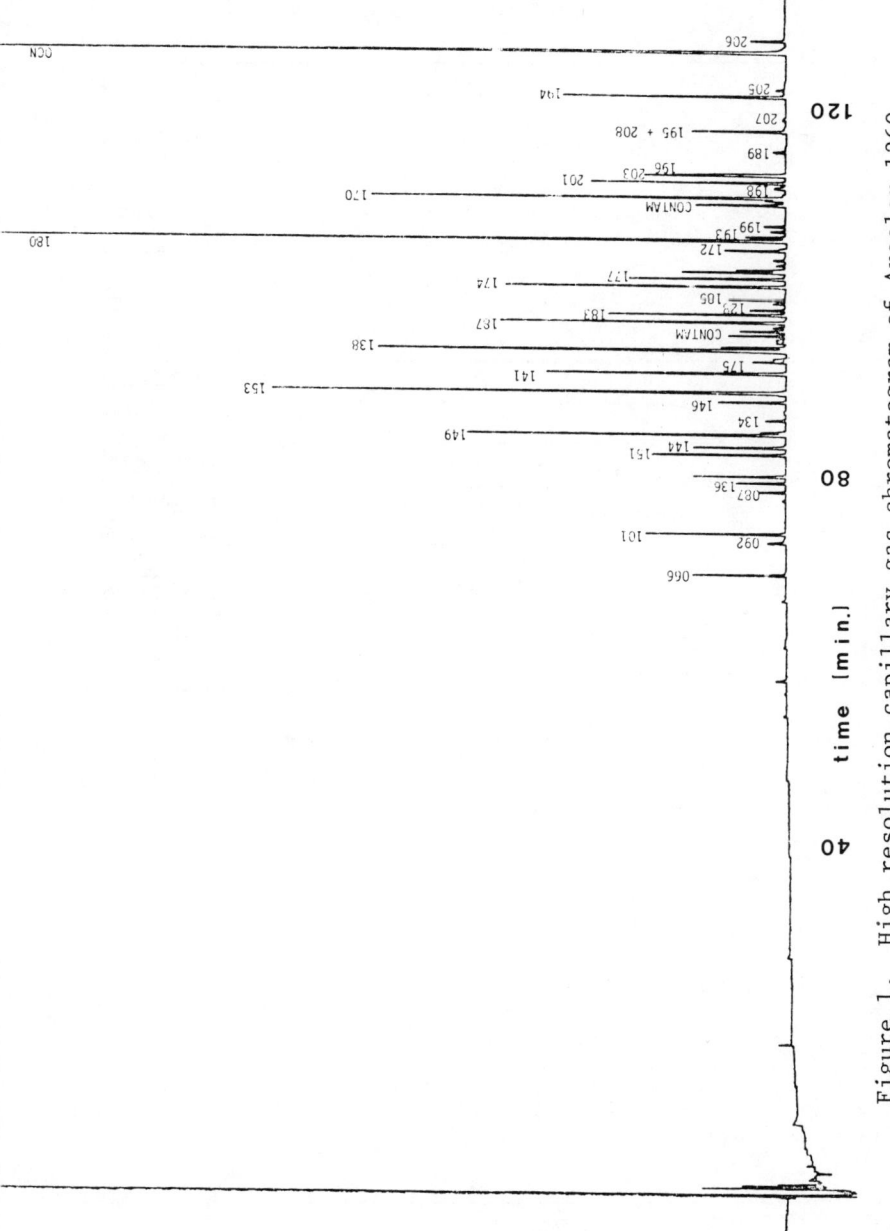

Figure 1. High resolution capillary gas chromatogram of Aroclor 1260

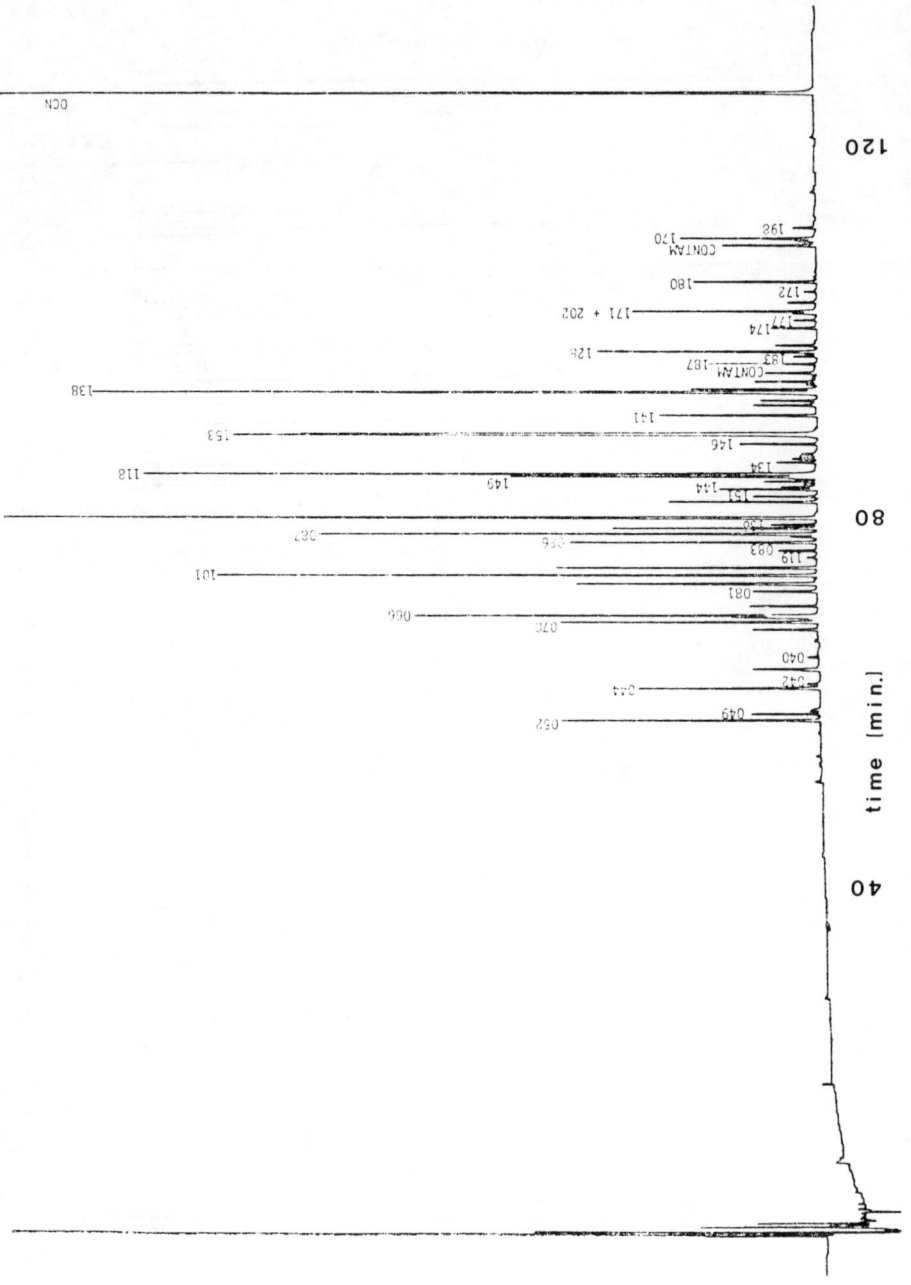

Figure 2. High resolution capillary gas chromatogram of Aroclor 1254

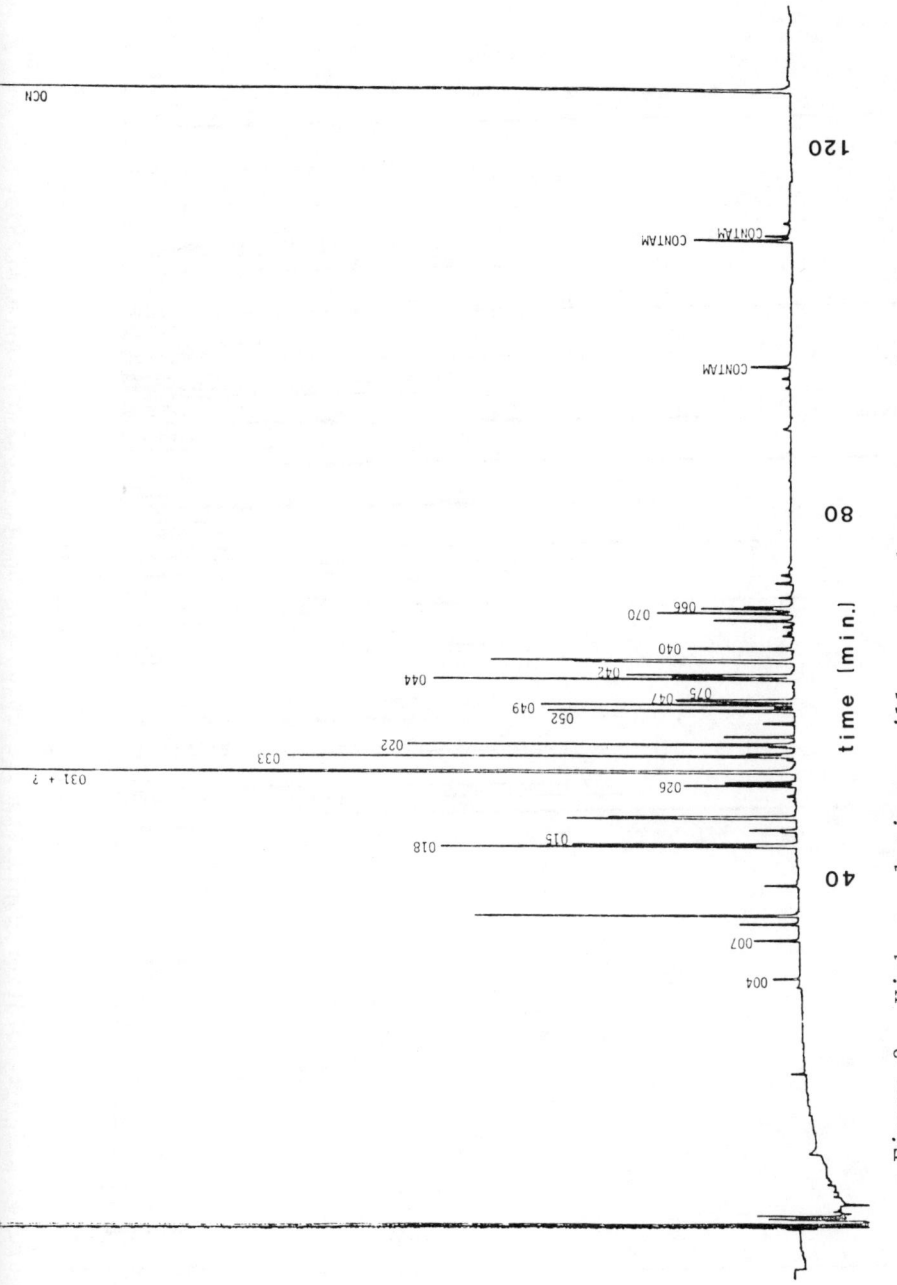

Figure 3. High resolution capillary gas chromatogram of Aroclor 1016

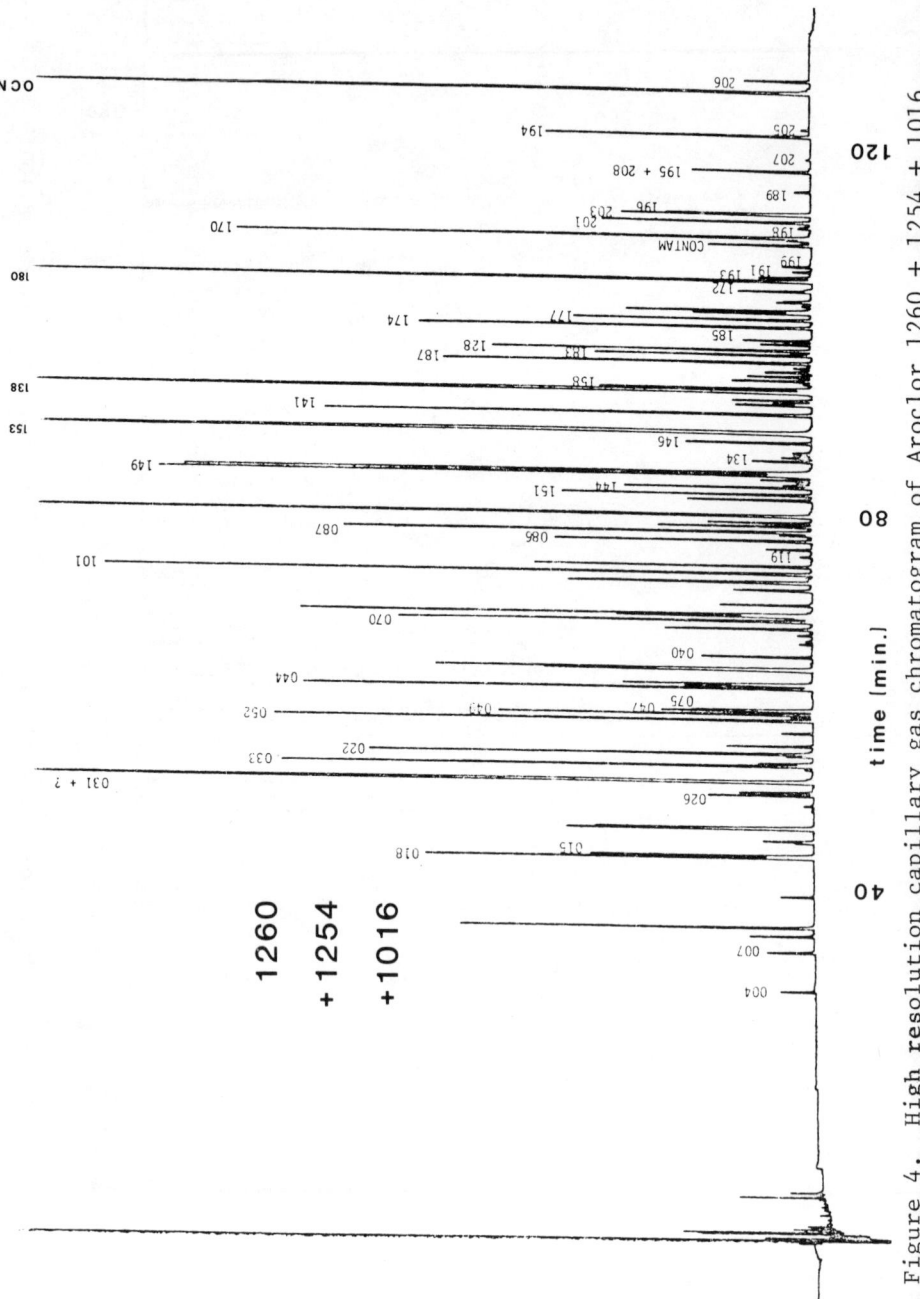

Figure 4. High resolution capillary gas chromatogram of Aroclor 1260 + 1254 + 1016

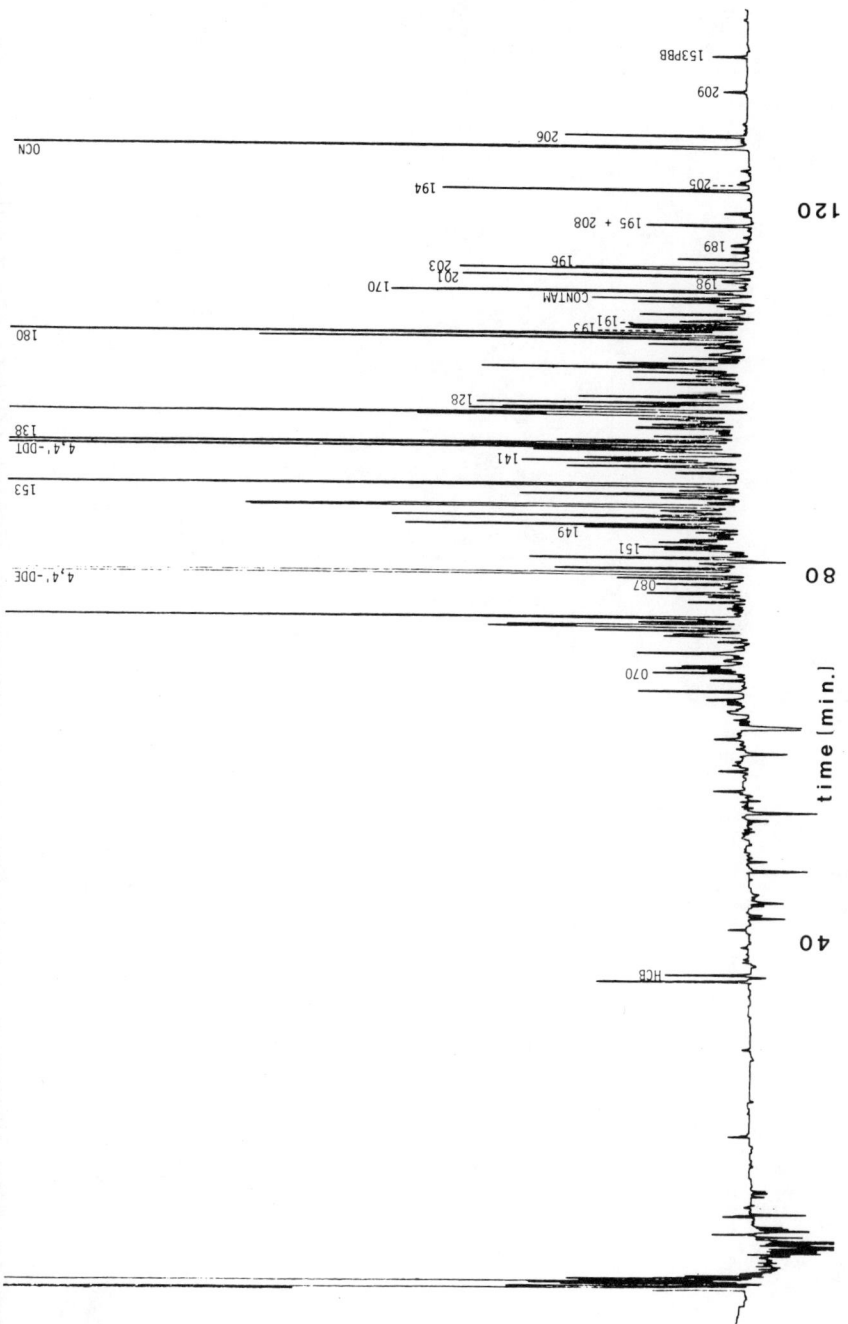

Figure 5. High resolution capillary gas chromatograph of a PCB fraction from a fish extract

CHAPTER 2

INFLUENCE OF VOLATILITY ON THE
COLLECTION OF AIRBORNE PCB AND
PESTICIDES WITH FILTER-SOLID
ADSORBENT SAMPLERS

Terry F. Bidleman, Nydia F. Burdick,
John W. Westcott, and W. Neil Billings
  Marine Science Program, Belle W. Baruch
  Institute for Marine Biology and Coastal
  Research, Department of Chemistry,
  University of South Carolina, Columbia, SC 29208

INTRODUCTION

   Solid adsorbents have gained wide acceptance within the last decade for low-and high-volume collection of trace organic vapors. An adsorbent bed functions as a frontal gas chromatographic system. Penetration of vapor fronts through the bed depends on the volatility of the sample and the total air volume pulled through the collector. Understanding this chromatographic behavior for different volatility solutes is important for designing collection systems that can retain compounds of interest without breakthrough.
   For several years we have been using solid adsorbents for high-volume collection of polychlorinated biphenyl (PCB) and chlorinated pesticide vapors. These pollutants have vapor pressures in the $10^{-4} - 10^{-7}$ torr range. Typical concentrations vary from a few nanograms per cubic meter in urban and rural air away from point sources, to picograms per cubic meter in remote continental areas and over the open ocean [1-9]. These levels are sufficiently low that hundreds to thousands of cubic meters of air must be sampled to obtain enough material for analysis, even using sensitive electron capture gas chromatography.
   A wide variety of adsorbents has been used to sample organic vapors. Reviews of vapor collection methods as well as the types and concentrations of airborne organics have been published by Lewis [10], Lamb et al. [11], and Simoneit and Mazurek [12]. Most of our work has been done using porous polyurethane foam (PPF), an adsorbent that is cheap, easy to handle, and retentive for many pesticides and PCB. We have also made comparison measurements using two other

solid adsorbents, Tenax-GC resin and Amberlite XAD-2 resin. In this report we discuss the influence of volatility on the collection of pesticides and PCB using a glass fiber filter-solid adsorbent sampler, and on the distribution of these organics between the apparent particle (filter-retained) and gaseous (adsorbent-retained) phases.

MEASUREMENT OF PCB VAPOR PRESSURES

Gas Saturation

Saturation of a flowing gas stream has been used to determine the vapor pressures of several pesticides as well as to estimate evaporation rates from soils. In a typical experiment, nitrogen is passed through a sand or soil column coated with pesticide and the effluent vapor is trapped in an organic solvent or on a solid adsorbent for analysis [13-15]. The vapor pressure is calculated from the measured vapor density using the ideal gas law:

$$p^o = dRT/M$$

where p^o is the vapor pressure, d is the saturation vapor density, M is the molecular weight, T is the absolute temperature, and R is the gas constant.

Vapor saturation experiments were carried out using pure PCB isomers obtained from Analabs, Inc. (North Haven, CT): 2',3,4-trichlorobiphenyl (2',3,4-TCB), 2,2',5,5'-tetrachlorobiphenyl (2,2',5,5'-TCB), and 2,2',4,5,5'-pentachlorobiphenyl (2,2',4,5,5'-PCB). Two compounds of the DDT family, o,p'-DDT and p,p'-DDE, obtained from the U.S. Environmental Protection Agency (Research Triangle Park, NC) were also selected to test our system, since their vapor pressures have been determined by Spencer and Cliath [14] using a gas saturation method. Their apparatus employed a large (6 x 43 cm) saturation column containing gram quantities of the pesticides. The high cost of pure PCB isomers made it necessary to scale down the saturation column to the point where 20 - 50 mg quantities of isomer could be used. The organochlorine compound was dissolved in pentane and coated onto 1.3-mm diameter glass beads packed into a U-shaped column 80 cm long x 0.6 cm i.d. After evaporating the solvent and conditioning the column for several days under a slow flow of dry air, gas saturation measurements were made by passing air at 0.1 - 0.4 ml/min through the column thermostated at 30°, 35°, or 40°C, and collecting the effluent vapor on a 3-g florisil trap. The trap was eluted with ethyl ether - petroleum ether, and the eluate was analyzed by electron capture gas chromatography using a Varian 3700 instrument and a 180-cm long x 0.4 cm i.d. glass column packed with 3% OV-225. Experimental details for packing and conditioning the saturation column and carrying out the vapor pressure determinations are given in Westcott et al. [16].

The results of the gas saturation experiments are summarized in Table I. Vapor pressures of p,p'-DDE (1.3 x 10^{-5} torr) and o,p'-DDT (8.8 x 10^{-6} torr) at 30°C agreed reasonably with Spencer and Cliath's values of 0.65 x 10^{-5} and 5.5 x 10^{-6} for the same compounds. For the three PCB isomers, vapor pressures were determined at 30°, 35° and 40°C, and Antoine plots of log p^o vs 1/T were extrapolated to yield vapor pressures at 25°C (Figure 1).

Table I. Organochlorine Vapor Pressures Determined by Gas Saturation

Compound	p^o, torr^a 30°C	25°C	Antoine Constants^b A	B
2',3,4-TCB	1.2 x 10^{-4}	1.0 x 10^{-4}	1.09	1.51 x 10^3
2,2',5,5'-TCB	3.6 x 10^{-5}	1.9 x 10^{-5}	11.8	4.92 x 10^3
2,2',4,5,5'-PCB	1.3 x 10^{-5}	7.2 x 10^{-6}	11.1	4.84 x 10^3
p,p'-DDE	1.3 x 10^{-5}			
o,p'-DDT	8.8 x 10^{-6}			

(a) The average coefficient of variation for determinations at 30°C was 18%. Vapor pressures were based on experiments for which gas-solid contact times in the coated bead column exceeded 35-40 minutes.

(b) Constants for the equation log p^o = A - B/T.

The gas saturation method is extremely time-consuming. Several days are required to prepare and condition the coated saturation columns. Also, solid-vapor equilibrium appeared to be established slowly. Constant vapor density values were obtained only for very slow airflow rates, such that the contact time of the air with the coated beads in the saturation column was 35 - 40 minutes. We have no explanation for this apparent slow approach to equilibrium in our semi-micro saturation columns. Spencer and Cliath [14] and Spencer et al. [15] obtained gas saturation for much shorter contact times in their large saturation columns. Because of the experimental difficulties involved with gas saturation on a semi-micro scale, we decided to explore gas chromatography as an alternate way to determine PCB vapor pressures.

Gas Chromatography

Gas chromatography has several advantages as a technique for determining vapor pressures: speed, tolerance to relatively impure compounds, the ability to determine vapor pressures of several compounds simultaneously, and small

Figure 1

Antoine plots for tri-, tetra-, and pentachlorobiphenyl isomers. The vertical lines represent standard deviations of a single determination.

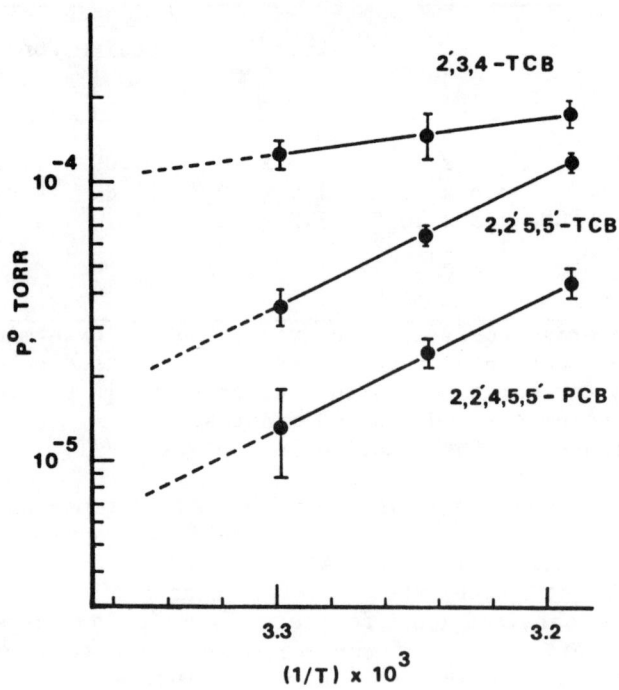

sample size requirements. Hamilton [17] determined the vapor pressures of 13 herbicide esters by comparing their retention volumes on a non-polar SE-30 column with that of a dibutyl phthalate reference. We used a similar technique to measure vapor pressures of five PCB isomers and two DDT compounds (o,p'-DDT and p,p'-DDE).

Vapor pressures for two substances at the same temperature are related through the equation:

$$(1) \quad \ln p_1^o = L_1/L_2 \ln p_2^o + C$$

where 1 and 2 refer to the test and reference compounds respectively, p^o is the vapor pressure, L is the heat of vaporization, and C is a constant. The ratio L_1/L_2 and the constant C may be calculated from the relative retention volume (V_R) or time of the test to the reference compound and the vapor pressure of the reference compound:

$$(2) \quad \ln (V_R)_1/(V_R)_2 = (1 - L_1/L_2) \ln p_2^o - C$$

A plot of $\ln (V_R)_1/(V_R)_2$ vs $\ln p_2^o$ should give a straight line with slope $(1 - L_1/L_2)$ and intercept -C. It is assumed that the ratio of the heats of vaporization of the two compounds is independent of temperature over the range of measurements. The derivation of equations 1 and 2 is presented in detail by Hamilton [17].

Selectivity in gas-liquid chromatography is influenced by the volatility of the solute and by chemical interactions between the solute and the stationary phase. If chemical effects can eliminated or at least minimized, partitioning between the stationary and mobile phases is controlled by solute volatility. We chose a 1-m glass capillary column coated with Apolane-87 for vapor pressure determinations. Use of a wall coated open tubular (WCOT) column eliminates adsorption effects due to the presence of a stationary phase support. Stationary phase selectivity can be minimized by choosing a stationary phase of as low a polarity as possible. Apolane-87 is an 87-unit hydrocarbon which is non-polar and non-chiral. McReynolds constants for Apolane-87 are lower than those for SE-30 and SF-96, both of which have been used as packed column stationary phases for vapor pressure work [17-19] (Table II).

Figure 2

Plots of $\ln (V_R)_1/(V_R)_2$ vs $\ln p_2^o$ (vapor pressure of 2,4,5-TIB at different temperatures) for PCB isomers and 2,4,5-trichlorophenoxyacetic acid, n-butyl ester (2,4,5-TNB).

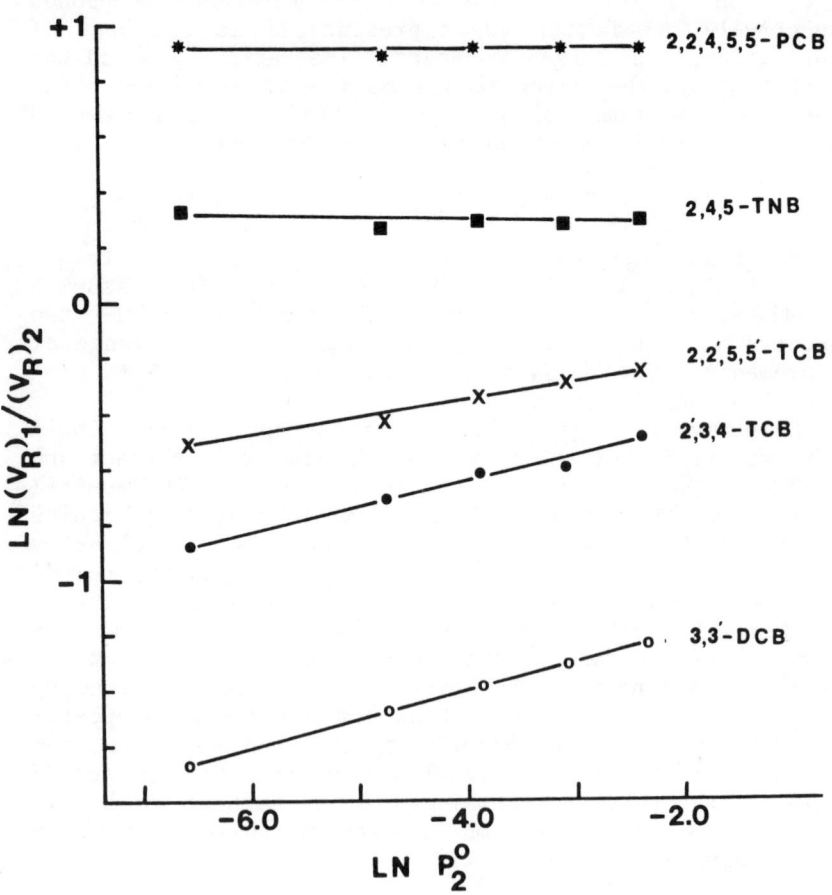

Table II. Apolane-87: Structure and McReynolds Constant

$$\begin{array}{c} H_{37}C_{18} \\ \diagdown \\ CH-(CH_2)_4-C-(CH_2)_4-CH \\ \diagup \\ H_{37}C_{18} \end{array} \begin{array}{c} C_2H_5 \\ | \\ | \\ C_2H_5 \end{array} \begin{array}{c} C_{18}H_{37} \\ \diagup \\ \diagdown \\ C_{18}H_{37} \end{array}$$

Phase	McReynolds Constant[*]
Apolane-87	71
SF-96	205
SE-30	217

[*] sum of five ΔI values, Applied Science Laboratories

Nanogram quantities of individual PCB isomers or pesticides in hexane were injected directly onto a 1-m x 0.25-mm i.d. WCOT glass capillary column installed in a Varian 3700 gas chromatograph. Use of pentane instead of hexane produced no difference in retention times. Hydrogen at a flow rate of 2-4 ml/min was used as a carrier gas, and nitrogen at 30 ml/min was used as a make-up gas for the ^{63}Ni electron capture detector. The column was operated isothermally at 60 - 130°C. The detector was maintained at 325°C, and the injector at 180°C. Retention times were measured in chart units from the injection point and were expressed relative to the retention time of 2,4,5-trichlorophenoxyacetic acid, isobutyl ester (2,4,5-TIB), which has a vapor pressure of 2.2 x 10^{-5} torr at 25°C [17].

Vapor pressures for the reference compound, 2,4,5-TIB, in the temperature range of our measurements were obtained from Hamilton's relative retention data for 2,4,5-TIB and dibutyl phthalate (DBP). Using his values for L_1/L_2 and C for the 2,4,5-TIB/DBP pair together with DBP vapor pressures at different temperatures (from the equation of Small et al., [20], quoted by Hamilton, [17]), p_2^o was calculated for 2,4,5-TIB in our temperature range using Equation 1. Regression lines were fitted to ln $(V_R)_1/(V_R)_2$ vs ln p_2^o plots (Figure 2) according to Equation 2, and the parameters L_1/L_2 and C were obtained for each of the test compound/2,4,5-TIB pairs. Two to seven of these plots were constructed for each test compound. Once values for L_1/L_2 and C were obtained [21], vapor pressures for the test compounds were calculated from the 2,4,5-TIB vapor pressure (2.2 x 10^{-5} torr at 25°C; 4.1 x 10^{-5} torr at 30°C) using Equation 1.

Vapor pressures for pesticides and PCB isomers as determined by capillary GC are given in Table III, along with our gas saturation and literature values for comparison. In general, the GC and gas saturation values at 25 - 30° agreed reasonably well. However we found a serious disagreement between the Antoine constants (for the equation $\log p^o = A - B/T$) determined by the two methods for 2',3,4-TCB. The constants as determined by GC are: $A = 10.92$, $B = 4480$; while the saturation method gave: $A = 1.09$, $B = 1510$. Equations using these two sets of constants give vapor pressures that agree well in the 25 - 30° range, but which deviate markedly at higher and lower temperatures. For example, predicted vapor pressures at 0°C are 3.2×10^{-6} torr using the GC equation, and 3.6×10^{-5} torr using the saturation equation.

Our conclusion is that more work should be done to critically compare GC and gas saturation methods for measuring vapor pressures for a variety of low volatility materials. GC methods have the advantage of being much more rapid than gas saturation techniques. The 1-m WCOT column used for these experiments was obviously of lower efficiency than the 30-50-m columns normally used for analytical separations, but the ability to resolve large numbers of compounds of similar volatility is not a major requirement of this technique. Figure 3 shows that the resolution of the column was sufficient to allow several compounds to be chromatographed at once. The use of a longer column would have been inconvenient because of the extremely long retention times for compounds of low vapor pressure at low column temperatures. On the 1-m column, 40 minutes were required to elute 2,4,5-TIB at 70°C and 3 ml/min carrier flow, while 1100 minutes were required to elute the same compound from Hamilton's 0.6-m packed column at 72°C and a carrier flow of 66 ml/min. On the WCOT column, 175 minutes were required to elute 2,2',4,5,5'-PCB at 60°C.

The solid PCB isomers 2',3,4-TCB and 2,2',5,5'-TCB are components of the commercial fluid Aroclor 1242, and 2,2', 4,5,5'-PCB is a component of Aroclor 1254 [22,23]. Reported vapor pressures for these fluids at 25°C are 4.06×10^{-4} torr and 7.71×10^{-5} torr, respectively [24]. Vapor pressures for the solid PCB isomers are thus approximately 5-10 times lower than for the commercial liquid mixtures.

COLLECTION OF ORGANOCHLORINE VAPORS BY SOLID ADSORBENT BEDS

Frontal Movement of Vapors Through Polyurethane Foam -- a Laboratory Study

Collection efficiency in the field is usually determined by observing the amount of vapor found on a backup

Table III. Vapor Pressures Determined by Gas Chromatography and Gas Saturation

Compound	Temperature, °C	GC	Vapor Pressure, torr Gas Saturation	Literature
p,p'-DDE	30	1.4×10^{-5}	1.3×10^{-5}	0.65×10^{-5} [14]
o,p'-DDT	30	8.4×10^{-6}	8.8×10^{-6}	5.5×10^{-6} [14]
2,4,5-T,n-butyl ester	25	1.6×10^{-5}		1.5×10^{-5} [17]
3,3'-DCB	25	1.9×10^{-4}		
2',3,4-TCB	25	7.7×10^{-5}	10.0×10^{-5}	
2,2',5,5'-TCB	25	5.3×10^{-5}	1.9×10^{-5}	
2,2',4,5,5'-PCB	25	9.0×10^{-6}	7.2×10^{-6}	
2,2',4,4',6,6'-HCB	25	1.3×10^{-5}		

Table III (cont.). Constants for $\log p^° = A - B/T$

Compound	GC		Gas Saturation	
	A	B	A	B
p,p'-DDE	11.37	4920		
o,p'-DDT	11.46	5010		
2,4,5-T,n-butyl ester	11.86	4970		
3,3'-DCB	10.68	4290		
2',3,4-TCB	10.92	4480	1.09	1510
2,2',5,5'-TCB	10.87	4510	11.8	4920
2,2',4,5,5'-PCB	11.28	4870	11.1	4840
2,2',4,4',6,6'-HCB	11.05	4750		

Figure 3

Chromatograms of 2',3,4-trichlorobiphenyl(tri-CB), 2,2',5,5'-tetrachlorobiphenyl (tetra-CB), and 2,2',4,5,5'-pentachlorobiphenyl (penta-CB) on a 1-m Apolane-87 capillary column at 80° and 100°C.

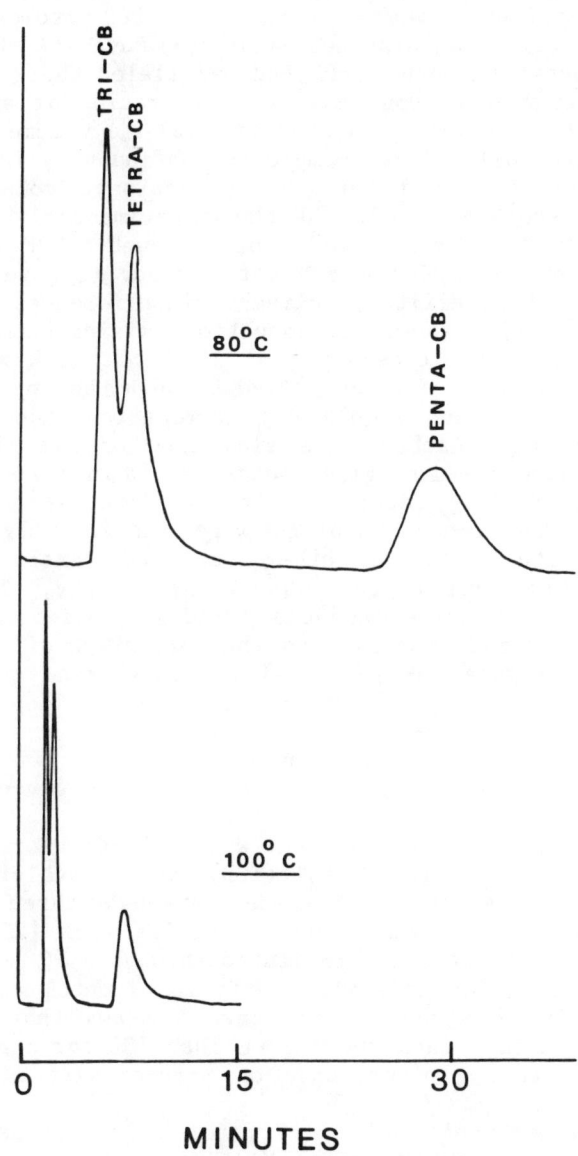

adsorbent trap. If the quantity on the backup trap is substantially less than the amount on the front trap, complete collection is assumed. While this approach is useful for checking breakthrough during monitoring, it offers little predictive capability. In earlier work, Simon and Bidleman [25] demonstrated that a PPF column performs as a chromatographic system. A single application of sample was introduced at the head of a 15-cm long x 7.6-cm diameter PPF column and eluted with prefiltered air. Analysis of 15 1-cm thick PPF slices revealed that: (a) PCB isomers moved through the column as distinct bands, (b) movement of Aroclor 1016 components through a PPF bed paralleled their order of elution from a GC column, and (c) penetration of an isomer band maximum was linearly related to total air volume.

In field collections, sample is continuously introduced and subsequently moved through an adsorbent column as a front. We therefore modified the experimental design to study the frontal movement of vapors through a PPF bed. The sampling train used for the frontal study (Figure 4) consisted of a PPF prefilter (to remove interferences from the laboratory air), a sample introduction/ mixing chamber, and a collection column consisting of 15 1-cm thick x 7.6 cm diameter PPF_3 plugs. Air was pulled through the apparatus at 0.3 - 0.5 m^3/min and sample vapors were continuously bled into the mixing chamber by a slow flow of air through a Pasteur pipet containing glass beads coated with about 50 mg of the compound of interest. At the termination of the experiment, the 1-cm PPF plugs were individually soxhlet extracted with petroleum ether, and the extracts were analyzed by electron capture gas chromatography. The quantity of organochlorine compound found on individual plugs was calculated and expressed as the percentage of the first plug value (Figures 5-7). Total organochlorine quantities in the 15-cm PPF trap ranged from 0.2 - 50 µg, depending on the volatility of the compound, the airflow through the coated bead column, and the sampling time. Details of PPF plug preparation and sampling train design are given in [25] and [26].

Experiments were carried out with 3,3'-dichlorobiphenyl (DCB), 2,4',5-trichlorobiphenyl (TCB), and hexachlorobenzene (HCB). Frontal movement of these compounds through a PPF bed is depicted in Figures 5-7. Earlier work [25] demonstrated that PCB isomers were eluted through a PPF column in order of decreasing volatility. In the frontal experiment this was also the case, and Figure 5 shows that DCB has penetrated further into the column than TCB for a given air volume. Increasing penetration of a front with air volume can be seen in Figure 6.

The breakthrough volume, V_B, for a frontal experiment may be taken as the point at which the concentration of a solute in the column effluent is half the concentration introduced to the column [27] This breakthrough volume should be identical with the retention volume, V_R, for an elution experiment carried out on the same column [28]. The thickness of foam (P-50) corresponding to 50% breakthrough was obtained for each frontal experiment (Figure 7) and plotted against total air volume. As was the case for previous elution experiments [25], linear relationships were obtained between vapor penetration and total air volume (Figure 8). Extrapolation of these plots yielded V_B values for a 7.6-cm diameter x 15 cm thick PPF column. The V_B values for DCB and TCB agreed well with V_R obtained in elution experiments (Table IV). The slightly greater V_B compared to V_R may be due to the fact that the frontal experiments were carried out at 20°C, whereas the elution experiments were done at 23-26°C.

Table IV.   Retention and Breakthrough Volumes for a 15-cm PPF Column

Compound	V_R, m^3 air elution	V_B, m^3 air frontal
HCB		250
3,3'-DCB	1250	1330
2,4',5-TCB	2700	2880

The vapor pressure of the 3,3'-DCB isomer is 1.1×10^{-4} torr at 20°C, as measured by capillary GC (see previous section). A few preliminary measurements indicate that the vapor pressure of the 2,4',5-TCB isomer is very close to that of the 2',3,4-TCB isomer, which has a vapor pressure of 4.3×10^{-5} torr at 20°C (GC). The elution characteristics of these two TCB isomers through a PPF column are also very similar [25]. Assuming 4.3×10^{-5} torr for the 2,4',5-TCB vapor pressure, the DCB/TCB volatility ratio is 2.6. This volatility ratio is reasonably close to the breakthrough volume ratio (2.2) for TCB/DCB. Hexachlorobenzene (HCB) is poorly retained by PPF, both under field sampling conditions [29,30] and in laboratory frontal studies. The breakthrough volume for HCB on a 15-cm PPF column was only 250 m^3 at 20°C (Figure 8).

Collection of Organochlorines in Urban Air Using Three Solid Adsorbents

Several solid adsorbents have been used to sample organic vapors, but few studies have been carried out to measure ambient concentration with two or more adsorbents over the same time period. We compared the collection characteristics of three adsorbents: Tenax-GC resin, Amer-

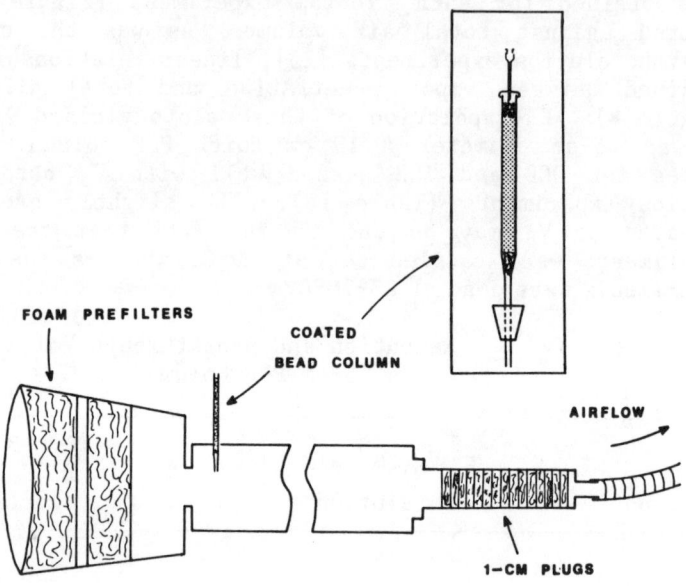

Figure 4. Apparatus used for studying frontal movement of PCB and HCB through a PPF column.

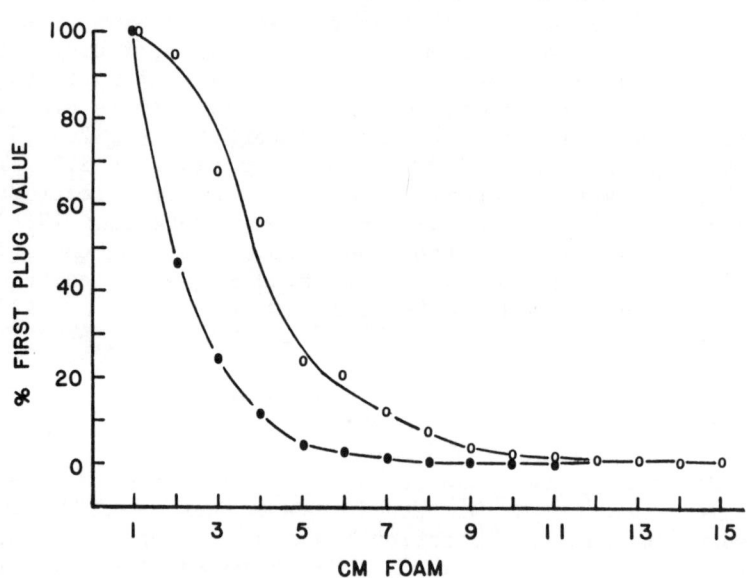

Figure 5. Frontal movement of ○ DCB and ● TCB isomers through PPF, 400 m^3 air.

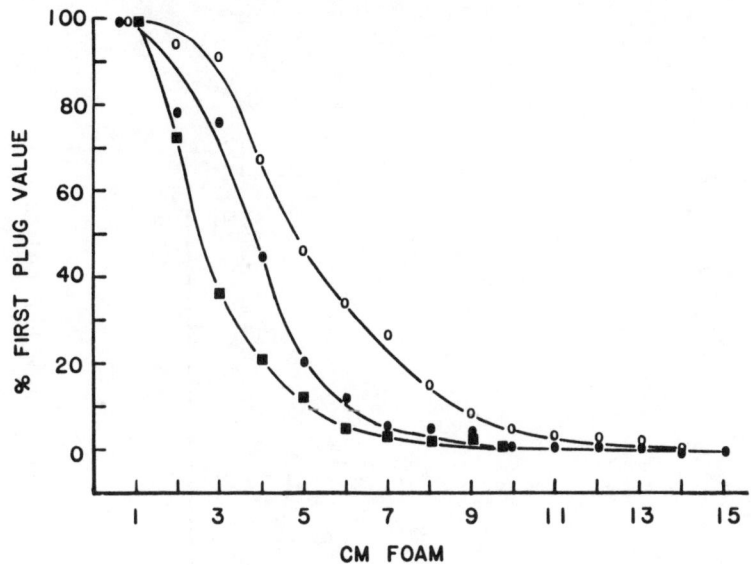

Figure 6. Frontal movement of TCB through a PPF column at ■ 500 ● 700 and O 900 m³ air.

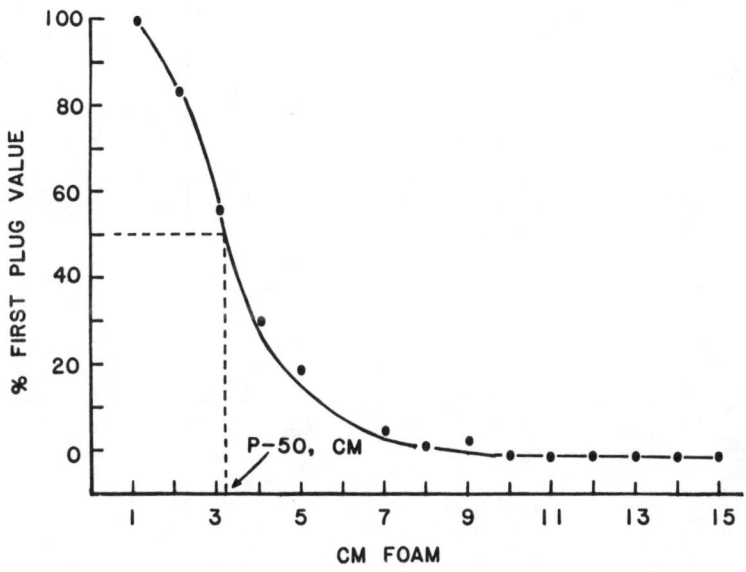

Figure 7. Frontal movement of HCB through a PPF column 50 m³ air. P-50 marks the point where the HCB concentration is 50% of the first plug value.

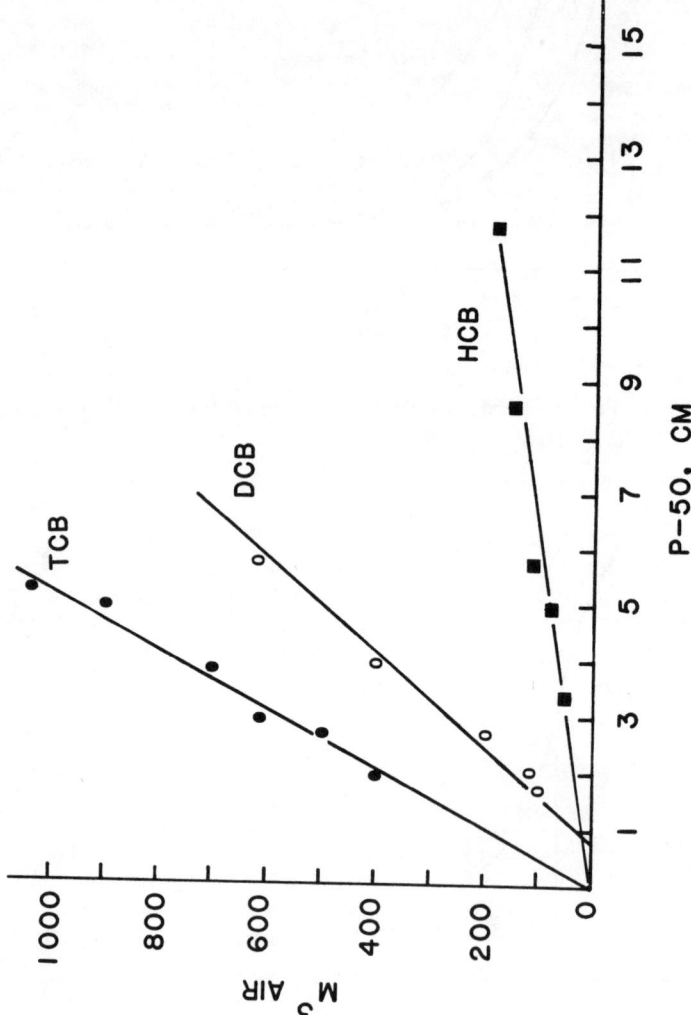

Figure 8. Linear movement of HCB, DCB, and TCB fronts through a PPF column with total air volume.

lite XAD-2 resin, and PPF for organochlorines in urban air with the intent of establishing (a) the adsorbent collection efficiencies for pesticide and PCB vapors at different sampling temperatures and total air volumes, (b) the reproducibility of replicate collections with a single adsorbent, and (c) the precision of measurements made by concurrent sampling with different adsorbents.

Three locations were chosen which have different climatic and air quality conditions. Columbia, SC is a medium-sized (pop. 380,000) southern city with mild temperatures. Concentrations of total suspended particles (TSP) are relatively low; during our 1977-80 sampling years geometric mean TSP levels ranged from 43 - 59 $\mu g/m^3$. Denver, CO was selected to test adsorbent collection characteristics and the distribution of organochlorines between the filter-retained and adsorbent-retained phases during cold weather. Samples in Denver were taken during January, 1980, at which time median daily temperatures ranged from -9 to +1°C, and TSP levels were 117-157 $\mu g/m^3$. Our third sampling site was the municipal landfill in New Bedford, MA, which received PCB waste from capacitor manufacturers in the area until 1975. Over the years an estimated 22,000 kg of PCB have been buried at the landfill [31]. Collections at the landfill were made in June, 1980. Median daily temperatures and TSP concentrations were 11-15°C and 140 - 220 $\mu g/m^3$.

Procedures for adsorbent cleanup, sample collection, and analysis have been given in detail elsewhere [2,29,32,33] and will only be described briefly here. Three adsorbents were selected for field comparison: PPF plugs 7.6 cm diameter x 7.6 cm thick (Olympic Products Corporation, Greensboro, NC, product no. 3014, density = 0.022 g/cm^3), Tenax-GC resin (Applied Science Laboratories, State College, PA, 35/60-mesh), and Amberlite XAD-2 resin (Rohm and Haas Company, Philadelphia, PA, 20/50-mesh). PPF and Tenax were pre-cleaned by soxhlet extraction with organic solvents and dried at 40-45°C in a vacuum desiccator under continuous water aspiration. XAD-2 was first vacuum-dried at 40-45°C overnight and then cleaned up as above.

Clean adsorbents were re-extracted with petroleum ether immediately or after several months' storage. The extracts were shaken with 7% fuming sulfuric acid, which removes most nitrogen- and oxygen-containing compounds but leaves most organochlorine pesticides and PCB intact, and analyzed by electron capture GC. All the peaks eluting in the range of Aroclors 1016 and 1254 were summed and quantified as PCB to establish blank values. A summary of blanks is given in Table V. More detailed information concerning adsorbent blanks before and after storage and after cleanup with different solvents is given in [30]. The adsorbent blanks are negligible compared to the quantity of PCB that would be extracted from 500 m^3 of urban air in a 24-hour sampling period, on the order of 1000 ng or more.

High volume air samples were taken by pulling air at 0.35 - 0.5 m^3/min through a 20 x 25-cm glass fiber filter followed by tandem traps containing the adsorbents. Adsorbent weights used in the front and backup traps were: 8.0 g PPF, 10.0 g Tenax, and 20.0 g XAD-2. The PPF plugs were contained in an aluminum cylinder, while the Tenax and XAD-2 resins were held in stacked metal sieves of 54-cm^2 cross section. The filter-adsorbent trap sampling trains were connected to Rotron DR-313 brushless pumps (Rotron Corporation, Woodstock, NY) by 4 m flexible hose. Flow rates were related to the pressure drop measured behind the sampler by using an orifice calibrator. At each location the samplers containing different adsorbents (or replicates of the same adsorbent) were placed the same height (1.5 - 2 m) above the roof or ground, and were spaced no more than 1.5 m apart. Concurrent sampling was carried out in Columbia, Denver, and at the New Bedford landfill using the three adsorbents, and in Columbia duplicate or triplicate collections were also made using the same adsorbent.

Filters and solid adsorbents were extracted with organic solvents. The extracts were cleaned up and fractionated on alumina and silicic acid columns, and analyzed by electron capture GC on at least two and usually three of the following 0.4 or 0.2 cm i.d. x 180 cm long glass columns: 4% SE-30/6% SP-2401, 1.5% SP-2250/1.95% SP-2401, 3% OV-225 (Supelco, Inc., Bellefonte, PA), mounted in a Tracor 222 or Varian 3700 chromatograph. Quantitative results were calculated from packed column chromatograms, using peak height. For multi-component organochlorines (PCB, toxaphene), the sum of peak heights in an Aroclor or toxaphene standard chromatogram were compared to the sum of matching peaks in sample chromatograms. Several samples from each location were also examined qualitatively by capillary electron capture GC on 30-m WCOT columns coated with SP-2100 or SE-54. Examples of capillary chromatograms obtained for Columbia, SC air and for some air samples collected over the open ocean during another study are shown in [2,3,29].

Average concentrations of PCB and organochlorine pesticides in air at the three locations are given in Table VI. Concentrations for each sampling date may be obtained from [29,30]. Distinct differences in organochlorine patterns were obtained at each location. Most of the organochlorine load in Columbia was chlorinated pesticides, especially toxaphene, chlordane, and hexachlorocyclohexane (HCH). Toxaphene showed the greatest variability. Typical concentrations were about 1-4 ng/m^3 for most of the year, but in 1977, 1978 and 1979 concentrations over an order of magnitude higher were encountered during late August through mid-September [29]. Organochlorines over the New Bedford landfill were dominated by PCB. Nevertheless total airborne PCB concentrations, averaging 44 ng/m^3, were far below the

Table V. Average Blank Values for Solid Adsorbents, Total Organochlorines as PCB

Adsorbent	Blank Before Storage, ng			Blank After Storage, ng[*]		
	A. 1016	A. 1254	Samples	A. 1016	A. 1254	Samples
PPF	9 ± 7	5 ± 3	14	19 ± 15	10 ± 5	6
Tenax	14 ± 16	9 ± 7	18	15 ± 12	14 ± 8	8
XAD-2	8 ± 5	7 ± 3	10	11 ± 5	14 ± 9	7

[*] Storage times ranged from 3 - 35 months for PPF, 8 - 11 months for Tenax, and 8 - 11 months for XAD-2.

400 - 1300 ng/m^3 levels measured by Stratton et al. [31] in June, 1977. Our samples were taken within the landfill perimeter, but upwind of the main work area (the sampling site was limited by the availability of electric power). This factor and/or changes in the amount of landfill cover since 1977 may account for our lower PCB concentrations. Concentrations of both PCB and pesticides were low in Denver, with light PCB (Aroclor 1016) accounting for most of the organochlorine load. HCB concentrations in all three cities were remarkably similar (Table VI).

Table VI. Arithmetic Mean Concentrations of Airborne Organochlorines at Columbia, Denver and the New Bedford Landfill[a]

	10^{-9} g/m^3		
	Columbia	Denver	New Bedford Landfill
Aroclor 1016	3.2	1.8	34.5
Aroclor 1254	1.5	0.45	9.3
HCB	0.29	0.24	0.18
p,p'-DDE	0.093	0.021	
Chlordane[b]	1.3	0.063	0.24
HCH[c]	1.1	0.30	1.0
Toxaphene	13.1		

(a) Sampling dates: Columbia, 1977-80; Denver, January, 1980; New Bedford, June, 1980.
(b) cis- + trans-isomers, may include trans-nonachlor in some cases
(c) α- + γ-HCH

Front and backup adsorbent traps were individually analyzed to check for sample breakthrough. PPF was a poor collector of HCB, with nearly equal quantities found on the front and back traps. The Denver samples were an exception. Because of the cold temperatures in Denver, very little penetration of HCB to a backup PPF plug occurred. In all three cities, Tenax and XAD-2 retained HCB effectively. HCB quantities on the backup traps averaged 10% or less of the front trap values for air volumes up to 1000 m^3. Aroclor 1254, chlordane (cis- and trans-isomers), p,p'-DDE, and toxaphene were effectively collected by all three adsorbents for air volumes up to 1600 m^3; less than 5% of these organochlorines was found on the backup traps even at the highest air volumes.

Collections in Columbia from 1977-80 as well as in Denver and New Bedford provided over 30 field experiments of breakthrough under a wide range of total air volumes and ambient temperatures. Aroclor 1016 was useful for evaluating field collection efficiencies. This mixture of di- and trichlorobiphenyls was effectively retained within a 2-plug PPF trap, but volatile enough that breakthrough from the first to the second trap could be studied in relation to sampling conditions. Laboratory investigations of PCB movement through PPF show that the penetration depth of vapor bands or fronts is directly related to total air volume ([25,26]; see previous section). However this simple relationship did not at first appear to hold for field sampling in urban air. In Figure 9 we have plotted a measure of Aroclor 1016 breakthrough --- the ratio of Aroclor 1016 found on the backup and front PPF traps --- against total air volume. The data are widely scattered and the correlation is very poor ($r^2 = 0.046$).

The air samples in Figure 9 were taken under temperature conditions ranging from summer heat in Columbia to winter cold in Denver. Vapor pressures for di- and trichlorobiphenyls increase by a factor of 1.7 - 1.8 for a 5° rise in temperature (vapor pressures determined by GC, [21] and Table III), and from gas chromatographic theory it is well known that the retention time of a solute is inversely related to its vapor pressure. For each field experiment we calculated a temperature weighting factor, P_T/P_{20}, the ratio of 2',3,4-TCB vapor pressures at the average ambient temperature for the run and at 20°C. The vapor pressures were calculated from the Antoine equation for 2',3,4-TCB given in Table III (GC method). The experimental air volume was then multiplied by the temperature weighting factor to give the air volume producing the same degree of breakthrough at 20°C. For example, vapor pressures of TCB at 20° and 25°C are 4.26×10^{-5} torr and 7.70×10^{-5} torr, and $P_T/P_{20} = 1.8$. A 500-m^3 air sample taken at 25°C should show as much breakthrough as a 500(1.8) = 906-m^3 sample taken at 20°C.

In Figure 10 the breakthrough data from Figure 9 have been replotted against the temperature-weighted air volume. While there is still considerable scatter, the improved correlation ($r^2 = 0.58$ vs. 0.046) is striking, and points out the great dependence of temperature on adsorbent collection efficiencies. The regression line in Figure 10 thus represents the average Aroclor 1016 penetration through a PPF column with total air volume at 20°C.

Precision data obtained in the three sampling locations are summarized in Table VII. We were interested in the reproducibility of duplicate or triplicate collections made with a given adsorbent (expressed as relative standard deviation, % RSD) as well as comparisons between airborne organochlorine measurements made with two or three different adsorbents. Table VII thus presents average % RSD's for adsorbent replicates and adsorbent comparisons.

Pesticide and PCB concentrations measured with two or three different adsorbents agreed within about 10-15%, or within the precision of our analytical methods [29]. In most cases HCB concentrations determined using PPF were much lower than those obtained using Tenax or XAD-2, the result of extensive HCB breakthrough on PPF (Denver was an exception, as noted earlier). However HCB concentrations measured with Tenax and XAD-2 agreed excellently, within 6% average RSD (Table VII).

From these field studies we conclude that because vapor pressures of several pesticides and PCB increase rapidly with temperature, a 5° rise in temperature causes nearly the same degree of sample penetration through a PPF bed as doubling the air volume. All the adsorbents tested in 10-20 g quantities effectively collected Aroclor 1254, p,p'-DDE, chlordane, and toxaphene, with negligible penetration to the backup trap. In a 24-hour period (500-700 m^3 air) at 20°C, losses of Aroclor 1016 to a backup PPF trap averaged about 10-12%, and the breakthrough was even less using Tenax or XAD-2. Any of the three adsorbents should thus be suitable for monitoring PCB in urban air, and in fact PCB levels measured using the three adsorbents agreed within our analytical precision.

OBSERVATIONS ON THE PARTICLE/GAS DISTRIBUTION OF ORGANOCHLORINES

The glass fiber filters used to precede the adsorbent traps effectively collect small particles. Collection efficiencies for particles of different radii are 99.9% for 0.3 μm and 98% for 0.015 μm [34]. Separate analysis of the filters and adsorbent traps provided an operational measure of the particle/gas (P/G) ratio; and the average P/G ratios in Columbia, Denver, and New Bedford are presented in Table VIII.

Junge [35] made some theoretical estimates of the fraction (ϕ) of particle-adsorbed vapors as a function of vapor pressure and particle surface area per unit volume of air. Figure 11, taken from his paper, shows that large variations in ϕ in urban air may be expected for compounds whose vapor pressures range from $10^{-5} - 10^{-8}$ torr. For example, the Aroclor 1254 component 2,2',4,5,5'-PCB has vapor pressures of 9 x 10^{-6} torr and 2.8 x 10^{-7} torr at 25°C and 0°C (Table III). Junge's calculations (Figure 11) indicate that in urban air having a median TSP concentration, ϕ for this isomer might range from about 10% to over 80% between these temperatures.

Figure 9. Breakthrough of Aroclor 1016 from first to second PPF traps at different total air volumes. ● Columbia ▲ Denver X New Bedford landfill.

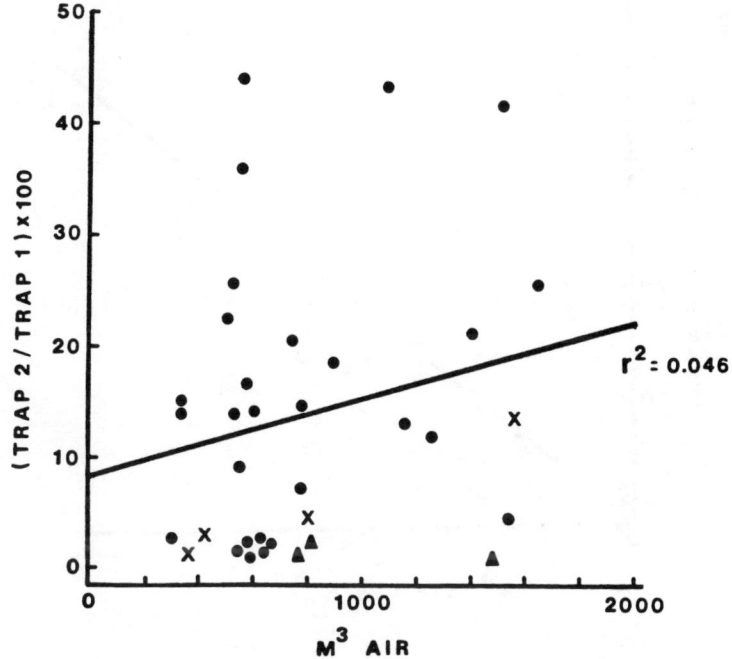

Figure 10. Breakthrough data from Figure 9 plotted vs temperature-weighted air volume, P_T and P_{20} are vapor pressures of a trichlorobiphenyl isomer at the average sampling temperature and 20°C. Symbols as in Figure 9.

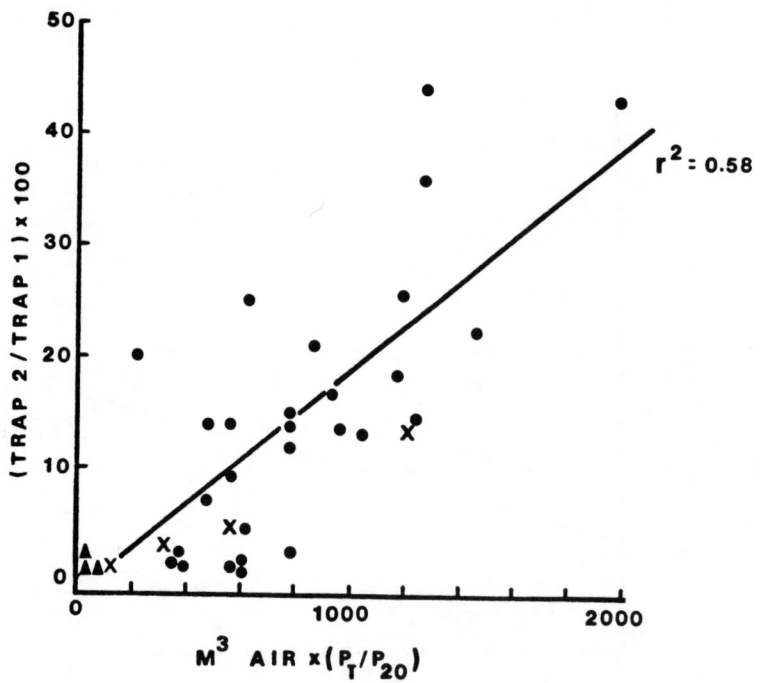

Table VII. Average Precisions for Single Adsorbent Replicates and Collections with Different Adsorbents

	Adsorbent Replicates [a] %RSD (no. of sample sets)			Adsorbent Comparisons [b] % RSD (no. of sample sets)			
	PPF	Tenax	XAD-2	PPF/Tenax	PPF/XAD-2	Tenax/XAD-2	PPF/Tenax/XAD-2
HCB	10(12)	22(4)	9(4)	c	c	6(8)	c
Aroclor 1016	12(9)	22(4)	6(4)	12(19)	12(11)	11(9)	14(9)
Aroclor 1254	13(11)	12(4)	11(4)	11(18)	14(11)	12(9)	15(9)
Chlordane	10(12)	8(4)	9(3)	10(16)	14(8)	7(7)	13(7)
p,p'-DDE	9(4)	—	10(4)	—	14(4)	16(2)	14(2)
Toxaphene	8(12)	12(4)	11(3)	10(12)	12(4)	5(3)	11(3)

(a) Average % RSD's for duplicate or triplicate collections with a single adsorbent.
(b) Average % RSD's for organochlorine levels measured with two or three different adsorbents.
(c) Not calculated due to poor retention of HCB by PPF.

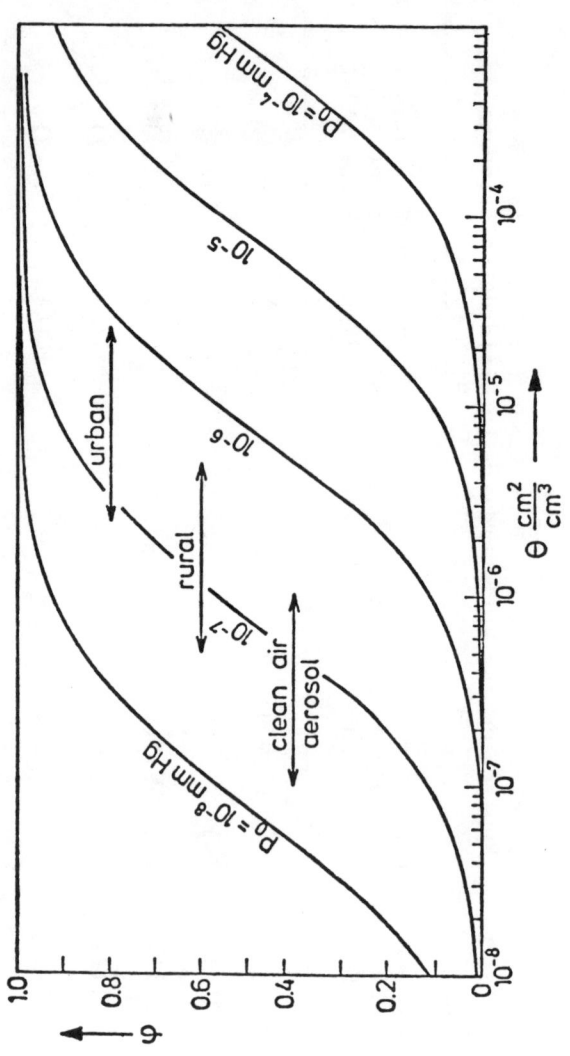

Figure 11. Fraction (Φ) of particulate organics as a function of vapor pressure and available particle surface area per cubic cm air (Θ). From Junge (35).

Table VIII. Filter-Retained Percentages of
Organochlorines in Urban Air

	Columbia	Denver	New Bedford Landfill
Aroclor 1016	< 2	9	3
Aroclor 1254	6	65	16
HCB		6	
p,p'-DDE	< 3	62	
Chlordane	< 2	53	

Very low percentages of PCB or pesticides in Columbia were filter-retained, probably a consequence of generally mild temperatures and low concentrations of TSP. In Denver, substantial proportions of the Aroclor 1254, p,p'-DDE, and chlordane were found on the filter. Others have reported filter-retained PCB in urban air, although the percentages were much lower than the Denver values. One study found 14-43% filter-retained PCB in Toronto [36], but in other cities filter-retained PCB appears to account for only about 10-20% of the total [7,8]. Cold temperatures and high TSP levels in Denver seem responsible for the large apparent P/G ratio. Denver TSP concentrations were about 2-3 times those in Columbia, providing more surface area for vapor adsorption, and temperatures during our sampling periods were usually below 0°C.

Filter-retained percentages in New Bedford were not high, despite the fact that TSP concentrations at the landfill (140-220 $\mu g/m^3$) were slightly higher than those in Denver (117 - 157 $\mu g/m^3$). The material collected on the filters appeared to be largely road dust, stirred up by trucks using the dirt roads leading to the landfill. This dust was probably of a large particle size and low in organic matter, and thus may not have been as retentive for organic vapors as urban aerosols. Also, the weather was warmer in New Bedford (11-25°C) than in Denver (-9 to +1°C).

As a final example of P/G ratios, we present some data for p,p'-DDT over the Northern Indian Ocean which were collected between December, 1976 - April, 1977 [3]. Concentrations of total DDT (particle plus vapor-phase) over the Arabian Sea, the Persian Gulf, and the Red Sea greatly exceeded levels found over the North Atlantic. The average DDT concentrations over the eastern seas were 25-40 times the 0.003 ng/m³ North Atlantic background value, measured in 1977-78 [2]. Concentrations of airborne DDT were lowest near the equator, and highest off the coast of India, in the Persian Gulf, and in the Red Sea (Figure 12). Tanabe and Tatsukawa [9] recently reported DDT levels even higher than ours in the Arabian Sea, the Bay of Bengal, and in the South China Sea. In 1980-81 Tanabe et al. [38] found 0.1 - 0.4 ng/m³ DDT between Australia and Antarctica. These elevated DDT concentrations are undoubtedly due to continued usage of DDT in countries bordering these seas. Projections of worldwide DDT use through 1981, made during the early 1970's, amounted to 116,000 metric tons/year --- about the same as DDT consumption in the 1960's. Most of this predicted use was in the tropical countries and in the southern hemisphere [37].

Figure 13 shows the percentage of filter-retained DDT for these samples, and it is interesting to note a qualitative correlation between the amount of DDT on the filter and filter color. Samples taken near the equator had white filters, and DDT on these filters was below 5-8% of the total. In the open Arabian Sea, the filters were light to medium grey, and filter-retained DDT averaged 5% of the total. Filters in the Persian Gulf were very dark grey to black, suggestive of large amounts of organic-rich particulate matter, and here the proportion of filter-retained DDT rose to 20-30%. The three samples taken in the Red Sea also had fairly heavy particle loadings (a qualitative observation; the filters were not weighed), but the material on the filters appeared to be light brown terriginous dust. For these samples, filter-retained DDT amounted to only 5% of the total.

These observations suggest that both the amount and organic content of filter-deposited particles is important in determining the extent of vapor adsorption.

ACKNOWLEDGEMENTS

This work was supported by the U.S. Environmental Protection Agency, Grant no. 807048. Contribution no. 454 of the Belle W. Baruch Institute.

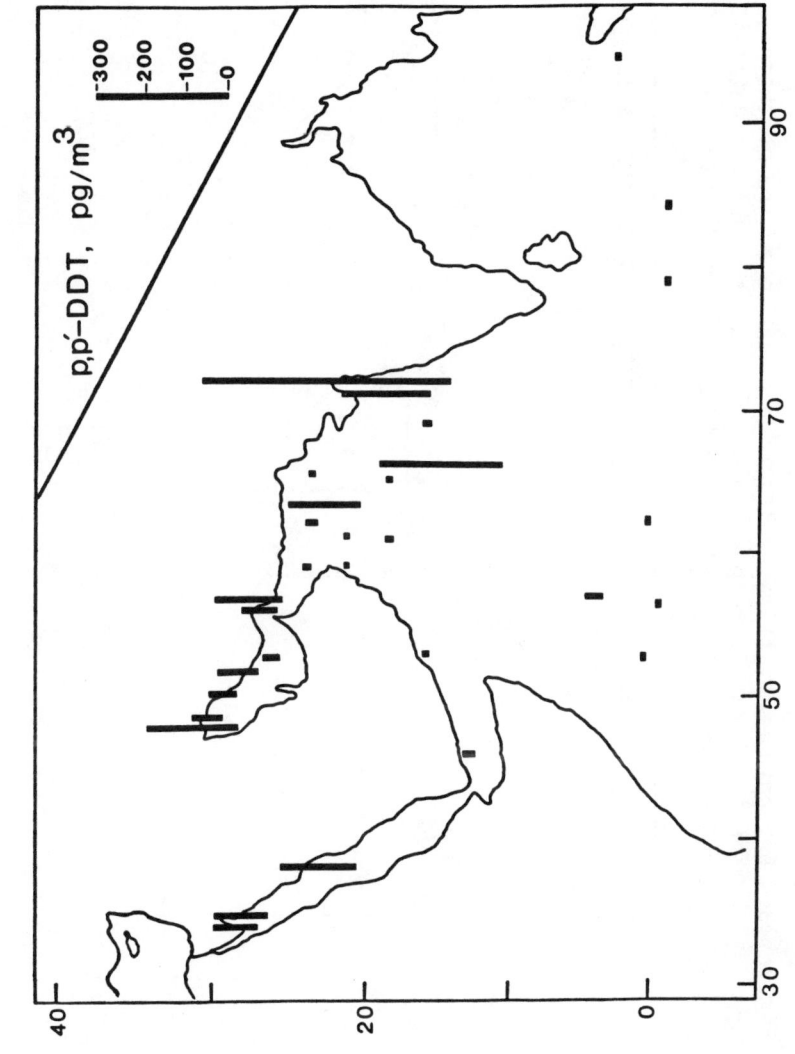

Figure 12. Concentrations of total p,p'-DDT (particulate + vapor-phase) over the northern Indian Ocean, Dec. 1976 - April, 1977; picograms/m^3.

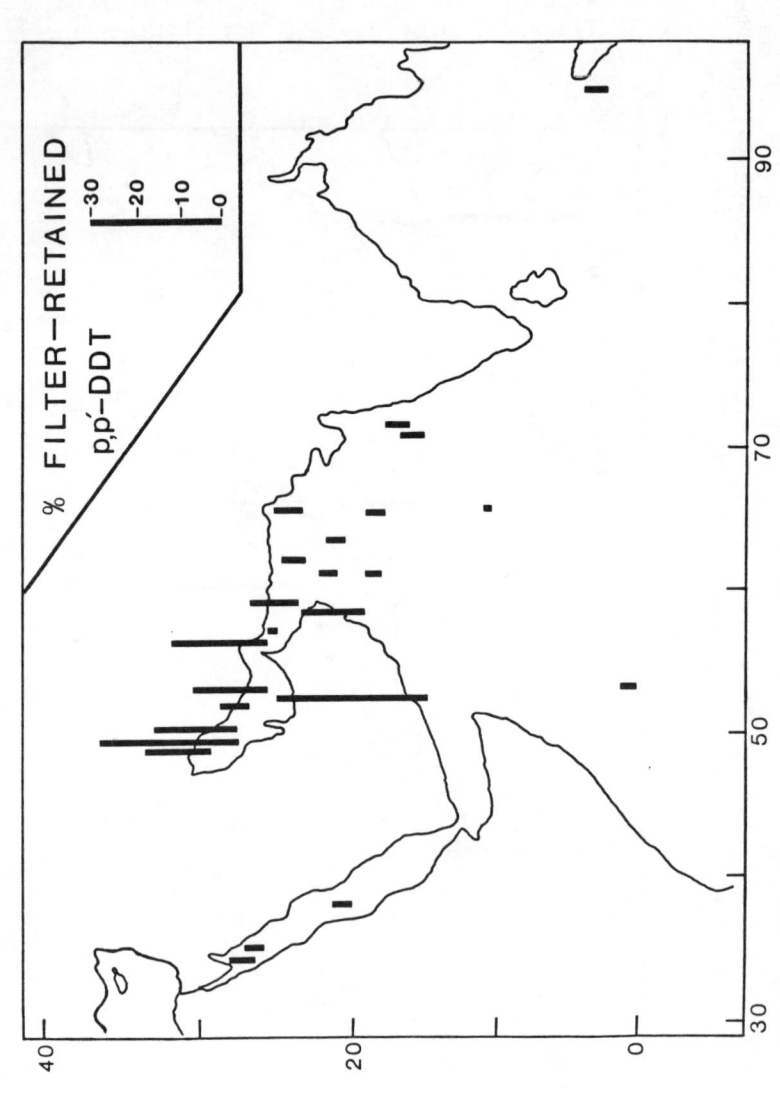

Figure 13. Percentages of filter-retained p,p'-DDT over the northern Indian Ocean. The two values for the equatorial samples represent detection limits, while measureable quantities were found on the other sample filters.

REFERENCES

1. Bidleman, T.F., Rice, C.P. and Olney, C.E., "High Molecular Weight Chlorinated Hydrocarbons in the Air and Sea: Rates and Mechanisms of Air/Sea Transfer". In: Windom, H.L. and Duce, R.A. (eds.) Marine Pollutant Transfer, D.C. Heath & Co., Lexington, MA, p. 323 (1976).

2. Bidleman, T.F., Christensen, E.J., Billings, W.N. and Leonard, R., "Atmospheric Transport of Organochlorines in the North Atlantic Gyre". J. Mar. Res., $\underline{39}$, 443 (1981).

3. Bidleman, T.F. and Leonard, R., "Aerial Transport of Pesticides Over the Northern Indian Ocean and Adjacent Seas". Atmos. Environ., $\underline{16}$, 1099 (1982).

4. Giam, C.S., Chan, H.S., Neff, G.S. and Atlas, E.L., "Phthalate Ester Plasticizers: A New Class of Marine Pollutant". Science, $\underline{199}$, 419.

5. Giam, C.S., Atlas, E., Chan, H.S. and Neff, G.S., "Phthalate Esters, PCB and DDT Residues in the Gulf of Mexico Atmosphere". Atmos. Environ., $\underline{14}$, 65 (1980).

6. Atlas, E.L. and Giam, C.S., "Global Transport of Organic Pollutants: Ambient Concentrations in the Remote Marine Atmosphere". Science, $\underline{211}$, 163 (1981).

7. Doskey, P.V. and Andren, A.W., "Concentrations of Airborne PCBs Over Lake Michigan". J. Gt. Lakes Res., $\underline{7}$, 15 (1981).

8. Eisenreich, S.J., Hollod, G.J. and Johnson, T.C., "Atmospheric Concentrations and Deposition of Polychlorinated Biphenyls to Lake Superior". In: Eisenreich, S.J. (ed.) Atomspheric Pollutants in Natural Waters, Ann Arbor Science, Ann Arbor, MI, p. 425 (1981).

9. Tanabe, S. and Tatsukawa, R., "Chlorinated Hydrocarbons in the North Pacific and Indian Oceans". J. Oceanog. Soc. Japan, $\underline{36}$, 217 (1980).

10. Lewis, R.G., "Sampling and Analysis of Airborne Pesticides". In: Lewis, R.G. and Lee, R.E., Air Pollution from Pesticides and Agricultural Processes, Chap. 3, CRC Press, Cleveland, OH, p. 51 (1976).

11. Lamb, S.E., Petrowski, C., Kaplan, I.R. and Simoneit, B.R.T., "Organic Compounds in Urban Atmospheres: A Review of Distribution, Collection, and Analysis". Air Pollut. Cont. Assoc. J., 30, 1098 (1980).

12. Simoneit, B.R.T. and Mazurek, M.A., "Air Pollution: The Organic Compounds". CRC Critical Reviews in Environmental Control, 11, 219 (1981).

13. Spencer, W. F. and Cliath, M.M., "The Vapor Density of Dieldrin". Environ. Sci. Technol., 3, 670 (1969).

14. Spencer, W.F. and Cliath, M.M., "Volatility of DDT and Related Compounds". J. Agric. Food Chem., 20, 645 (1972).

15. Spencer, W.F., Shoup, T.D., Cliath, M.M., Farmer, W.J. and Haque, R., "Vapor Pressures and Relative Volatility of Ethyl and Methyl Parathion". J. Agric. Food Chem., 27, 273 (1979).

16. Westcott, J.W., Simon, C.G. and Bidleman, T.F., "Determination of Polychlorinated Biphenyl Vapor Pressures by a Semi-micro Gas Saturation Method". Environ. Sci. Technol., 15, 1375 (1981).

17. Hamilton, D., "Gas Chromatographic Measurement of Volatility of Herbicide Esters". J. Chromatog., 195, 75 (1980).

18. Castello, G. and D'Amato, G., "Influence of Vapor Pressure and Activity Coefficients on the Retention Volumes of Branched-Chain Alkanes". J. Chromatog., 107, 1 (1975).

19. Castello, G. and D'Amato, G., "Influence of Vapor Pressure, Activity Coefficient and Structure on the Retention Volumes of Branched-Chain Nonanes". J. Chromatog., 116, 249 (1976).

20. Small, P.A., Small, K.W. and Cowley, P., "The Vapor Pressures of Some High Boiling Esters". Trans. Faraday Soc., 44, 810 (1948).

21. Westcott, J.W. and Bidleman, T.F., "Determination of Polychlorinated Biphenyl Vapor Pressures by Capillary Gas Chromatography". J. Chromatog., 210, 331 (1981).

22. Webb, R.G. and McCall, A.C., "Identification of PCB Isomers in Aroclors". J. Assoc. Offic. Anal. Chem., 55, 746 (1972).

23. Sissons, D. and Welti, D., "Structural Identification of Polychlorinated Biphenyls in Commercial Mixtures by Gas-Liquid Chromatography, Nuclear Magnetic Resonance, and Mass Spectrometry". J. Chromatog., 60, 15 (1971).

24. MacKay, D. and Wolkoff, A.W., "Rate of Evaporation of Low-Solubility Contaminants from Water Bodies to Atmosphere". Environ. Sci. Technol., 7, 611 (1973).

25. Simon, C.G. and Bidleman, T.F., "Sampling Airborne Polychlorinated Biphenyls with Polyurethane Foam--A Chromatographic Approach to Determining Retention Efficiencies". Anal. Chem., 51, 1110 (1979).

26. Burdick, N.F. and Bidleman, T.F., "Frontal Movement of Hexachlorobenzene and Polychlorinated Biphenyl Vapors Through Polyurethane Foam". Anal. Chem., 53, 1926 (1981).

27. Pellizzari, E.D., Bunch, J.E., Berkley, R.E. and McRae, J., "Collection and Analysis of Trace Organic Vapor Pollutants in Ambient Atmospheres. The Performance of a Tenax-GC Cartridge Sampler for Hazardous Vapors". Anal. Lett., 9, 45 (1976).

28. Reilley, C.N., Hildebrand, F.P. and Ashley, Jr., J.W., "Gas Chromatographic Response as a Function of Sample Input Profile". Anal. Chem., 34, 1198 (1962).

29. Billings, W.N. and Bidleman, T.F., "Field Comparison of Polyurethane Foam and Tenax-GC Resin for High Volume Air Sampling of Chlorinated Hydrocarbons". Environ. Sci. Technol., 14, 679 (1980).

30. Billings, W.N. and Bidleman, T.F., "High Volume Collection of Chlorinated Hydrocarbons in Urban Air Using Three Adsorbents". Atmos. Environ. (in press, 1982).

31. Stratton, C.L., Whitlock, S.A. and Allan, J.M., "A Method for the Sampling and Analysis of Polychlorinated Biphenyls (PCBs) in Ambient Air". EPA Research Report EPA-600/4-78-048, U.S. Environmental Protection Agency, Research Triangle Park, N.C. 27711.

32. Bidleman, T.F. and Olney, C.E., "High Volume Collection of Atmospheric Polychlorinated Biphenyls". Bull. Environ. Contam. Toxicol., 11, 442 (1974).

33. Bidleman, T.F., Matthews, J., Olney, C.E. and Rice, C.P., "Separation of Polychlorinated Biphenyls, Chlordane and p,p'-DDT from Toxaphene by Silicic Acid Column Chromatography". J. Assoc. Offic. Anal. Chem., 61, 820 (1978).

34. Butcher, S. S. and Charleson, R.J., "An Introduction to Air Chemistry", Academic Press, NY, p. 49 (1972).

35. Junge, C.E., "Basic Considerations About Trace Constituents in the Atmosphere as Related to the Fate of Global Pollutants". In: Suffet, I.H. (ed.) "Fate of Pollutants in the Air and Water Environments", Advances in Environmental Science and Technology Series, Vol. 8, Wiley-Interscience, p. 7 (1977).

36. Gilbertson, M., "Background to the Regulation of PCBs in Canada, Task Force on PCBs". Report to the Environmental Contamination Committee of Environment Canada and Health and Welfare, Toronto, Ontario.

37. Goldberg, E.D., "Synthetic Organohalides in the Sea". Proc. Roy. Soc. Lond. B189, 277 (1975).

38. Tanabe, S., Kawano, M. Tatsukawa, R., "Chlorinated Hydrocarbons in the Antarctic, Western Pacific, and Eastern Indian Oceans". Transact. Tokyo Univ. Fisheries, 5, 97 (1982).

CHAPTER 3
VAPOR EXCHANGE OF PCBs
WITH LAKE MICHIGAN: THE
ATMOSPHERE AS A SINK FOR PCBs

Thomas J. Murphy
Jean C. Pokojowczyk
    DePaul University
    Chicago, IL

Michael D. Mullin
    U.S. Environmental Protection Agency
    Grosse Ile, MI

INTRODUCTION

One of the major problems remaining in determining the PCB budgets of the Great Lakes is the question of whether there is net transport of PCB vapor into or out of the water. Conceptually, the problem is a simple one. Given the concentration of PCBs in the atmosphere, lake water, and the equilibrium distribution coefficient (the Henry's Law constant, HLC), the direction of net transfer of PCBs through the air/water interface can be determined. The amount of PCBs transferred may then be estimated through the use of an available mass transfer model which relates the differences in fugacity to a mass transfer rate. The problem is that the necessary HLCs and water concentrations have been difficult to determine and reasonable estimates of them are not yet available.

Rather than discussing and comparing the concentrations and amounts of PCBs in the air and water phases, it is conceptually much easier compare their fugacities. The concept of fugacity, developed by G. N. Lewis, recently has been clarified and promulgated by Mackay (1) and by Mackay and Paterson (2). Fugacity is a measure of the escaping tendency of a material in a particular phase. The attraction for the use of fugacity is that for a material distributed between two or more phases, and at equilibrium, the fugacity of the material in each phase is the same. In contrast, the concentrations or amounts of the material in each of the phases at equilibrium will be different, with the amounts present related to the fugacity capacity (Z) for the material in that phase. For a system not at equilibrium, transport will be in the direction to equalize the fugacities.

The fugacity of a material in a phase is proportional to the concentration of the material divided by the capacity factor (Z) for that material (1). The fugacity of vapor phase materials is relatively easy to determine. It is equal to the vapor pressure of the material (atm). The fugacity of a material in the liquid phase is equal to the amount in solution (mol/m^3) times the Henry's Law Constant ($atm-m^3/mol$), for that material.

In this paper, preliminary results of the application of an equilibrium method for determining the Henry's Law Constants (HLCs) for all the individual chlorobiphenyl compounds in the Aroclor^R mixtures are reported. Most of the individual compounds in Aroclor 1242 and 1254 have HLCs in the range of 2 to 7×10^{-4} $atm-m^3/mol$. A method is also described which permits the fugacity of the PCBs in natural waters to be determined. Preliminary results of this method with Lake Michigan water indicates that about 60% of the PCBs in the water samples tested were in true solution.

The implication of these results to the sources and sinks of PCBs in natural waters is discussed.

HLC's for Chlorinated Biphenyls and Aroclor Mixtures

The HLCs are required in order to calculate the fugacity of the PCBs in the water phase. Also, as will be seen, they are needed to determine the percent of the total PCBs in a water sample that are in solution. Unfortuately, the HLCs for PCBs are not well known. A recent article (3) discussing this problem shows a range of 10^4 for the partition coefficient of Aroclor 1242. Some of the reasons for this are that the low vapor pressure and solubility of the individual compounds at room temperature makes their determination difficult. The fact that essentially all of

the pure chlorinated biphenyl compounds are solids at room
temperature while the commercial mixtures are liquids also
complicates this determination. Finally, since more than
100 of the 209 possible different compounds do occur in the
commercial mixtures, the need to determine all the
individual HLCs is a formidable task.

The method used in this project to determine the HLCs
for PCBs involves the preparation of saturated solutions of
some of the individual chlorinated biphenyl compounds and
commercial PCB mixtures. The solutions were prepared by the
method of Haque and Schmedding (4). The material to be
dissolved was placed on the bottom of a clean, 4-liter glass
bottle, about 3 liters of clean, organic-free water was
added [the water was prepared with a Millipore Super-Q^R
system and then passed through XAD-2 resin, or deionized
water was refluxed for 24 hours with acid dichromate and
then distilled through an all-glass, packed column], and the
solution stirred for one to two months. The solution was
then permitted to stand undisturbed for at least two
additional months to come to equilibrium.

The water phase was sampled by taking 2.0 ml of water
from beneath the surface with a pipet which had been
equilibrated with some of the solution. The water was added
to a vial containing 0.200 ml of isooctane. The vial was
shaken to extract the water, and the amount of PCBs in the
isooctane was determined by gas chromatography with an
electron capture detector (GC/ECD).

The container with the solution of the PCBs was covered
with aluminum foil and permitted to stand undisturbed for
several days. A sample of the vapor was collected by
puncturing the foil with a glass, disposable pipet
containing about 0.5 gm of Florisil^R. A measured volume of
air (usually 50 or 100 ml) was drawn through the pipet. The
chlorobiphenyl compounds collected on the Florisil were
eluted with hexane, and the amount present determined with
GC/ECD. The volume of air collected was corrected for the
dilution which occurred as air was drawn into the container
being sampled.

The use of capillary gas chromatography to determine
the amounts of each of the chlorinated biphenyl isomers
present in the vapor and water phases of an equilibration of
one of the Aroclor mixtures, permits the HLCs for ALL of the
chlorinated biphenyl isomers present in measurable amounts
in both phases to be determined simultaneously. Also, since
excess Aroclor is present, the solution is saturated with

respect to the Aroclor. If the Aroclor composition is known, then the vapor pressure and water solubility of each of the chlorinated biphenyl isomers may also be calculated.

This determination assumes that all of the PCBs present in the water phase are in solution. Low HLC values would result if any PCBs are present on particles. To minimize problems with particulates, clean, organic-free water was used to prepare the solutions. As a check, the concentrations of PCBs found in all of the solutions were in close agreement with the reported water solubility of the PCB used.

The results of the capillary gas chromatographic (GC) determination of the HLCs of the chlorinated biphenyl compounds in Aroclor mixtures are shown in Table 1. HLCs with two different units are shown. While the HLCs reported in Table 1 for the components of the Aroclors may have an error of a factor of 2 to 3 associated with them at this time, the relative HLCs of the individual compounds should be reasonably accurate. It is remarkable how similar the HLCs are for the individual compounds. The differences between most of them are not statistically significant. It is also interesting to note that the HLCs tend to increase for the higher molecular weight compounds. This indicates that the solubility of the compounds decreases faster as more chlorines are substituted on the biphenyl molecule, than does the vapor pressure.

Henry's Law Constants for three chlorinated biphenyl compounds were determined both on solutions of the single isomers and on solutions of the Aroclor mixtures. These results are summarized on Table 2 along with some information on particular batches of two Aroclors and information from the literature on individual chlorobiphenyl compounds. Most of the measurements on the solutions of the single isomers were made with packed column GC, and the results were in close agreement with those made with capillary GC. There is very good agreement between the literature values for the 2,2',5,5'-tetrachlorobiphenyl and the values found in this project.

There is a difference in all three cases studied, however, between the HLC for a specific compound determined on a solution of the individual compound versus a determination made on a solution of an Aroclor mixture containing that compound (Table 2). This difference could imply that the activity of a PCB compound is different in a solution of the pure compound than in a solution prepared from the Aroclor. More probably, however, it reflects the experimental error in these determinations at this time.

Table 1. Henry's Law Constants for PCBs

Compound	Compound Identity Number(10)	HLC (atm-m^3/mol) X 10^4	HLC (ng/m^3) $\overline{\text{ng/l}}$
22'	4	2.2	9
(23')	(6)	2.1	8.8
(24')	(8)	2.2	9
(22'6)	(19)	2.1	8.8
22'5	18	2	8.3
44'+22'3	15+16	3	12.3
Unknown		1.4	5.8
Unknown		2.4	10
(23'4)	(25)	1.6	6.5
24'5	31	2	8
2'34	33	1.5	6.2
22'56'	53	3	12.8
234'	22	1.8	7.6
Unknown		2	8.3
Unknown		2.2	9
22'55'	52	2.2	9
22'45'	49	2	8.4
22'44'+ 244'6	47+75	3.7	15
22'35'	44	2.4	10
23'4'5	70	2	7.6
22'345'	87	3.3	13.5
(233'4'6)	(110)	3.7	15
22'33'4	82	2	7.6
22'355'6	151	3	12
22'345'6	144	6	25
Unknown		5	21
22'34'5'6'	149	3	12
23'44'5	118	4	15
22'33'56	134	5.7	23
22'44'55'	153	3.5	14
22'3455'	141	4	16
22'344'5	138	4.8	20
Unknown		6	24
22'344'5'	138	4.8	20
233'44'6	158	6.4	26
Unknown		11	46
Unknown		7.3	30
22'344'5'6	183	6.3	26
22'33'44'	128	5	20.6
22'33'4'56	177	22	90
233'44'5	156	8.7	36
233'44'5'	157	5.8	24

Table 2. Henry's Law Constants for Chlorobiphenyl Compounds and Aroclor Mixtures from Equilibrium Measurements

2,2',5-Trichlorobiphenyl

Source	HLC (atm-m^3/mol) X10^4	HLC (Pa-m^3/mol)	HLC (ng/m^3) ng/l
Single Isomer (3)*	3.4	35	14.3 ± 2.7
Aroclor 1242 (3)	2	20	8.3 ± 2.7

2,2',5,5'-Tetrachlorobiphenyl

Single Isomer (3)	2.6	27	10.8 ± .54
" "	3.1-5.3	30-55	13-22 ref. (5)
" "	2.8	28	11.5 ref. (6)
Aro 1242 & 1254 (3)	2	21	9 ± 5.7

2,2',4,4',5,5'-Hexachlorobiphenyl

| Single Isomer (3) | 2.7 | 27 | 11 ± 1.7 |
| Aroclor 1254 | 3.5 | 35 | 14 |

Aroclors

| Aroclor 1242 (3) | 2.2 | 23 | 9 ± 4.5 |
| Aroclor 1254 | 2 | 21 | 8.4 |

*Number of determinations

Table 3. Determination of the Fugacity of PCBs in Lake Michigan

2,2',5-Trichlorobiphenyl (TCB); HLC = 3.5 X 10^{-4} atm-m^3/mol

1111 ng added to 4.5 liters of Lake Michigan water in a 20 liter container

Air concentration determined = 2220 ng/m^3
= 2.1 X 10^{-10} atm

Total measured water TCB concentration = 260 ng/liter
= 1 X 10^{-6} mol/m^3

TCB concentration in the water in equilibrium with the vapor = $\dfrac{2.1 \times 10^{-10} \text{ atm}}{3.5 \times 10^{-4} \text{ atm-m}^3/\text{mol}}$ = 0.6 X 10^{-6} mol/m^3

TCB in solution = $\dfrac{0.6 \times 10^{-6}}{1 \times 10^{-6}}$ X 100 = 60 percent of total

Water Phase Fugacity of PCBs in Lake Michigan

The difficulty in determining the water phase PCB fugacity arises for several reasons. The major one is that PCBs are mixtures of a large number of compounds, and thus PCB mixtures do not have a true HLC, but rather each PCB compound has an HLC, and the particular PCB mixture present has some "apparent HLC". This apparent HLC presumably varies through time due to preferential gain or loss of component compounds. In addition, the low solubility and vapor pressures of the many of the individual PCB compounds makes the determination of these properties for these compounds difficult. Finally, the fugacity of PCBs in water depends on the amount in true solution, rather than the total amount present in the water. Since much of the PCBs in the water may be associated with small organic or inorganic particulates, it is difficult to determine the amount in true solution.

The method used to determine the fugacity of the PCBs in a water sample involves the use of a solution of an individual chlorinated biphenyl compound whose HLC is known. The procedure is as follows:

1) The total amount of PCBs in a water sample is measured. A solution of a single chlorinated biphenyl compound is added to the water sample. The sample is stirred until equilibrium is attained between the compound in the water and vapor phases.

2) The concentration of this compound in the vapor is measured. This gives its vapor fugacity and also its water fugacity (they are equal at equilibrium). The concentration of the compound in solution in the water is calculated (fugacity/HLC).

3) The total concentration of the chlorinated biphenyl compound in the water is measured, and the ratio of the concentration of the compound in solution to the total amount in the water is calculated.

4) This ratio is multiplied by the total amount of PCBs originally found in the water sample to determine the amount in solution in the water.

These determinations were made by adding a solution of 2,2',5-trichlorobiphenyl (TCB) to about 5 liters of lake water in a 20 liter container. After several days, a sample (~10 liters) of vapor was collected along with a sample of the water (~10 ml) and their TCB concentrations determined. The fugacity of TCB in the vapor phase is equal to its vapor pressure in the atmosphere. The concentration of the PCBs in true solution in the water then is its fugacity (which is equal to the fugacity of the vapor) divided by the HLC for TCB (3.5 X 10^{-4} atm-m^3/mol). This amount when divided by the total TCB in solution, gives the fraction of the TCB in true solution in the water. This fraction is assumed to hold for the PCBs originally present in the solution. From the total concentration of PCBs determined on the original sample, and this fraction, the fugacity of PCBs in the water sample can be calculated. The fugacity of PCBs in three samples of inshore Lake Michigan water was 59% of the total PCB concentration in the water. The calculations for one of these samples is shown in Table 3.

The Direction of PCB Movement Across the Air/Water Interface

For a total PCB concentration in Lake Michigan of 2 X 10^{-8} mol/m^3 (5 ng/l), a HLC of 2 X 10^{-4} atm-m^3/mol and 59% of the total PCBs being dissolved, the water phase fugacity of PCBs is 2.4 X 10^{-12} atm. This is an order of magnitude higher than the fugacity of PCBs in the atmosphere over Lake Michigan of 1.6 X 10^{-13} atm (1 ng/m^3)(3). These results indicate that the net movement of PCB vapor is from the water to the atmosphere. Estimates of the amounts transferred indicate that it could be substantial (3).

While the concentration of the chlorobiphenyl compound used to determine the fugacity of PCBs in the aqueous phase (Table 3) is a factor of 50-100 higher than the concentrations normally found in the Great Lakes, it should still be low enough that the nature of the interactions between the dissolved and particulate phases is not substantially changed.

It should be pointed out that the presence or absence of a surface film and its chemical nature, or loss to the walls of the container should have no effect on these measurements if care is taken to exclude the surface layer when the bulk water is sampled. This is because the system is sampled at equilibrium when the fugacity of PCBs in all of the phases present is equal. While the concentrations of PCBs may be considerably higher in a surface film, this just means that the fugacity capacity (Z) for PCBs of this phase is higher than that of water. This, in fact, would be expected since the microlayer is known to be less polar in nature than is water, and it is different enough chemically to be physically separate from the water phase.

Determinations of the HLCs for Aroclors and chlorobiphenyl compounds reported here are mostly in the range of 2 to 7 X 10^{-4} atm-m^3/mol. The ramifications of this with respect to atmospheric deposition are as follows. A water concentration of 1 ng/l of a PCB with an HLC of 3.3×10^{-4} atm-m^3/mol will support a vapor PCB concentration of 13.3 ng/m^3. Thus, precipitation in equilibrium with an atmospheric vapor concentration of 2 ng/m^3 of PCB, will have a dissolved PCB concentration of 0.15 ng/l. This is a concentration two orders of magnitude or more lower than is normally found in precipitation (7), and indicates that vapor scavenging can account for less than 1% of the wet deposition PCB inputs. This is in agreement with experimental findings (8,9) that the PCBs scavenged by precipitation come mostly from particulates.

The results reported here for the HLCs were made on experiments where the temperature of the water was not carefully controlled. Also, only a few replicate determination were made. More careful measurements are now being made on solutions at constant temperature; on solutions at different temperatures; and on solutions at concentrations other than at saturation. The results of these experiments should give better data on the HLCs of concern for environmental transport needs.

ACKNOWLEDGEMENTS

The support of this work by The U. S. Environmental Protection Agency under Co-operative Agreement CR-807412, and by DePaul University is gratefully acknowledged. We are grateful to Mr. Nelson Thomas who fomented this work and has encouraged it through the years it took to learn to work with a few nanograms of PCB vapor.

REFERENCES

1. Mackay, D. Finding Fugacity Feasible. Envir. Sci. & Tech., 13, 1218 (1979).

2. Mackay, D. and S. Paterson. Calculating Fugacity. Envir. Sci. & Tech., 15, 1006 (1981).

3. Doskey, P. V. and A. W. Andren. Modeling the Flux of Atmospheric Polychlorinated Biphenyls across the Air/Water Interface. Envir. Sci. & Tech., 15, 705-11 (1981).

4. Haque, R. and D. Schmedding. A Method of Measuring the Water Solubility of Hydrophobic Chemicals: Solubility of Five Polychlorinated Biphenyls. Bull. Envir Tox. & Contam., 14, 13 (1975).

5. Westcott, J. W., C. G. Simon and T. F. Bidleman. Determination of PCB Vapor Pressures by a Semimicro Gas Saturation Technique. Envir. Sci. & Tech., 15, 1375-78 (1981).

6. Andren. A. W. Personal Communication (1981).

7. Murphy, T. J., A. Schinsky, G. Paolucci and C. P. Rzeszutko. Atmospheric Inputs of PCBs to the Great Lakes. In: Atmospheric Inputs of Pollutants to Natural Waters, S. J. Eisenreich, Ed., Ann Arbor Press (1981).

8. Murphy, T. J. and C. P. Rzeszutko. Precipitation Inputs of PCBs to Lake Michigan. J. of Great Lakes Research, 3, 305-312 (1977).

9. Strachan, W. M. and H. Huneault. Polychlorinated Biphenyls and organochlorine pesticides in Great Lakes precipiation. J. of Great Lakes Research, 5, 61-68 (1979).

10. Ballschmiter, K. and M. Zell. Analysis of PCBs by glass capillary gas chromatography. Composition of technical Aroclor- and Clophen-PCB mixtures. Fres. Z. fur Anal. Chim., 302, 20 (1980).

CHAPTER 4

PHYSICAL CHEMICAL PROPERTIES
OF POLYCHLORINATED BIPHENYLS

D. Mackay, W.Y. Shiu

J. Billington, G.L. Huang

Department of Chemical Engineering
and Applied Chemistry
University of Toronto
Toronto
Ontario, Canada.

INTRODUCTION

For predicting and interpreting the environmental behaviour of PCBs, it is essential to have reliable data for their physical-chemical properties, notably those properties such as vapor pressure P^S (Pa), aqueous solubility C^S (mol/m^3), Henry's law constant H (Pa m^3/mol) and the various partition coefficients such as those for octanol-water (K_{ow}), organic carbon-water (K_{oc}) and biota-water (K_B). It is not economically or practically feasible to measure these data for all 209 congeners; thus, it is necessary to infer values for one compound from another.

Here we review the thermodynamic relationships between these environmentally relevant physical chemical partitioning properties. Available data for PCBs are compiled and correlated as a function of chlorine number to provide order-of-magnitude estimates of these properties for any congener. Some environmental implications are discussed, especially for air-water exchange, and the need for more accurate property determinations is illustrated. Finally, we believe that the approach suggested here may be applicable to other homologous series such as the dibenzodioxins or naphthalenes.

Thermodynamic Basis

There are several basic properties of each compound which individually or collectively determine the partitioning behaviour. These include vapor pressure (or fugacity), water phase activity coefficient (γ_w) and octanol phase activity coefficient (γ_o). Also important are the phase transition temperatures of melting (T_M) and boiling (T_B).

Vapor Pressure or Fugacity

At low pressures, it can be assumed that these quantities are equal. The values are temperature-dependent and are also influenced by melting point since solids have lower vapor pressures than the subcooled liquids, the ratio of these values being the fugacity ratio ϕ which can be approximated as

$$\phi = \exp(-\Delta S_F(T_M/T - 1)/R)$$

where ΔS_F is the entropy of fusion (J/mol K), T is the system temperature (K) and R is the gas constant (8.314 J/mol K). Walden's Rule suggests that ($\Delta S_F/R$) has a value of approximately 6.8, as discussed by Yalkowsky (1). It follows that the solid vapor pressure P_S^S is related to the sub-cooled liquid vapor pressure P_L^S, as $P_S^S = \phi P_L^S$, ϕ being less than unity, except at the melting point. For liquid compounds (i.e., T exceeds T_M), there is no need to define ϕ.

When correlating PCB vapor pressure versus chlorine number, it is thus essential to convert each solid P_S^S value to P_L^S and correlate P_L^S, since variations due to T_M and ϕ are eliminated. But information may also be available about boiling point T_B. Recently Mackay et al. (2) have proposed an equation relating T_B to P_L^S for hydrophobic organics which is a variation of the Rankine Equation, namely

$$\ln(P_L^S/P_B) = -(4.4 + \ln T_B)\{1.803(T_B/T-1)-0.803\ln(T_B/T)\}$$

where P_B is the vapor pressure at the boiling point of a pressure of 1 atm or 101325 Pa.

Using this equation, it is possible to estimate P_L^S from T_B and vice-versa, thus incorporating boiling point data for mixtures (Aroclors) into a correlation. The correlating technique is then to convert all available P_S^S data to P_L^S, combine with any P_L^S data, convert each P_L^S to T_B and combine these calculated T_B^L values with any T_B data as one plot of T_B versus chlorine number N. A regression can be obtained of T_B as a factor of N. The smoothed T_B values can then be converted back to P_L values, and when necessary to P_S using the melting point. The advantage of this procedure is that it combines all available information. Errors in the T_B - P_L^S equation affect only the T_B data, since the P_L experimental data are subjected to the equation and its inverse, thus introducing no net error.

Solubility

Yalkowsky and Valvani (3) have convincingly demonstrated that for a homologous series of hydrophobic organic compounds aqueous solubility C^S correlates well with calculated total molecular surface area (TSA). For sparingly soluble liquid substances C_L^S can be assumed to equal $1/\gamma_w v_w$ where v_w is the molar volume of water (18×10^{-6} m^3/mol). For solids C_S^S is ϕC_L^S. This form of correlation was applied to PCBs by Mackay et al. (4), yielding an equation for γ_w and C_S in terms of TSA. Billington (5) has shown that γ_w is almost as well correlated with chlorine number N; thus, a direct correlation is possible between N and C_L^S, and through T_M to C_S^S.

Henry's Law Constant

H is by definition P_L^S/C_L^S or P_S^S/C_S^S at saturation conditions, the fugacity ratio ϕ cancelling. Normally it is assumed that H does not vary with concentration; thus, the same value applies at lower (subsaturated) concentrations. It is worthy of emphasis that H can be calculated as indicated above from the solid or liquid saturation values but not from a mixture, eg. P_L^S/C_S^S. Although H is usually calculated from a vapor pressure and a solubility it can be determined directly by a stripping procedure (Mackay et al., (6)). A recent review in this area has been published by Mackay and Shiu (7).

The dimensionless air-water partition coefficient H' is H/RT where R is the gas constant (8.314 Pa m^3/mol K) and T is the temperature (K), P/RT being a concentration (mol/m^3).

Octanol-Water Partition Coefficient

It can be shown that K_{ow} is given by

$$K_{ow} = \gamma_{wo} v_{wo} / \gamma_{ow} v_{ow}$$

where γ_{wo} is the activity coefficient of the solute in water saturated with octanol, v_{wo} is the molar volume of water saturated with octanol and γ_{ow} and v_{ow} are the corresponding values in octanol saturated with water. There is some doubt about the effect of octanol on γ_w, i.e., if γ_w and γ_{wo} differ and if so, to what extent the difference depends on the solute concentration. The primary determinant of K_{ow} is clearly γ_{wo} or γ_w; thus, a good correlation is apparent with C_L^S as discussed by Mackay, et al. (8) who developed the correlation

$$K_{ow} = 1797/C_L^S = 1797\phi/C_S^S$$

This equation can thus be applied to liquid and solid solutes. It was developed primarily for relatively low K_{ow} solutes, ie. $K_{ow} < 10^5$ and it appears that it may break down at higher values of K_{ow} or lower values of C_L^S. This is possibly due to an increase in γ_{ow} with molecular size; thus, it seems reasonable to postulate that K_{ow} could be correlated by an expression of the form

$$K_{ow} = 1797/\{C_L^S f(N)\}$$

where N is chlorine number, the f(N) term, of as yet unknown form, corresponding to the increase in γ_{ow} with molecular size.

An attractive rapid method of estimating K_{ow} is to measure retention time on a HPLC column and interpolate between known values.

Bioconcentration Factor

Mackay (9) has recently suggested that the bioconcentration factor K_B for fish can be simply related to K_{ow} as

$$K_B = 0.048 \, K_{ow}$$

thus, K_B can be estimated from N using the K_{ow} equation.

Organic Carbon Sorption Coefficient

Karickhoff (10) has suggested that this partition coefficient K_{oc} can be related to K_{ow} by

$$K_{oc} = 0.414 \ K_{ow}$$

thus, K_{oc} can also be estimated from N.

In summary, vapor pressure, solubility, Henry's law constant and the partition coefficients between octanol and water, biota and water and organic carbon and water can be related to each other and to chlorine number by thermodynamically reasonable equations. It must be emphasized that differences are expected between equal chlorine number congeners over and above those differences attributable to melting point. For example, ortho substitution render the molecule non-planar and imparts different properties both physically and toxicologically. But such differences are expected to be of relatively small, or at least predictable, magnitude.

DATA ANALYSIS

Available data for solubility, vapor pressure, melting and boiling point, Henry's law constants and octanol/water partition coefficients for congeners and Aroclors were gathered and analysed. A set of consistent and reasonable values of these quantities were selected using judgement as to accuracy. Details of the selection process and the source data are described elsewhere (11) but the process involved selection of a C_L^S value (based primarily on total surface area), then a P_L^S value (based on vapor pressure and boiling point data) such that C_L^S and P_L^S were consistent with literature H values. K_{ow} values for the congeners were taken from the recent report of Bruggeman et al (12), averages being calculated for each chlorine number. Values of f(N) were then calculated. The entire data set are given in Table 1.

It must be emphasised that many values are speculative especially those in parentheses. It is not possible to assign confidence limits in detail but it is believed that the following 95% error limit factors may apply to C_L^S and P_L^S with somewhat higher factors applying to H, when

N=0 to 3 value is within a factor of 2.5 of stated value
N=4 to 6 value is within a factor of 5 of stated value
N > 6   value is within a factor of 10 of stated value

Table 1. Literature and Selected Values for Properties of Congeners and Aroclors. Data are for 25°C.

Compound	N	MW	T_B	ϕ	Literature Values (ranges)			Suggested Values				log K_{ow}	f(N)
					C_S^S	P_S^S	H	C_L^S	P_L^S	H			
Biphenyl	0	154.1	529	.35	7.5	1.33	30-66	21.4	3.8	27		4.10	1.03
Mono CBP	1	189	558	0.3-1.0	0.9-5.9	1.1-5.6	58-74	7.2	2.3	60		4.66	1.03
Di CBP	2	223	585	.06-1.0	.06-1.5	.03-.36	97	2.2	0.6	60		5.19	1.18
Tri CBP	3	257	610	.24-.65	.015-.64	.01-.27	82-102	0.67	0.2	77		5.76	1.20
TetraCBP	4	292	633	.1-.68	.0008-.17	.003-.104	75-94	.23	0.06	76		(6.35)*	(1.00)
PentaCBP	5	326	654	.1-.31	.004-.03	.0009-.001	–	.072	.015	68		(6.85)*	(1.15)
HexaCBP	6	361	673	.06-.25	.0004-.01	.0016	–	.021	.005	86		(7.33)*	(1.44)
HeptaCBP	7	395	690	.06	.0005	–	–	.006	(.0015)	(100)		(7.93)*	(1.39)
OctaCBP	8	430	705	.04	.0002-.007	–	–	.002	(.0005)	(100)		(8.55)*	(1.09)
NonaCBP	9	464	718	.016	.0001	–	–	.0007	(.00015)	(100)		(9.14)*	(0.86)
DecaCBP	10	499	729	.002	.00002	–	–	.0002	(.00004)	(100)		(9.60)*	(1.13)
1016	3	257	598-629		.4-.91	.06	.3-1300	(.67)	(0.2)	(77)		4.4-5.8	
1221	1.15	192	548-593		3.5-15	.89	.75	(6.5)	(2.0)	(60)		4.1-4.7	
1232	2.04	221	563-593		1.45	.54	–	(2.0)	(0.55)	(60)		4.5-5.2	
1242	3.10	261	598-639		.13-.70	.054	.03-770	(0.40)	(0.10)	(65)		4.5-5.8	
1248	3.90	288	613-648		.054	.066	–	(.20)	(.06)	(86)		5.8-6.3	
1254	4.96	327	638-663		.01-.07	.01	.01	(.06)	(.015)	(82)		6.1-6.8	
1260	6.30	372	658-693		.003	.005	–	(.017)	(.004)	(88)		6.3-7.5	

1) Units are as follows: Molecular Weight MW g/mol. Boiling Point T_BK. Solubility C_S^S (solid) and C_L^S (liquid) g/m^3. Vapor Pressure P_S^S (solid) and P_L^S (liquid)Pa. Henry's law constant H Pa m^3/mol. The others are dimensionless.
2) In some cases the literature solubilities for congeners are for liquids. The Aroclor values are for liquid mixtures.
3) H is P_L^S MW/C_L^S and f(N) is 1797 MW/C_{Low}^{SK}. Values in parentheses () are speculative, * are calculated

This factor should be applied to K_{ow} not log K_{ow}.

It should be noted that only a single value is given for each chlorine number but differences occur between isomers, thus the value given is an average of the isomer values. It is known for example that K_{ow} values depend on the substitution position (12).

When selecting property values for one N, the values for congeners of (N+1) and (N-1) were used to influence the selection process in order that a fairly smooth dependence on N was obtained.

The Aroclor data are more difficult to analyse because no single property value can be assigned to the mixture. There is a wide variation in the reported values and there are some inconsistencies between the Aroclor values and the corresponding congener values. For example Aroclor 1248 is predominantly tetrachlorobiphenyl (N=4) (40%) with 20% N< 4 and 40% N_3>4 (13). The reported liquid solubility is 0.054 g/m^3 (13) which is a factor of four lower than that of liquid tetrachlorobiphenyl and in the range of the solid isomer values. Reported K_{ow} values range from 5.75 to 6.11 but Bruggeman et al (12) report a higher value of 6.35 for N=4. It appears that the Aroclors display lower-than-expected solubilities and K_{ow} values.

The explanation may be that the effective values measured are for different mixtures. If a large excess of PCB is present in the determination the values obtained reflect those of the lower, more water-soluble congeners, thus in a K_{ow} determination with excess PCB in the octanol, the value obtained for 1248 is essentially that for N=3, i.e. approximately 5.8. If in a solubility determination only a small excess of PCB is present the value obtained may be influenced by the N=5 congeners. For these reasons, the values suggested for Aroclors in Table 1 are regarded as quite speculative and imprecise. Care must be exercised when selecting values since the "correct" value depends on the circumstances being considered.

Discussion

The data in Table 1 suggest that addition of each chlorine causes the following changes in properties.

The boiling point T_B rises by some 29K for the first chlorine; then by increments progressively 2K smaller for subsequent chlorine addition.

The subcooled liquid vapor pressures generally fall by a factor of 2 to 3 per chlorine added.

The subcooled liquid solubilities also fall by a similar factor.

The Henry's law constants remain remarkably constant with chlorine number because of the fairly equal fall in P_L^S and C_L^S. The values obtained range from 20 to 100 Pa m^3/mol. The dimensionless partition coefficient H/RT is thus in the range 0.008 to 0.040. As chlorines are added, both H and H/RT may tend to increase slightly; thus, the higher PCBs tend to partition more into the air.

Log K_{ow} increases consistently by 0.55 units i.e. K_{ow} increases by a factor of 3.5, for each chlorine added, but there are differences depending on position as discussed by Bruggeman et al (11).

It is striking that the H values show no definite trend with N. This has also been observed for PNAs (7). The implication is that PCBs in air and water may have similar congener distributions, there being no marked enhancement of for example low N congeners in the vapor phase.

The K_{ow} values in Table 1 can be used to estimate K_{oc} and bioconcentration factors as discussed earlier.

The f(N) values are remarkably constant showing no trend to increase or decrease, the mean being 1.14. The implication is that for all PCBs the product of K_{ow} and C_L^S (mol/m^3) is 1580, and there is a negligible change in the activity coefficient of PCBs in octanol with N. This could be confirmed by measuring the solubility of PCBs in octanol saturated with water. An interpretation of this observation is that the solubility of liquid PCBs in octanol is constant at 1580 mol/m^3.

It must be emphasised that K_{ow} values of the higher congeners in Table 1 are calculated, not measured and that the calculation implies a near-constant f(N). It is possible, even likely, that f(N) increases as N increases and that the K_{ow} values are lower than indicated.

Using these likely values it is interesting to examine the implications for PCB volatilization. The usual approach for calculating this rate is to use the two resistance model (14).

$$\text{Flux} = K_{OL} (C_W - C_A) \text{ mol/m}^2\text{s}$$
where $1/K_{OL} = 1/K_L + RT/HK_G$

Here K_{OL} is the overall mass transfer coefficient (MTC) (m/s) K_L is the water phase MTC (m/s) K_G is the air phase MTC (m/s) C_W is the water concentration mol/m^3 C_A is the air concentration mol/m^3.

Typically K_L is 3 to 5 cm/h or 10^{-5} m/s while K_G is 1000 to 2000 cm/h or 400x10^{-5} m/s. It follows that the gas and liquid resistances (RT/HK$_G$) and 1/K$_L$ will be equal when RT/H is of the order of 400. Since RT is typically 2500 Pa m^3/mol this corresponds to H of 6 Pa m^3/mol. In practice when H ranges from 1.0 to 10 both resistances are important. If H exceeds 10 liquid phase diffusion dominates and volatilization is relatively "fast". If H is less than 1.0 gas phase diffusion controls and volatilization is relatively "slow".

The H values in Table 1 are consistently greater than 10. It is almost inconceivable that they have values below 10, thus there is a compelling case for concluding that PCB volatilization is liquid phase controlled and should thus be "fast".

Environmental observations do not however support this conclusion, in that to reconcile mass balances a very low value of K_L must be invoked. This issue is discussed in the chapter by Richardson elsewhere in these proceedings. Several explanations are possible for the low volatilization rate

(i) H is in error. If H is of the order of 1.0 $P_a m^3$/ mol there will be an appreciable gas phase resistance which will slow volatilization. It is difficult to accept such a low value.

(ii) K_L is lower than expected. Certainly the low water diffusivity of PCBs retards volatilization but the magnitude of the effect is predictable. It may only account for a factor of 2 in K_L. Another possibility is that K_L is smaller in highly dilute systems. No K_L measurements have been reported for organics at the ng/L concentration range. Surface films may also retard volatilization but again the effect should be relatively small.

(iii) PCBs are appreciably sorbed in solution. If K_p is of the order of 10^6 in a solution of 1 mg/L suspended matter then half the PCB may be sorbed. O'Connor and Connolly (15) have suggested that such values are possible, the issue being discussed elsewhere in these proceedings by DiToro.

(iv) PCBs are close to equilibrium in air and water. It is apparent that average air concentrations are known reliably and are a few ng/m^3 while water concentrations are a few ng/L, i.e. a difference of a factor of 1000. If these are close to equilibrium H/RT must be approximately 0.001 or H must be 2.4. It is difficult to reconcile such a low value with the data in Table 1. It is concluded that a substantial fugacity driving force exists from water to air.

On the basis of this data analysis it is likely that the second and third explanations, alone or in combination, account for the discrepancy, but further study is obviously required.

Conclusions

It is clear that to make progress in understanding the environmental fate of PCBs or indeed any toxic substance in the Great Lakes and elsewhere, it is essential to have reliable physical chemical data, especially for solubility, vapor pressure and octanol-water partition coefficient. It is regrettable that so much doubt remains about these quantities, especially for the higher congeners. By compiling the available data and analysing it using known thermodynamic relationships and empirical correlations a set of consistent and "most likely" values has been obtained. To improve on this set requires further careful measurement of C^S, P^S, H and K_{ow} and the development of a more statistically rigorous procedure for selecting the most likely values.

The values obtained suggest that addition of one chlorine reduces solubility and vapor pressure by approximately a factor of 3 thus the Henry's law constant is little changed, and K_{ow} increases by a similar factor.

This analysis suggests that PCBs should be volatilizing relatively rapidly from the Great Lakes and that the unexpectedly low volatilization rates deduced from environmental monitoring and modeling must be due to the liquid phase transfer coefficient being lower than expected or the PCBs being more sorbed in the water column than expected. It is unlikely that explanations based on H being in error are valid.

Finally, it is hoped that this analysis will stimulate similar approaches for other series of organic contaminants such as the PNAs, chloronaphthalenes, and dioxins.

REFERENCES

1. Yalkowsky, S. H., Ind. Eng. Chem. Fundam., 18 108 (1979).

2. Mackay, D., "Equations for the vapor pressure of organic compounds", manuscript submitted for publication.

3. Yalkowsky, S. H., and Valvani, S. L., J. Chem. Eng. Data, 24, 127 (1979).

4. Mackay, D., Mascarenhas, R., Shiu, W. Y., and Yalkowsky, S. H., Chemosphere, 9, 257 (1980).

5. Billington, J., MASc. Thesis, Dept. of Chemical Engineering and Applied Chemistry, University of Toronto, (1982).

6. Mackay, D., Shiu, W. Y. and Sutherland, R. P., Envir. Sci. and Technol, 13, 333 (1979).

7. Mackay, D. and Shiu, W.Y., J. Phys. Chem. Ref. Data, 10, 1175 (1981).

8. Mackay, D., Bobra, A., Shiu, W. Y. and Yalkowsky, S. H., Chemosphere, 9, 701 (1980).

9. Mackay, D., Envir. Sci. and Technol., 16, 274 (1982).

10. Karickhoff, S. W., Chemosphere, 10, 833 (1981).

11. Mackay, D. and Shiu, W. Y., "A compilation and analysis of the physical chemical properties of PCBs" Unpublished manuscript.

12. Bruggeman, W. A., Van Der Steen, J., Hutzinger, O., J. Chromatog., 238, 335 (1982).

13. National Academy of Sciences "Polychlorinated Biphenyls" Washington, DC (1979).

14. Mackay, D. and Leinonen, P. J., Envir. Sci. and Technol, 9, 1178 (1975).

15. O'Connor, D. J., Connolly, J. P., Water Research, 14, 1517 (1980).

CHAPTER 5

REACTIVITY AND ENVIRONMENTAL
PERSISTENCE OF PCB ISOMERS*

W. Brock Neely
Environmental Sciences Research
Dow Chemical U.S.A.
Midland, MI 48640

INTRODUCTION

The acronym "PCB" was originally associated with the product resulting from the chlorination of the biphenyl molecule. Accordingly, there were many PCB's containing a variety of congeners ranging from the unsubstituted biphenyl to the fully chlorinated homologue. As the scientific community began to analyze environmental samples many of the chlorinated species began to show up in locations far removed from where the original fluid was used [1]. Unfortunately, the mixtures and even the single components that were seen were labeled PCB. In actual fact the only material that should be called PCB is the original item of commerce that was prepared by chlorinating the biphenyl molecule.

This paper will examine the individual components in a typical fluid in an effort to identify the properties that cause certain congeners to accumulate in environmental systems. By examining the properties of the individual isomers instead of the mixture known by the acronym PCB more confidence will be gained in our ability to prevent similar events from occurring in the future. As Jonathan Livingston Seagull said, "...learn nothing and the next world is the same as this one, all the same limitations and lead weights to overcome...". Hopefully we can learn from the large-scale environmental experiment that has taken place through the use of fluids known as PCB.

*The Dow Chemical Company   B-600-002-82

PROPERTIES

The two key physical properties that are required for assessing the movement of chemicals are water solubility and vapor pressure. In the case of the chlorinated biphenyls both of these properties decrease in magnitude with increasing degree of chlorination. As the values become smaller the degree of difficulty in making accurate experimental measurements increases. This has created a situation where the uncertainty in the number reported also increases with the degree of chlorination. Consequently, it becomes impossible to say what is the "true" water solubility and vapor pressure. At the present time a 100% error in the reported number is not uncommon and this becomes one of the variables that must be included in any prediction of environmental fate. Each of these properties will now be presented.

Water Solubility

Mackay et al. [2] have recently summarized the available data. Table I gives the average and the standard deviation for the reported results of a few homologues. Included in this tabulation are some recent experimental numbers [3]. From the water solubility a value for the octanol water partition coefficient may be estimated using the equation of Banerjee et al. [4]. These calculations are reported in Table I.

$$\log K_{ow} = 6.5 - 0.89 \left(\log \frac{S \times 1000}{M}\right) - 0.015 \, (mp) \quad (1)$$

where K_{ow} = octanol-water partition coefficient
S = solubility in mg/L
mp = melting point in °C
M = molecular weight

Vapor Pressure

Most experimentally determined vapor pressure measurements are made at temperatures where pressures of 1 mm Hg or greater are seen. For the chlorinated biphenyls this is much higher than room temperature. Consequently, a method is required to extrapolate the measured vapor pressure to ambient temperature. Such extrapolations usually take the form of the Antoine equation

Table I. Water Solubility, Melting Point and Octanol Water Partition Coefficient for Various Homologues of Chlorinated Biphenyls

Chemical	MW	mp^1 °C	$\log^2 K_{ow}$	Water Solubility mg/L 25°C Mackay[1]	Bailey et al[3]
Biphenyl	154	71	3.98	6.6(1.2)[4]	6.8
2-CBP[5]	189	34	4.72	5.0(0.8)	7.8
4-CBP	189	78	4.34	2.4(1.1)	1.2
2,2'-CBP	223	61	4.96	1.1(0.3)	
4,4'-CBP	223	149	4.77	0.06(0.01)	
2,5,2'-CBP	257	44	5.63	0.44(0.2)	
2',3,4-CBP	257	60	6.1	.078	
2,5,2',5'-CBP	292	87	6.12	0.026(0.02)	0.027-0.046[6]
3,4,3',4'-CBP	292	180	4.3	.087(0.08)	
2,4,2',4'-CBP	292	41	6.44	0.068	0.22(0.08)
2,3,4,5,2'-CBP	326	100	6.38	0.009	

[1]Reported in Reference [2].
[2]Estimated using equation 1.
[3]Reference [3].
[4]Number in parentheses is the standard deviation.
[5]CBP = chlorobiphenyl; the number indicates where on the ring the chlorine is located.
[6]Reference [5].

$$\log P = A - [B/(T + C)] \qquad (2)$$

where A, B, & C are constants
T is the temperature of interest
P is the estimated pressure

For chemicals where such equations have not been developed a different approach is required. Grain [6] has recently summarized the available techniques. The one that will be used here is suitable for chemicals where the pressure is as low as 10^{-7} mm Hg, a range that includes the higher chlorinated biphenyls. The method requires a knowledge of the boiling point which may be estimated [7]. Table II is a tabulation of the results. It should be noted that the experimental values have been corrected for the liquid solid transition that occurs for these materials in extrapolating to 25°C. In other words the reported vapor pressure is the pressure exerted by the solid as opposed to the liquid.

There are a few interesting observations that can be made from the data in these two tables. The spread in vapor pressure between the 2,2' and 4,4' CBP is noteworthy. A similar divergence is noted in the melting point of these two isomers. One hypothesis is that the 2,2' isomer is sterically prevented from assuming a planar configuration. This in turn will prevent close packing and hence the melting point will be lower and the vapor pressure will be greater than what is exhibited by a more planar molecule.

There are two pieces of information in the recent literature that help to substantiate this speculation [3, 5]. Bailey et al [3] experimentally determined that the air/water partition coefficient for 2,4,2',4'-CBP was 2×10^{-2}. Using the data in Table II and the equation of Dilling [10].

$$H = \frac{16.04 \text{ PM}}{TS} \qquad (3)$$

where H = air/water partition coefficient
P = vapor pressure (mm Hg)
M = molecular weight
T = temperature in °K
S = solubility (mg/L)

a vapor pressure of 8.6×10^{-5} mm Hg can be estimated. The second piece of data is the vapor pressure data for the 2,5,2',5'-CBP [5] reported in Table II. Both of these numbers are higher than the calculated value of 2.3×10^{-6} mm Hg. If it is assumed that the calculated value is for

Table II. Vapor Pressure and Boiling Points for Several Homologues of Chlorinated Biphenyl

Chemical	Boiling Point Exp. °K	Boiling Point Calc. °K	Vapor Pressure (mm Hg) Exp. 25°C	Vapor Pressure (mm Hg) Calc.[1]
Biphenyl	528	526	[2]9.5 x 10^{-3}	3.86 x 10^{-3}
2-CBP	547	547	[2]8.4 x 10^{-3}	1.26 x 10^{-3}
4-CBP	564		[2]4.6 x 10^{-3}	
2,2'-CBP		567	[3]1 x 10^{-3}	2.10 x 10^{-4}
4,4'-CBP			[3]1.9 x 10^{-5}	
2,5,2'-CBP		585	[4]9.0 x 10^{-5}	6.58 x 10^{-5}
2',3,4-CBP				
2,5,2',5'-CBP			[4]3.7 x 10^{-5}	
3,4,3',4'-CBP		604		2.30 x 10^{-6}
2,4,2',4'-CBP			[2]8.6 x 10^{-5}	
2,3,4,5,2'-CBP		621		5.76 x 10^{-7}

[1] The calculations do not distinguish between isomers.
[2] Reference [8].
[3] Reference [9].
[4] Reference [5].
[5] Based on data in Reference [3] and calculated as described in text.

the molecule possessing no internal strain (i.e. the planar 3,4,3',4'-CBP conformation) then these observations would support the suggestion that the sterically hindered isomers have greater vapor pressures. What is needed is the experimental vapor pressure for 3,4,3',4'-CBP.

Once these chemicals enter the environment they are subjected to various degradation reactions, such as photolysis and biodegradation. The latter is one of the most important dissipating reactions for PCBs and the microbial data of Furukawa et al. [11] and Wong et al. [12] will be used to quantitate these reactions. Pseudo first order rate constants were derived and are shown in Table III. The rate constants are based on the microbial concentrations of 10^6 cells/mL in the sediment and 10^5 cells/mL in the water column for an eutrophic lake. These values were adopted by the National Academy of Sciences [13] based on the assumptions of Baughman and Lassiter [14].

Table III. Rate Constants for Microbial Degradation of Various Chlorinated Biphenyls

Chemical	Pseudo-First Order Rate Constants[1] for a Lake (1/Day)		Ref.
	Water Column	Sediment	
Biphenyl	0.29	2.98	12
2-CBP	0.17	1.7	12
4-CBP	0.10	1.0	12
2,2'-CBP	1.83×10^{-3}	1.83×10^{-2}	11
2,5,2'-CBP	5.7×10^{-4}	5.7×10^{-3}	11
2,5,2',5'-CBP	2.74×10^{-4}	2.7×10^{-3}	11
3,4,3',4'-CBP	----- Too slow to measure -----		11
2,4,2',4'-CBP	----- Too slow to measure -----		3
2,3,4,5,2'-CBP	1.36×10^{-5}	1.36×10^{-4}	11

[1]First order constants derived by multiplying second order rate constant by concentration of cells in the water column and sediment [14].

Atmospheric photodegradation is another avenue for degradation. In this case the only quantitative data available are those of Appleby [15] who studied the photodecomposition of 2-CBP under simulated sunlight in a 50% relative humidity environment. Correcting his data for the dark reaction a rate constant of 0.53 day^{-1} was estimated. It is known that the higher chlorinated materials are more resistant to photodegradation but experimental rate data is not available [16].

APPLICATION TO LAKE MICHIGAN

The implications of these results to a large body of water such as Lake Michigan can be examined with the aid of Figure 1. The scenario will be to add a mixture of chlorinated biphenyls to the lake and to observe the distribution of the various isomers and homologues in the water and biota at steady state. The mass balance equation for this ecosystem is given by

$$V_w \frac{dC_w}{dt} = k_o - [k_1 A + (k_2 + k_3) K_{sw} M_s] C_w \qquad (4)$$

where V_w = volume of water (mL)
M_s = mass of sediment (g)
K_{sw} = soil water sorption coefficient (mL/g)
A = area (cm^2)
k_o = input to lake (g/day)
k_1 = rate of volatility (cm/day)
k_2 = rate of biodegradation (1/day)
k_3 = rate of burial (1/day)
C_w = weight fraction in water

The burial constant (k_3) measures the speed that material absorbed to the soil particulates becomes buried in the bottom sediments. A value of 2×10^{-4} day^{-1} will be assigned to k_3 [17].

Integration of equation 4 yields

$$C_w = [k_o/(V_w K)][1 - e^{-Kt}] \qquad (5)$$

where $K = \dfrac{k_1}{d} + \dfrac{(k_2 + k_3) K_{sw} M_s}{V_w} \qquad (6)$

d = depth = $\dfrac{V_w}{A}$

In order to use equation 5 for estimating water concentrations the various parameters in equation 6 need to be evaluated.

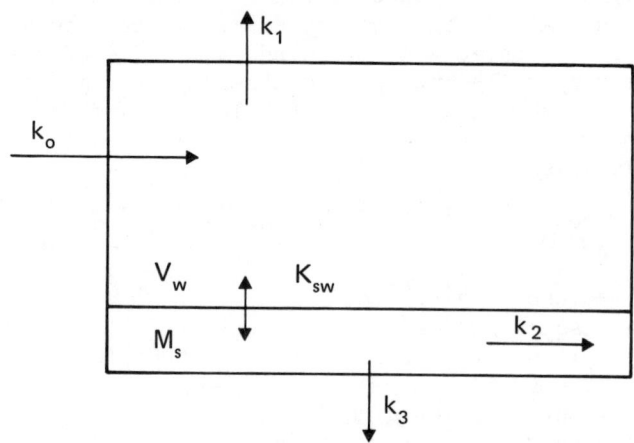

Figure 1. Diagram Illustrating the Transport and Transformation of Chlorinated Biphenyls in Lake Michigan

Lake Michigan

The volume, depth, area and weight of the active sediment layer were taken from a previous publication [18]. These are summarized in Table IV.

Table IV. Physical Parameters Used to Describe Lake Michigan

Parameter	Value
Volume (V_w)	4.9×10^{18} mL
Area (A)	5.7×10^{14} cm^2
Depth (d)	8,515 cm
Sediment Layer (M_s)	2.2×10^{15} gms

k_o (Input)

The input will be equated to a typical commercial PCB mixture. The composition was taken from the National Academy of Sciences [13] and is shown in Table V. The addition of a thousand pounds per day to the lake was simulated and the system allowed to reach steady state. The water concentration is given by

$$C_w = k_o/V_w K \qquad (7)$$

k_1 (Volatility)

The rate constant for loss by volatility is estimated using the theory of Liss and Slater [19] and the equation established by Dilling [10].

Table V. Mixtures of Chlorinated Biphenyls for Simulated Experiment in Lake Michigan (Mixture Equivalent to Aroclor 1242®)

Chemical	% Composition	Input, Pounds (kg/day)
Monochlorobiphenyl		
2-CBP	0.5	5 (2.3)
4-CBP	0.5	5 (2.3)
Dichlorobiphenyl		
2,2'-CBP	8	80 (36.2)
4,4'-CBP	8	80 (36.2)
Trichlorobiphenyl		
2,5,2'-CBP	49	500 (226.5)
Tetrachlorobiphenyl		
2,5,2',5'-CBP	8	80 (36.2)
2,4,2',4'-CBP	8	80 (36.2)
3,4,3',4'-CBP	8	80 (36.2)
Pentachlorobiphenyl		
2,3,4,5,2'-CBP	10	100 (45.3)
	Total Input	1,000 (453.)

$$k \text{ (cm/day)} = \frac{3.18 \times 10^5}{(\frac{1.042}{H} + 100)M^{1/2}} \qquad (8)$$

where H = Henry's Constant (from equation 3)
M = molecular weight

Using the data in Tables I and II, the rate constants were calculated and are shown in Table VI.

Table VI. Derived Parameters to Be Used on the Components of the PCB Mixture Added to Lake Michigan

Chemical	k_1^1 (1/day)	k_2^2 (1/day)	K_{sw}	k_4^3 (1/day)
2-CBP	1.3×10^{-2}	1.7	629	4.9×10^{-1}
4-CBP	2.1×10^{-2}	1.0	262	1.4×10^{-1}
2,2'-CBP	1.2×10^{-2}	1.7×10^{-2}	1094	2.0×10^{-2}
4,4'-CBP	6.9×10^{-3}	1.7×10^{-2}	707	1.2×10^{-2}
2,5,2'-CBP	3.7×10^{-3}	5.7×10^{-3}	5118	1.6×10^{-2}
2,5,2',5'-CBP	1.48×10^{-2}	2.74×10^{-3}	15819	3.5×10^{-2}
2,4,2',4'-CBP	1.4×10^{-2}	ND	33050	1.4×10^{-2}
3,4,3',4'-CBP	8.37×10^{-4}	ND	240	8.57×10^{-4}
2,3,4,5,2'-CBP	2.0×10^{-3}	1.4×10^{-4}	28785	6.39×10^{-3}

[1] Rate constant for evaporation includes the depth
[2] Rate constant for biodegradation in the benthic layer
[3] $k_4 = [k_1 + (k_2 + k_3) K_{sw} M_s/V]$. This is the overall rate constant for the particular chemical in Lake Michigan, where k_3 = burial constant (2×10^{-4} day^{-1}).

k_2 (Rate Constant for Biodegradation)

The numbers developed in Table III are for an eutrophic lake. Since Lake Michigan is oligotrophic it was decided to ignore the small amount of degradation that would take place in the water column. Consequently, the data in Table VI refers to the biodegradation taking place in the sediment.

K_{sw} (Sorption Coefficient)

The sorption coefficient between the chemical and the sediment is given in Table VI. The estimation was generated with the aid of equation 9 developed by Karickhoff <u>et al.</u> [20].

$$K_{sw} = [K_{ow} \times 0.6] \times 0.02 \tag{9}$$

where K_{ow} = octanol/water partition coefficient

0.02 = % organic content assumed to be present in the bottom sediments of Lake Michigan

RESULTS

The results of adding 1,000 pounds/day of the PCB mixture to Lake Michigan are shown in Table VII. The steady state concentration of the chemicals in both the water column and the fish are illustrated. In addition, the half-life, or the time it would take for the water concentration to be reduced by half (once the input had been terminated) is also given.

The fractionation of the original PCB mixture (Table VI) by water and finally by the fish is shown graphically in Figure 2. The phenomena illustrated by the penta series are the phenomena that need to be restricted. These isomers because of their long time constants eventually dominate the final composition. In the example cited the pentachlorobiphenyl represents 10% of the original fluid whereas it is 48% of the chlorinated biphenyl in the fish. All of the lower halogenated homologues show a decreasing percentage of the total composition in moving from the fluid to water to fish. The one exception is the 2,4,2',4'-CBP isomer which also shows an increased concentration in the fish compared to the original fluid. The reason for this build-up is due to the lack of microbial

Table VII. Results of Simulating the Addition of the Mixture of Chlorinated Biphenyls to Lake Michigan

Chemical	Water Concentration[1] ng/L (ppt)	Fish Concentration[2] g/kg (ppb)	Half-Life[3] Days
2-CBP	.0009	.0018	1.4
4-CBP	.0034	.0033	4.9
2,2'-CBP	.036	1.18	34.5
4,4'-CBP	0.62	1.40	57.5
2,5,2'-CBP	2.88	35.07	43.1
2,5,2',5'-CBP	0.21	6.67	19.7
2,4,2',4'-CBP	0.52	30.90	49.2
3,4,3',4'-CBP	8.6	7.75	805
2,3,4,5,2'-CBP	1.45	76.62	108

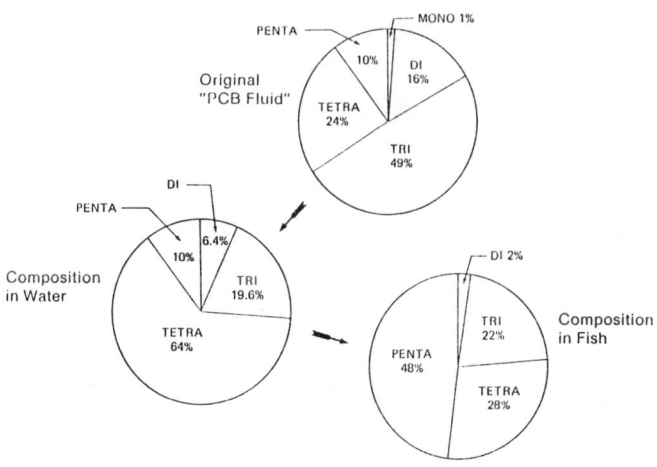

Figure 2. Fractionation of a Hypothetical Commercial PCB as It Moves from Water to Fish

degradation and the high bioconcentration factor associated with the 2,4,2',4'-CBP.

The 3,4,3',4'-CBP is an interesting isomer. Because of the low vapor pressure and the high water solubility (compared to the penta) the rate constant for loss by evaporation is small. When this low rate constant is combined with no observable loss by biodegradation a long residence time and a high water concentration are the result. The low partition coefficient, compared to the other two tetra isomers, prevents the 3,4,3',4'-CBP from building up in the fish.

Unfortunately, the actual experiment described in this simulation could not be found in the literature. However, Stalling [20] in examining the composition of PCB residues in lake trout from Lake Michigan noted that the isomers containing 5, 6, and 7 chlorine atoms dominated the pattern. For example, the ratio of a hexa to a tri isomer in a 1:1 mixture of 1248-1254 Aroclor® was 1 while in the lake trout there was three times as much of the particular hexa isomer. This would tend to substantiate the pattern shown in Figure 2 where the higher chlorinated biphenyls are predicted to be the major components present in the fish.

CONCLUSIONS

1. The criteria associated with the fraction of the PCB mixture that accumulates in fish are as follows:

    a. vapor pressure ~10^{-7} mm Hg

    b. water solubility ~10 µg/L

    c. log K_{ow} ~6.0

    d. molecular weight ~350

    It is interesting to note that these criteria are very close to DDT a classic example of a material that persists and bioaccumulates. For comparison the reported values for DDT are 10^{-7} mm Hg, 1.2 µg/L, 6.5 and 352, respectively.

2. In addition to the above criteria, microbial degradation must be sufficiently slow in order that the chemical can build up in the various compartments.

From the scenario presented for Lake Michigan the criteria for microbial degradation to be considered insignificant would be if the rate constant is the same order of magnitude or slower than the burial constant (2×10^{-4} day^{-1}).

3. The different physical properties exhibited by the isomers that have the ortho position blocked or unblocked, i.e., the 2,4,2',4'-CBP and 3,4,3',4'-CBP parallels the biological differences that have been observed. For example, Stalling et al [22] have demonstrated that the 3,4,3',4'-CBP is much more toxic to rainbow trout than the isomers where the chlorine is on the ortho position. In a similar manner Goldstein [23] established that of the five hexachlorobiphenyl congeners only the 3,4,5,3',4,5' was toxic. The conclusion was that all the other isomers were sterically hindered at the ortho position and were incapable of eliciting the toxic response.

4. The acronym "PCB" should be restricted to describing the commercial mixture resulting from the chlorination of the biphenyl molecule. In this way a great deal of confusion would be eliminated in talking about the problems that have been created by the presence of these materials in environmental samples. Thus if the problem being studied is the pharmacology of 2,4,5,2',4',5 then it should not be referred to as a study of PCB. If it were then it might be concluded that the PCB studied did not exhibit toxic symptoms. This would follow from the previously cited study [23] which demonstrated that this particular hexachloro isomer did not show the classic pattern of PCB poisoning.

REFERENCES

1. Gustafson, C. G., "PCB's -- Prevalent and Persistent". Env. Sci. and Technol., 4, 814 (1970).

2. Mackay, D., Mascarenhas, R. and Shiu, W. Y., "Aqueous Solubility of Polychlorinated Biphenyls". Chemosphere, 9, 257 (1980).

3. Bailey, R. E., Rhinehart, W. L., Gonsior, S. J., Batchelder, T. L, Mendoza, C. G. and Neely, W. B.,

"Hazard Assessment of Monochloro Biphenyl in the Aquatic Environment. A Case History". Presented at the Society of Env. Toxicol., meeting in Washington, D.C. (1981).

4. Banerjee, S., Yalkowski, S. H. and Valvani, S. C., "Water Solubility and Octanol/Water Partition Coefficients of Organics. Limitations of the Solubility Partition Coefficient". Env. Sci. and Technol., 14, 1227 (1980).

5. Westcott, J. W., Simon, C. G. and Bidleman, T. F. "Determination of Polychlorinated Biphenyl Vapor Pressures by a Semimicro Gas Saturation Method". Env. Sci. and Technol., 15, 1375 (1981).

6. Grain, C. F., in "Research and Development Methods for Estimating Physiochemical Properties of Organic Compounds of Environmental Concern". Edited by Lyman, W. J. Published by A. D. Little, Inc., Cambridge, Massachusetts. June, 1981. Chapter 14.

7. Rechsteiner, C. E., Jr., in "Research and Development Methods for Estimating Physiochemical Properties of Organic Compounds of Environmental Concern." Edited by Lyman, W. J. Published by A. D. Little, Inc., Cambridge, Massachusetts. June, 1981. Chapter 12.

8. McDonald, R. A., Thermal Laboratory, The Dow Chemical Company, Midland, MI 48640, unpublished data.

9. Smith, N. K., Gorin, G., Good, W. D. and McCullough, J. P. "The Heats of Combustion, of Sublimation and Formation of Four Dihalo Biphenyls". J. Phys. Chem. 68, 940 (1964).

10. Dilling, W. L., "Interphase Transfer Processes. II. Evaporation Rates of Chloromethanes, Ethanes, Ethylene Propanes and Propylenes from Dilute Aqueous Solutions". Env. Sci. and Technol., 11, 405 (1977).

11. Furukawa, K., Tonomura, K. and Kamibayashi, A. "Effect of Chlorine Substitution on the Biodegradability of Polychlorinated Biphenyls". Appl. Environ. Technol., 35, 223 (1978).

12. Wong, P. T. S. and Kaiser, K. L. E. "Bacterial Degradation of Polychlorinated Biphenyls. II. Rate Studies". Bull. Env. Contam. Toxicol., $\underline{13}$, 249 (1975).

13. National Academy of Sciences. "Polychlorinated Biphenyls" published by the Nat. Acad. of Sci. Washington, D.C. (1979).

14. Baughman, G. L. and Lassiter, R. R. "Prediction of Environmental Pollutant Concentration" in Estimating the Hazard of Chemical Substances to Aquatic Life. Edited by Cairns, J., Dickson, K. L. and Maki, A. W., Am. Soc. for Testing and Materials, Philadelphia, Pennsylvania (1978).

15. Appleby, A. "Atmospheric Freons and Halogenated Compounds". EPA 600/3-76-108, November, 1976.

16. Hendry, D. G. and Kenley, R. A. "Atmospheric Reaction Products of Organic Compounds". EPA 560/12-70-001, June, 1979.

17. Neely, W. B. and Mackay, D. "An Evaluation Model for Estimating Environmental Fate" in Hazard Assessment. Edited by Dickson, K. L., Maki, A. L. and Cairns, J. To be published by Ann Arbor Press, Ann Arbor, Michigan. 1982.

18. Neely, W. B. "A Material Balance Study of Polychlorinated Biphenyls in Lake Michigan". Science of the Total Environ., $\underline{7}$, 117 (1977).

19. Liss, P. S. and Slater, P. G. "Flux of Gases across the Air-Sea Interface". Nature, $\underline{247}$, 181 (1974).

20. Karickhoff, S. W., Brown, D. S., and Scott, T. A. "Sorption of Hydrophobic Pollutants in Natural Sediments". Water Res., $\underline{13}$, 241 (1979).

21. Stalling, D. C. "An Investigation of Chlorinated Hydrocarbon Residues in Fish from Lake Michigan". Second Int. Congress of Pesticide Chem. Tel Aviv, Israel, February (1971).

22. Stalling, D. L., Huckins, J. N., Petty, J. D. Johnson, J. L. and Sanders, H. O. "An Expanded Approach to the Study and Measurement of PCBs and Selected Planar Halogenated Aromatic Environmental Pollutants". Anal. New York Acad. Sciences, 320, 48 (1979).

23. Goldstein, J. A. "The Structure-Activity Relationships of Halogenated Biphenyls as Enzyme Inducers". Anal. New York Acad. Sciences, 320, 164 (1979).

CHAPTER 6
REVERSIBLE AND RESISTANT
COMPONENT MODEL OF HEXACHLORO-
BIPHENYL ADSORPTION-DESORPTION
RESUSPENSION AND DILUTION

Dominic M. Di Toro
  Environmental Engineering and Science Program
  Manhattan College, Bronx, New York 10471

Lewis M. Horzempa
  Envirosphere Corporation
  2 World Trade Center, New York 10048

INTRODUCTION

The purpose of this paper is to summarize the experimental data and theoretical framework for a model of adsorption and desorption of hexachlorobiphenyl from suspended particles. In addition, the results of a series of experiments are presented that test the model predictions of the extent of reversibility and the effects of varying suspended sediment concentrations under conditions that might be encountered in field situations.

In contrast to the desorption reaction, a large body of information already exists for the adsorption reaction. Relationships have been developed that relate the extent of adsorption of organic chemicals to their characteristics such as aqueous solubility (1) and adsorbent properties such as specific surface area (2) and organic carbon content (3). This is not the case, however, for the desorption reaction. The available information indicates that for a great many organic chemical-adsorbent systems the desorption reaction is not completely or even moderately reversible (4-16). As a consequence the assumption of reversible behavior is neither justified nor realistic, and it is not possible to directly apply the large body of adsorption theory and data to describe desorption since, for nonreversible systems, it is not the same reaction.

This is unfortunate since it is not clear how to incorporate nonreversible behavior into modeling frameworks that have been, and are being, developed for the computation of the fate of toxic chemicals in natural waters (17-20).

If adsorption were either completely reversible, or completely irreversible so that no desorption occurred, then it would be straightforward to include such behavior in a fate model. What has been found experimentally, however, is that some desorption takes place. The amount is variable and depends on the details of the situation such as the mass of adsorbent and the adsorbate-adsorbent pair involved.

The use of a desorption "partition coefficient" in a way that is analogous to the use of the adsorption partition coefficient in fate computations is not a solution to the problem since the actual quantity of chemical desorbed when exposed to lower aqueous concentration is not directly related to only the desorption partition coefficient but also to the quantity of chemical previously adsorbed. As it happens, for the desorption model described below, the desorption partition coefficient does have a specific meaning, which the model clarifies, but it cannot be used directly in fate computations. Without a specific model for nonreversible desorption, it is not surprising that this mechanism has not been explicitly included in fate computations.

The nonreversible behavior of adsorption and desorption can have important consequences for the fate of chemicals in natural waters. As inputs of toxic chemicals are reduced, the desorption of already existing toxic chemical from suspended solids and sediments will constitute the major inputs of dissolved toxicants into the water column. The magnitude and extent of this reaction can control the environmental distribution and the exposure level for the biota. If the quantity of chemical desorbed is much less than the quantity initially adsorbed then assuming completely reversible behavior can significantly overestimate the dissolved chemical in the water column. This overestimate may translate into an underestimation of the impact of remedial measures such as discharge reductions via treatment of effluents. Hence, a quantitative understanding of the factors that influence the behavior of the desorption reaction is an essential component for understanding the fate of toxic chemicals in natural waters and the consequences of remedial actions.

SUMMARY

As part of the effort at Manhattan College to formulate and test mathematical models of the fate of PCBs in the Great Lakes (18) a series of experiments has been conducted using tritiated hexachlorobiphenyl (abbreviated as HCBP) as the adsorbate and natural sediments and inorganic clays as the adsorbents. The experiments concentrated on the desorption behavior as well as conventional adsorption tests. Nonreversible desorption occurred and an effort was made to formulate a model which explained the data.

It is assumed that the adsorbed HCBP is made up of two components: a reversible component which readily and reversibly desorbs and readsorbs depending upon aqueous phase concentration, and a second component, which is termed resistant, which resists desorption until exposed to very low (or possibly zero) aqueous concentrations. This idea is often used to explain nonreversible behavior in qualitative terms, e.g. physical adsorption versus chemisorption. Methods were developed for calculating the quantity of the reversible and resistant components from the experimental adsorption and desorption data. This is the unique feature of the model since it gives quantitative estimates of the magnitudes of these components. An analysis of the individual component behavior suggested that each is describable in terms of (distinct) linear isotherms. This regular behavior, for both natural sediments and inorganic clays, is a significant simplification and provides a codification of a large quantity of HCBP adsorption and desorption data in terms of distinct partition coefficients for the reversible and resistant components.

The basic idea is illustrated in Figure 1. A conventional adsorption-desorption data set (assuming three adsorption points and three desorption points for clarity) is shown in Figure 1a. The fact that two distinct isotherms are found for the adsorption and desorption data indicates that the desorption is not completely reversible. Consider a single pair of points corresponding to a single adsorption-desorption experiment, Figure 1b. If it is assumed that continued desorption cycles follow a straight line, then the intersection of this line and the ordinate defines the sediment bound HCBP concentration which is resisting desorption initially and may, in fact, be completely nonexchangeable. This resistant component concentration, r_o, is illustrated in Figure 1b. Once the resistant component concentration, r_o, has been found, the differences between this concentration and that found at adsorption and desorption equilibria must be the reversible component since two components are assumed to be present. Note that two reversible component data points result: at adsorption equilibrium, r_{xa}, and at desorption equilibrium, r_{xd}. These correspond to the two aqueous concentrations c_a and c_d, respectively. Thus, each adsorption-desorption data point

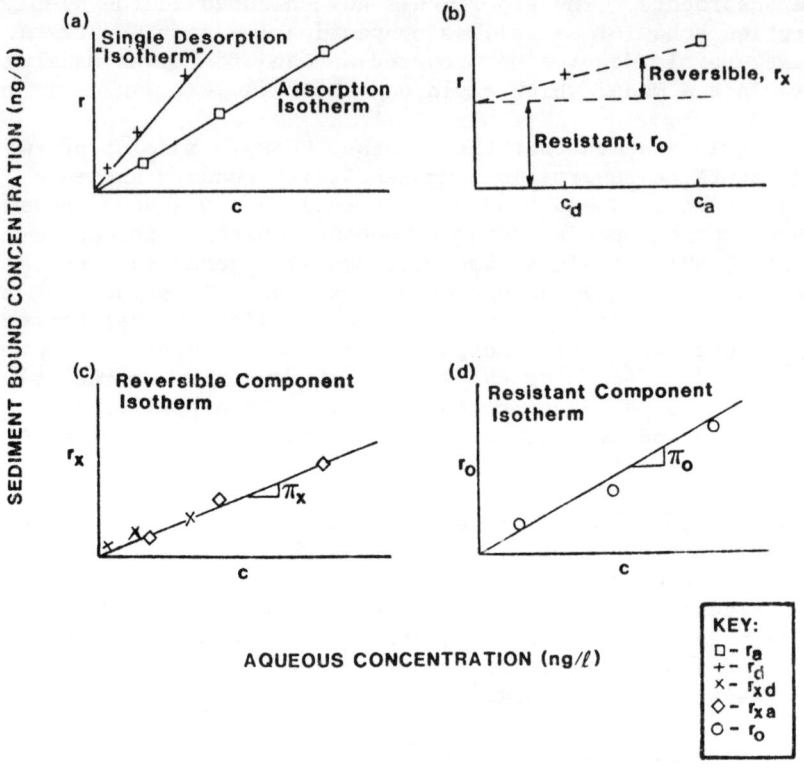

Figure 1. Reversible-Resistant Component Model of Adsorption-Desorption

pair provide an estimate of the resistant component concentration for that experimental condition and two estimates of the reversible component concentrations. If this analysis is repeated for the remaining two adsorption-desorption data pairs in the Figure 1a illustration, then a total of three resistant component and six reversible component concentration data result.

The validity of this analysis depends upon the observation that all the reversible component data conform to a single isotherm, Figure 1c, regardless of whether they correspond to the quantity of reversible component that is present at adsorption, r_{xa}, in equilibrium with aqueous concentration, c_a, or at desorption, r_{xd}, in equilibrium with aqueous concentration, c_d. That is, the reversible component is behaving in accordance with classical reversible adsorption-desorption theory.

The three resistant component concentrations calculated from the data analysis have also been found to follow one isotherm, illustrated in Figure 1d. They have been found to be a linear function of the adsorption aqueous concentration as illustrated.

An actual data set for an HCBP adsorption-desorption isotherm experiment, corresponding to the illustrations in Figure 1. are shown in Figure 2. Additional results of the isotherm analysis for HCBP and a full discussion of the development of the adsorption-desorption model is available (21).

A second focus of the experiments conducted with HCBP was the effect of the concentration of adsorbent on the partition coefficients. It had been observed from an

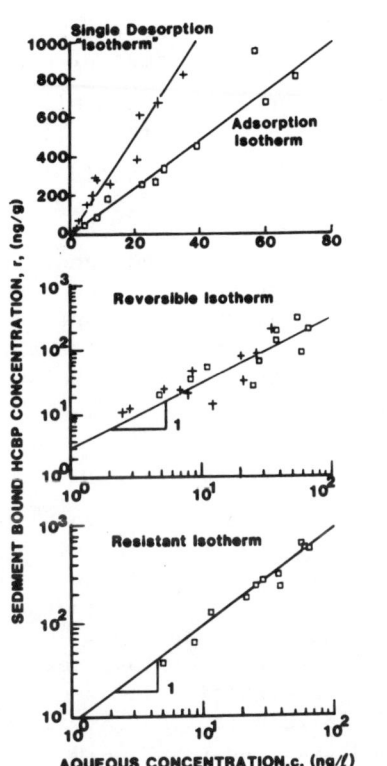

Figure 2. HCBP Isotherm Analysis. Saginaw Bay Station # 50.

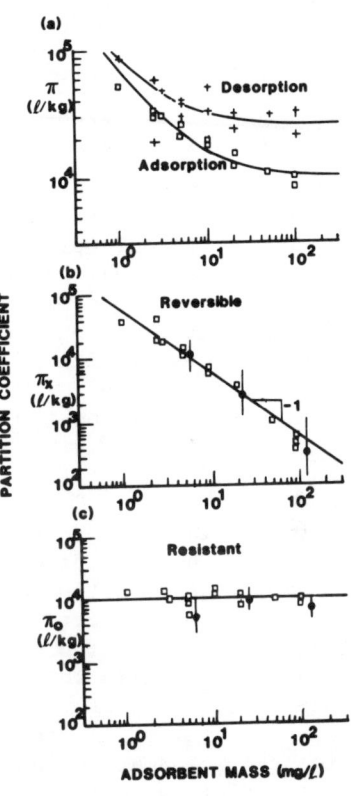

Figure 3. HCBP Partition Coefficient versus Adsorbent Concentration. Saginaw Bay Station #50.

analysis of published data (22) that adsorption partition coefficients decrease as adsorbent concentration increases. This phenomena was investigated for HCBP adsorption and also for desorption. It was found to occur for both reactions. If the data is interpreted in terms of reversible and resistant components, it was found that the resistant component partition coefficient is essentially independent of adsorbent concentration whereas the reversible partition coefficient is inversely proportional to adsorbent concentration. An example is presented in Figure 3. The adsorption and desorption partition coefficients are seen to decrease as adsorbent concentration, m, increases (Figure 3a). Note that the extent of irreversibility increases as the sediment concentration increases. That is, the desorption partition coefficient becomes increasingly larger relative to the adsorption partition coefficient as m increases. The reversible partition coefficient, π_x, is seen to be inversely proportional to m (Figure 3b) whereas the resistant partition coefficient, π_o, is independent of m (Figure 3c). This suggests a definition of a reversible component distribution coefficient: $\nu_x = \pi_x m$ which is itself independent of sediment concentration. The solid lines in Figure 3a are the variations of adsorption and desorption partition coefficients predicted by the reversible-resistant desorption model. The variation of these mass-independent parameters as functions of sediment properties has been examined. The details, together with additional data and a more complete description of the adsorbent mass effects are available (23). Table 1 in the Appendix lists the notations and equations describing the model.

IMPLICATIONS

The formulation and validation of a desorption model are guided by the use for which this model is being developed: namely as a portion of the computation of toxic chemical fate. An important process in such models is the settling and resuspension of the adsorbent particles. Consider the situation in which particles settle into a region near the sediment-water interface. The concentration of adsorbent particles will increase in this region. Since the experimental information indicates that the reversible partition coefficient decreases as sediment concentration increases (Figure 3b), it is possible that some desorption from the particles will take place. This mechanism could liberate PCBs to the dissolved phase and thereby increase the exposure concentration for the biota.

Conversely, decreases in suspended solids concentrations can occur if influents, which carry both PCBs and suspended solids, are dispersed and diluted in the receiving water. For example, a tributary with a high concentration of both PCBs and suspended solids enters a deeper, more quiescent, region of a lake. The mass-dependent partitioning for the reversible components would interact to produce a redistribution of the total quantity of PCB among the adsorbed components and the soluble fraction. Both of these situations are of practical significance and are discussed below.

RESUSPENSION EXPERIMENT

Consider an experiment that is designed to investigate the effect of increased sediment concentration. Following adsorption equilibrium and adsorbent separation by centrifugation, a portion of the contaminated aqueous phase is removed. No new aqueous phase is added. Rather the adsorbent is then resuspended into the previously contaminated aqueous phase and equilibrated. The only change has been to increase the concentration of the adsorbent, since it is now resuspended into only a portion of the aqueous phase volume.

If the reversible-resistant model of desorption is valid, then it is expected that the reversible partition coefficient should decrease since it is inversely proportional to adsorbent mass (Figure 3b). This should cause an increase of dissolved chemical in the aqueous phase. Hence, the first prediction is that the aqueous phase HCBP concentration at resuspension equilibrium should increase.

This is a somewhat surprising prediction since during resuspension equilibrium the only change that has occurred is in the number of particles per unit volume. The aqueous phase is the same as that which resulted from adsorption equilibrium. Thus each particle is exposed to the same aqueous phase concentration as it was previously exposed to, and with which it was in equilibrium. The only difference is that the particle-particle interaction is increased: each particle is exposed to more neighboring particles than at adsorption equilibria. Accordingly, one would expect no change if the particle concentration has no effect on this desorption. But the experiments at varying sediment concentrations indicate that increasing particle concentration does have an effect on the reversible component partition coefficient. Hence the resuspension experiment is a direct check on the validity of the sediment concentration effects and the formulation of the desorption model.

Three experiments were performed: for the first two experiments, the sediment concentration was increased four-fold from the adsorption sediment concentration of (1) m_a = 55 mg/ℓ to the resuspension sediment concentration of m_{rs} = 220 mg/ℓ; and (2) from m_a = 220 mg/ℓ to m_{rs} = 1100 mg/ℓ. For the third experiment a twenty-fold change was examined: (3) from m_a = 55 mg/ℓ to m_{rs} = 1100 mg/ℓ. Saginaw Bay #50 sediment was employed. The results are

TABLE 2. Resuspension Experiment - Saginaw Bay #50

Experiment No.	Adsorption			Resuspension			Dissolved Concentration Ratio
	m_a (mg/ℓ)	c_a (ng/ℓ)	r_a (ng/g)	m_{rs} (mg/ℓ)	c_{rs} (ng/ℓ)	r_{rs} (ng/g)	c_{rs}/c_a
(1)	55	41.8	692.	220	104.	345.	2.49
	55	37.1	647.	220	58.3	575.	1.57
	55	38.1	663.	220	70.9	443.	1.86
(2)	220	22.1	300.	880	32.2	286.	1.46
	220	23.3	269.	880	42.3	298.	1.82
	220	21.5	259.	880	27.5	301.	1.28
(3)	55	29.8	497.	1100	84.8	488.	2.85
	55	29.3	480.	1100	74.4	458.	2.54
	55	27.6	511.	1100	63.6	487.	2.30

shown in Table 2 where each replicate is listed. In all cases the aqueous concentration at resuspension equilibrium, c_{rs}, is larger than at adsorption equilibrium, c_a, in conformity with the rather unexpected model prediction.

The experiment is a clear confirmation of the effect of sediment concentration on partitioning that was observed in the conventional adsorption-desorption experiments at different solids concentrations (Figure 3). What is unexpected is that by simply resuspending the solids into a reduced volume of supernatant aqueous phase but with the same HCBP concentration as at adsorption equilibrium, causes any change at all. The only difference is that the solids concentration has increased. What seems to be occurring is that the particle-particle interactions are increased and that either the quantity of available reversible sites are decreased, causing desorption, or, more likely, that the binding strength of all the reversible sites is decreased by the particle-particle interactions, causing desorption from all these sites.

Note that the effect is more pronounced for adsorption at

(1) $m_a = 55$ mg/ℓ and resuspension at $m_{rs} = 220$ mg/ℓ for which $c_{rs}/c_a \cong 2.0$ then at (2) $m_a = 220$ mg/ℓ and $m_{rs} = 1100$ mg/ℓ for which $c_{rs}/c_a \cong 1.5$. This is due to the fact that the quantity of the reversible component is larger at the lower sediment concentration $m_a = 55$ mg/ℓ than at $m_a = 220$ mg/ℓ. To see this, consider the fraction of adsorbed HCBP that is reversible:

$$\frac{r_x}{r_a} = \frac{\pi_x c_a}{\pi_a c_a} = \frac{\nu_x/m}{\pi_o + \nu_x/m} = \frac{\nu_x}{\nu_x + m\pi_o} \quad (1)$$

so that for $\pi_o \cong 10{,}000$ ℓ/kg and $\nu_x = 0.5$, $r_x/r_a \cong 0.5$ at $m_a = 55$ mg/ℓ, whereas at $m_a = 220$ mg/ℓ, $r_x/r_a \cong 0.2$.

The largest change in dissolved HCBP occurs for the twentyfold increase in sediment concentration (3) $m_a = 55$ mg/ℓ and $m_{rs} = 1100$ mg/ℓ for which $c_{rs}/c_a \cong 2.6$, which is the result of the initially large fraction of reversible component present and the large change in sediment concentration.

The quantitative model predictions are derived as follows. Consider the ratio of the particulate fraction of total HCBP present at adsorption equilibrium:

$$f_p = m_a r_a / c_{Ta} \quad (2)$$

to the dissolved fraction:

$$f_d = c_a / c_T \quad (3)$$

That is:

$$f_p/f_d = m_a r_a / c_a = \nu_x + m_a \pi_o \quad (4)$$

where the second equality is predicted from the reversible-resistant component model (Table 1).

Now consider the effect of resuspension. The aqueous concentration at resuspension equilibrium c_{rs}, is increased, $c_{rs} > c_a$, due to increased sediment concentration $m_{rs} > m_a$.

Hence additional HCBP should adsorb to the resistant and reversible sites. Thus the total HCBP concentration at resuspension equilibrium, c_{Trs}, is given by:

$$c_{Trs} = c_{rs} + m_{rs}r_{rs} = c_{rs} + m_{rs}(r_o + r_x) \quad (5)$$

since r_{rs} is the sum of the reversible and resistant com-components. The resistant component concentration should be given by the resistant component isotherm: $r_o = \pi_o c_{rs}$, since the aqueous concentration increased. The reversible component is always determined by the reversible isotherm: $r_x = \pi_x c_{rs}$. Therefore Equation 5 becomes:

$$c_{Trs} = c_{rs} + m_{rs}(\pi_o + \pi_x)c_{rs} = c_{rs}(1 + \nu_x + m_{rs}\pi_o) \quad (6)$$

where the second equality follows from the reversible partition coefficient mass dependent relationship: $\pi_x = m/\nu_x$. The dissolved fraction is:

$$f_d = \frac{c_{rs}}{c_{Trs}} = \frac{1}{1 + \nu_x + m_{rs}\pi_o} \quad (7)$$

the particulate fraction is $f_p = 1 - f_d$ as before, and their ratio is:

$$f_p/f_d = \nu_x + m_{rs}\pi_o \quad (8)$$

which is the same form as the particulate to dissolved ratio at adsorption equilibrium (Equation 4) except that it is evaluated at the sediment concentration at resuspension, m_{rs}. Figure 4 presents the results for both the adsorption

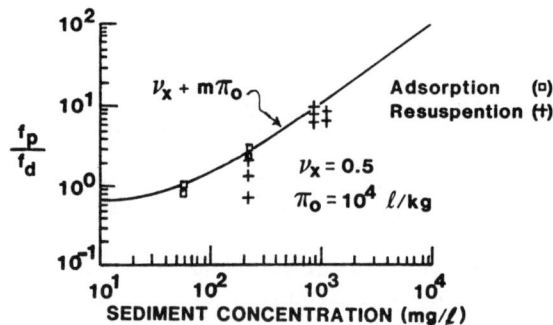

Figure 4. Resuspension Experiment - Model Predictions and Observations

and resuspension ratios, and the resistant-reversible model prediction for this experiment. The ratio of the fraction of particulate to dissolved chemical is predicted to increase as sediment mass is increased. As shown the experimental results are in conformity with the quantitative prediction. The data, although somewhat scattered, is consistent with $\nu_x = 0.5$ and $\pi_o = 10,000$ ℓ/kg, which are in close agreement to the results of the adsorption- desorption experiments at varying sediment concentrations (Figure 3, $\pi_o = 10,000$ ℓ/kg and $\nu_x = 0.624$).

DILUTION EXPERIMENT

A dilution experiment is, in a sense, the reverse of a resuspension experiment. The concentration of sediment is reduced by adding uncontaminated aqueous phase to the vessel after adsorption equilibrium is achieved. This procedure is sometimes used to construct adsorption isotherms without an intervening solids separation step. It has been suggested (15) that a possible cause of the nonsingularity of adsorption and desorption isotherm is the solids separation step, which involves centrifuging and subsequent resuspension into adsorbate-free aqueous phase. This mechanical procedure can conceivably alter the particle aggregations so that the available sites and, therefore, the observed desorption partition coefficient increases.

The dilution isotherm avoids the solids separation step: following adsorption equilibrium, a fraction of the mixed aqueous-solids phase is removed and replaced by adsorbate-adsorbent-free aqueous phase. This step lowers the aqueous phase adsorbate concentration and causes desorption to occur. Unfortunately the dilution step also lowers the solids concentration so that the analysis of this experiment must account for that effect as well. If sediment concentrations affect the partitioning then the dilution "isotherm" is not a direct experimental determination of the desorption isotherm because of the sediment concentration reduction which occurs at dilution. However if the sediment concentration effect has been quantified, then it is possible to account for this effect.

An analysis of the dilution experimental data can be made using the predicted ratio of particulate to dissolved fractions at adsorption and dilution equilibrium in the same way as the resuspension experiment. At adsorption equilibrium, the ratio of the particulate, f_p, to the dissolved fraction: $f_d = c_a/c_{Ta}$, is given by Equation (4). The sediment concentration effect accounts for the difference between this expression and the conventional expression: $f_p/f_d = m_a \pi_a$.

Consider the situation at dilution equilibrium. The total concentration is given by:

$$c_{Td\ell} = c_{d\ell} + m_{d\ell} r_{d\ell} \qquad (9)$$

where $c_{d\ell}$ and $r_{d\ell}$ are the aqueous and particulate concentrations and $m_{d\ell}$ is the sediment concentration at dilution equilibrium. The particulate concentration is made up of the sum of the resistant, $r_{od\ell}$, and reversible, $r_{xd\ell}$, components, that is:

$$r_{d\ell} = r_{od\ell} + r_{xd\ell} \qquad (10)$$

The reversible component has reacted to the lowered aqueous concentration caused by the dilution and it is given by the reversible isotherm:

$$r_{xd\ell} = \pi_x c_{d\ell} \qquad (11)$$

However, assume that the resistant component has not reacted, since it is resisting desorption. Rather it remains at the concentration that is determined by the resistant component isotherm at adsorption equilibrium:

$$r_{od\ell} = r_o = \pi_o c_a \qquad (12)$$

Hence the particulate concentration at dilution equilibrium is, from Equations (10, 11 and 12):

$$r_{d\ell} = \pi_o c_a + \pi_x c_{d\ell} \qquad (13)$$

Using this expression in Equation 9 yields:

$$c_{Td\ell} = c_{d\ell} + m_{d\ell}(\pi_o c_a + \pi_x c_{d\ell}) \qquad (14)$$

and since $m_{d\ell} \pi_x = \nu_x$ from the reversible partition coefficient sediment concentration relationship, the result is that:

$$c_{Td\ell} = c_{d\ell}(1 + \nu_x + m_{d\ell} \pi_o c_a / c_{d\ell}) \qquad (15)$$

and the dissolved fraction at dilution equilibrium is given by:

$$f_d = \frac{c_{d\ell}}{c_{Td\ell}} = \frac{1}{1 + \nu_x + m_{d\ell}\pi_o(c_a/c_{d\ell})} \qquad (16)$$

and finally, the particulate to dissolved fraction ratio is:

$$f_p/f_d = \nu_x + m_{d\ell}\pi_o \, c_a/c_{d\ell} \qquad (17)$$

Note the difference between this formula, which applies at dilution equilibrium, and Equation 4 which applies at adsorption equilibrium. The ratio is larger for dilution equilibrium due to the presence of the factor: $c_a/c_{d\ell}$, which is due to the assumption that the resistant component concentration is determined by the aqueous concentration at adsorption equilibrium: $r_o = \pi_o c_a$ and that it remains constant at dilution equilibrium and does not decrease in response to the decreased aqueous concentration that is: $c_{d\ell} < c_a$, whereas: $r_{od\ell} = r_o$, Equation 12.

The dilution experimental data is compared to the (incorrect) expression for the particulate to dissolved ratio that applies only at adsorption equilibrium:

$$f_p/f_d = \nu_x + m\pi_o \quad \text{(Adsorption)} \qquad (18)$$

in Figure 5. Note that the ratio found for the dilution experiment is significantly larger than that predicted from the adsorption sediment concentration effect alone. This suggests that indeed dilution equilibration is not simply another technique to produce adsorption isotherms even if the sediment concentration effect is taken into account. In fact the nonreversibility persists and the particulate fraction is larger than expected. It is not linearly related to sediment concentration as is predicted by the adsorption equilibrium expression Equation 18.

By contrast, if the nonreversibility is taken into account, then the particulate-dissolved concentration ratio is given by:

(Adsorption and Dilution)

$$f_p/f_d = \nu_x + m\pi_o \, c_a/c_{d\ell} \qquad (19)$$

This expression predicts a straight line if the ratio is

plotted against sediment concentration, m, modified by the ratio of aqueous concentration of adsorption equilibrium, c_a, to aqueous concentration at dilution equilibrium, $c_{d\ell}$. Since the former concentration is not measured in the

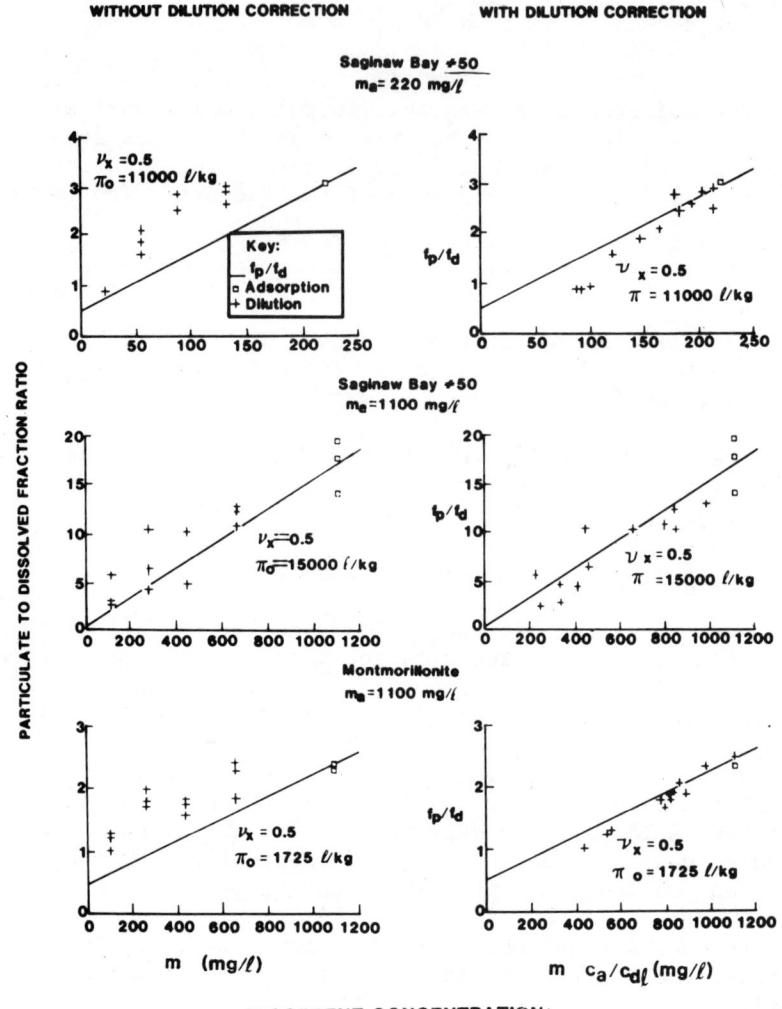

Figure 5. Comparison of Observed f_p/f_d Adsorbent Concentration Relation Without Dilution Correction.

Figure 6. Comparison of Observed f_p/f_d to Adsorbent Concentration, modified by $c_a/c_{d\ell}$.

dilution experiment, it is estimated from the total concentration at adsorption equilibrium, c_{Ta}, which is measured, and the adsorption partition coefficient, π_a, which is available from the parallel adsorption experiment, as described previously. The results are shown in Figure 6 where f_p/f_d are plotted against $mc_a/c_{d\ell}$. The straight line corresponds to Equation 19, the prediction of the reversible-resistant component model. The values of π_o and ν_x (from which $\pi_a = \pi_o + \nu_x/m_a$ is also calculated) are shown in the figure. The value of the reversible distribution coefficient is $\nu_x = 0.5$ in all cases, in conformity with the experiments for which sediment concentration was varied. The resistant component partition coefficient, π_o, is chosen to conform with the parallel adsorption experiment as shown in Figure 6.

Note that the dilution experimental data are in conformity with the prediction except at the largest dilution, corresponding to the lowest aqueous concentrations. This suggests that, although the reversible-resistant model is certainly more successful than the model which accounts for only the sediment concentration variations (compare Figure 5 to Figure 6), it is not fully descriptive of the behavior of the experiments at large dilutions. This failure was also noted at low aqueous concentrations in the analysis of consecutive desorption isotherms, where for m_a = 220 mg/ℓ, it appeared that eventually it may be possible to desorb the resistant component (24). This point is discussed below in more detail.

COMPONENT BEHAVIOR AT DILUTION EQUILIBRIUM

The utility of the division of adsorbed HCBP into reversible and resistant components depends upon the applicability of the isotherms $r_o = \pi_a c_a$ and $mr_x = \nu_x c_x$ to all situations that may be encountered during simulations of mixing phenomena in natural waters. In particular the dilution experiment can be thought of as the experimental analog of the mixing that would occur if a tributary stream carrying a high HCBP and suspended solids concentration enters a relatively uncontaminated, suspended solids free receiving water. The question is: are the isotherms capable of predicting the results. More specifically, is each component behaving as expected. The analysis of the ratio of particulate to dissolved fractions suggest that it is. However a more detailed answer to this question involves a direct measurement of the component concentrations at dilution equilibrium. This can be determined if a desorption step follows the dilution step in the experiment. The final desorption step generates the desorption points (c_d, r_d), which when extrapolated, can be used to calculate the reversible and resistant components at dilution equilibrium in a way that is directly analogous to the method employed in the conventional adsorption-desorption experiment with the exception that the dilution concentrations replace the adsorption concentrations.

That is the resistant particulate concentration is estimated using:

$$r_{od} = \frac{r_d - \beta_d r_{d\ell}}{1 - \beta_d} \qquad (20)$$

where $\beta_d = c_d/c_{d\ell}$, and the reversible partition coefficient is given by:

$$\pi_{xd} = (r_{od} - r_{d\ell})/c_{d\ell} \qquad (21)$$

The analysis of the dilution experiments is based upon the hypothesis that the resistant component concentration achieved at adsorption equilibrium, r_o, shows no change during dilution since the aqueous concentration decreases. By performing the subsequent desorption step an estimate, r_{od}, is made of the resistant component concentration. If it were entirely nonexchangeable it should remain constant as aqueous concentrations decrease. The experimentally estimated resistant concentrations (Equation 20) are shown in Figure 7 for the three dilution experiments. The parallel adsorption-desorption experimental estimates are shown as well. They are plotted against c_a and $c_{d\ell}$ respectively. It can be seen that for the lower initial sediment concentration experiment (Saginaw Bay, m_a = 220 mg/ℓ) the resistant component concentration appears to decrease with decreasing aqueous concentrations in violation of the assumption of complete nonexchangeability. For the higher initial sediment concentrations however, the nonexchangeable hypothesis is supported.

By contrast, the experimentally estimated reversible partition coefficients appear to conform to an inverse relationship to sediment concentration in all cases as shown

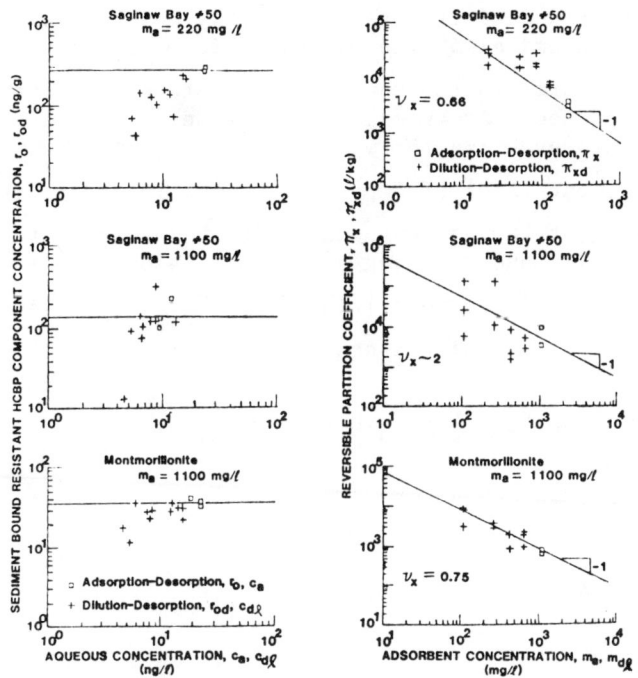

Figure 7. Resistant Component Concentration at Adsorption and Dilution Equilibrium

Figure 8. Reversible Partition Coefficient at Adsorption and Dilution Equilibrium

in Figure 8 although the scatter in the data is substantial. The lines have a slope of minus one in conformity with the relationship: $\pi_x \sim m^{-1}$ and $\nu_x \sim 0.5$ except for the scattered data at m_a = 1100 mg/ℓ for Saginaw Bay. Hence, while the reversible component behavior appears to be in agreement with the model hypothesis, the resistant component appears to be more exchangeable at lower sediment concentrations than is assumed in the model formulation.

APPLICATION TO RECEIVING WATER FATE MODELS

The use of models for the computation of toxic chemicals exposure levels in natural waters is currently an important component of rational toxic chemical regulation and control. These models have a common approach in dealing with the adsorption-desorption reaction. The mass balance equations are written in terms of total chemical, c_T, with the transport and kinetic terms suitably modified with the fraction of chemical in the dissolved form, f_d, depending on whether the terms in the equation apply to particulate or dissolved phases. As an example, consider a two layer segmentation representing the water column of depth H_1, and an active sediment layer of depth H_2. These interact via vertical mixing of the aqueous phases, with mass transfer coefficient K_L; and settling and resuspension of the particulate phases, with velocities w_a and w_{rs} respectively. The governing mass balance equations are (19):

$$H_1 \frac{dc_{T1}}{dt} = K_L(f_{d2}c_{T2} - f_{d1}c_{T1}) - w_a f_{p1} c_{T1}$$
$$+ w_{rs} f_{p2} c_{T2} + W \qquad (22)$$

$$H_2 \frac{dc_{T2}}{dt} = K_L(f_{d1}c_{T1} - f_{d2}c_{T2}) + w_a f_{p1} c_{T1}$$
$$- w_{rs} f_{p2} c_{T2} \qquad (23)$$

where c_{T1} and c_{T2} are the total chemical concentrations in the water column and sediment layers respectively, and W is the input mass loading rate ($M/L^2/T$). Note the central role of the dissolved (f_{d1} and f_{d2}) and particulate (f_{p1} and f_{p2}) fractions, in the water column and sediment segments respectively. They directly affect the magnitudes of the mass transfer coefficients and therefore the fate of the chemical. A more complex fate computation would include terms for outflow, the various appropriate decay mechanisms, and sedimentation losses. However, the principle is still the same. Once the total concentration is computed, the dissolved water column concentration is given by: $c_{d1} = f_{d1} c_{T1}$, with analogous expressions for the particulate concentration. Again the particulate and dissolved fractions play a central role, and these fractions are a direct result of the adsorption-desorption model employed.

For completely reversible adsorption-desorption and a linear isotherm, the dissolved and particulate fractions are given by:

$$f_d = 1/(1 + m\pi) \tag{24}$$

$$f_p = m\pi/(1 + m\pi) \tag{25}$$

where π is the reversible partition coefficient and m is the adsorbent concentration. The subscripts 1 and 2 in Equations 22 and 23 refer to evaluating these fractions using the appropriate adsorbent concentration in segments 1 and 2.

For the HCBP reversible-resistant component model of adsorption-desorption, these fractions depend upon the model parameters: π_o, the partition coefficient for the resistant component; and ν_x, the distribution coefficient for the reversible component; and the maximum dissolved aqueous concentration to which the particle has been exposed: c_{max}. This latter concentration sets the magnitude of the resistant component. It is shown in the previous section that the dissolved and particulate fractions are given by the expressions:

$$f_d = \frac{1}{1 + \nu_x + m\pi_o(c_{max}/c_d)} \tag{26}$$

$$f_p = \frac{\nu_x + m\pi_o(c_{max}/c_d)}{1 + \nu_x + m\pi_o(c_{max}/c_d)} \tag{27}$$

where c_d is the current dissolved aqueous phase concentration. The particulate fraction as a function of adsorbent solids concentration, m, is shown in Figure 9. The conventional expression, assuming reversible behavior is also shown. There is a significant difference between the conventional reversible formulation and the reversible-resistant model. The particulate fraction is always a substantial portion of the total HCBP concentration, even at low suspended solids concentrations that are characteristic of most receiving waters (10-100 mg/ℓ). This suggests that fate computations using the reversible-resistant model will give quite different results which emphasize the importance of particle transport, even at low particle concentrations.

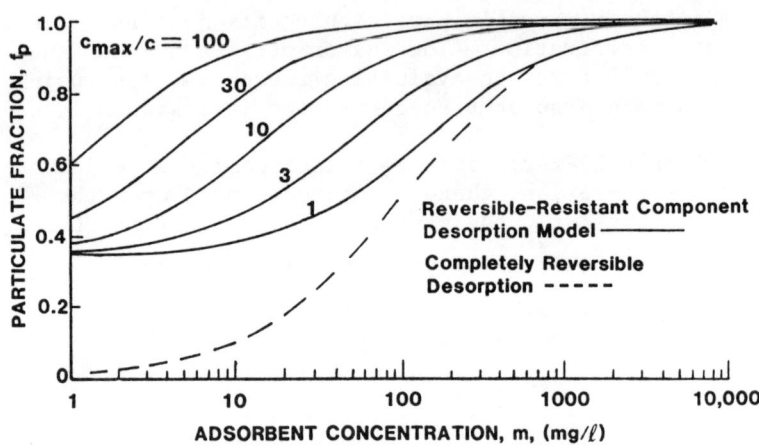

Figure 9. Particulate HCBP fraction versus adsorbent concentration for reversible desorption, Equation (25); and reversible-resistant component model, Equation (27), with $\pi_o = 10^4$ ℓ/kg and $\nu_x = 0.5$.

ACKNOWLEDGEMENT

The authors are pleased to acknowledge the assistance and cooperation of our colleagues: the members of the EPA Large Lakes Research Station, Grosse Ile, Michigan: Nelson Thomas, William Richardson, Michael Mullin and John Filkins, and our group at Manhattan College: John Jeris, Robert Thomann, Donald O'Connor, John Mancini, Maureen Casey, Joanne Guerriero and Margaret Cafarella. The research described in this paper was supported by EPA Grant R805229 and Cooperative Agreement CR807853.

TABLE I - Reversible-Resistant Component Model

1. Notation

c_T = total HCBP concentration (dissolved + sediment bound) (ng/ℓ)
c = dissolved HCBP concentration (ng/ℓ)
r = sediment bound HCBP (ng HCBP/gsediment)
π = r/c = partition coefficient (ℓ/g-sediment)

Table I (continued)

m = concentration of adsorbent (sediment) (mg/ℓ)
f_d = fraction dissolved = c/c_T
f_p = fraction particulate = $1 - f_d$ = mr/c_T

2. Subscripts

   a = adsorption equilibrium
   d = desorption equilibrium
   rs = resuspension equilibrium
   $dℓ$ = dilution equilibrium
   o = resistant component
   x = reversible component

3. Mass Balance Equation

   $c_T = c + mr$

4. Component Definitions

   $r_a = r_o + r_{xa}$
   $r_d = r_o + r_{xd}$

   By linear extrapolation to $c = o$:

   $$r_o = (r_d - \beta r_a)/(1 - \beta)$$

   where:

   $$\beta = c_d/c_a$$

5. Isotherms

   Adsorption $r_a = \pi_a c_a$
   Desorption $r_d = \pi_d c_d$
   Resistant $r_o = \pi_o c_a$
   Reversible $r_x = \pi_x c$

6. Partition Coefficient Relationships

   $$\pi_o = \frac{m \pi_a (\pi_d - \pi_a)}{1 + m(\pi_d - \pi_a)}$$

$$\pi_x = \frac{\pi_a}{1 + m(\pi_d - \pi_a)}$$

where the desorption experiment is such that $c_{Td} = mr_a$ and $m_a = m_d$

7. Adsorbent Concentration Dependence

π_o = constant

$\pi_x = \nu_x/m$; ν_x = constant

8. Reversibility Assumptions

$r_o = \pi_o c_{max}$

where c_{max} is the largest dissolved concentration to which the particles are exposed.

$r_x = \pi_x c$

which responds reversibly to any change in c.

REFERENCES

1. Chiou, C.T., Peters, L.J., Freed, V.H. A Physical Concept of Soil-Water Equilibria for Nonionic Organic Compounds. Science 206(16) p. 831. (1979).

2. Hiraizumi, Y.M., Takahashi, M., Nishimura, H. "Adsorption of Polychlorinated Biphenyl onto Sea Bed Sediment, Marine Plankton, and other Adsorbing Agents." Environ. Sci. Tech., Vol. 13(5), p. 580. (1970).

3. Karickhoff, S.W., Brown, D.S., Scott, T.A. Sorption of Hydrophobic Pollutants on Natural Sediments. Water Research, Vol. 13, p. 241. (1979).

4. Huang, Ju-C, Liao, C. Adsorption of Pesticides by Clay Minerals. J. Sanit. Engr. Div., ASCE, Vol. 96, SA5, pp. 1057-1078. (1970).

5. Pierce, R.H., Jr., Olney, C.E., Felbeck, G.T., Jr. pp'-DDT Adsorption to Suspended Particulate Matter in Sea Water. Geochim. Cosmochim. Acta, Vol. 38, pp. 1061-1073. (1974).

6. Felsot, A., Dahm, P.A. Sorption of Organophosphorus and Carbamate Insecticides by Soil, Agricul. Food Chem., Vol. 27, pp. 557-563. (1979).

7. Wildish, D.J., Metcalfe, C.D., Akagi, H.M., McLeese, D.W. Flux of Aroclor 1254 Between Estuarine Sediments and Water. Bull. Environ. Contam. Toxicol. Vol. 24, pp. 20-26. (1980).

8. Swanson, R.A., Dutt, G.R. Chemical and Physical Processes that Affect Atrazine Distribution in Soil Systems. Soil Sci. Soc. Am. Proc., Vol. 37, pp. 872-876. (1973).

9. van Genuchten, M.Th., Davidson, J.M., Wierenga, P.J. An Evaluation of Kinetic and Equilibrium Equations for the Prediction of Pesticide Movement Through Porous Media. Soil Sci. Soc. Amer. Proc., Vol. 38, pp. 29-35. (1974).

10. Savage, K.E., Wauchoye, R.D. Fluometuron Adsorption-Desorption Equilibria in Soil. Weed Sci., Vol. 22, pp. 106-110. (1974).

11. van Genuchten, M.Th. Wierenga, P.J., O'Connor, G.A. Mass Transfer Studies in Sorbing Porous Media: III. Experimental Evaluation with 2,4,5-T. Soil Sci. Soc. Am. J., Vol. 41, pp. 278-285. (1977).

12. Bowman, B.T. Method of Repeated Additions for Generating Pesticide Adsorption-Desorption Isotherm Data. Can. J. Soil Sci., Vol. 59, pp. 435-437. (Nov. 1979).

13. Bowman, B.T., Sans, W.W. Adsorption of Parathion, Fenitroghion, Methyl Parathion, Aminoparathion and Paraoxon by Na, Ca, and Fe Montmorillonite Suspension. Soil Soc. Am. J., Vol. 41, pp. 514-519. (1977).

14. Koskinen, W.C., O'Connor, G.A., Cheng, H.H. Characterization of Hystersis in

the Desorption of 2,4,5-T in Soils.
Soil Sci. Soc. Am. J., Vol. 43, pp.
871-874. (1979).

15. Rao, P.S.C., Davidson, J.M. Estimation
of Pesticide Retention and Transformation
Parameters Required in Nonpoint Source
Models in <u>Environmental Impact of Nonpoint Source Pollution</u>, ed. M.R. Overcash,
J.M. Davidson. Ann Arbor Science Publishers, Inc., pp. 23-67. (1980).

16. Peck, D.E., Corwin, D.L., Farmer, W.J.
Adsorption-Desorption of Diuron by
Freshwater Sediments. J. Environ.
Qual., Vol. 9, pp. 1010-106. (1980).

17. Baughman, G.L., Lassiter, R.R. Prediction
of Environmental Pollutant Concentration
in <u>Estimating the Hazard of Chemical Substances to Aquatic Life</u>. Cairns, J.,
Dickson, K.L., Maki, A.W. eds. ASTM
STP 657. pp. 35-70. (1978).

18. Thomann, R.V., Di Toro, D.M. Preliminary
Model of Recovery of the Great Lakes
Following Toxic Substances Pollution
Abatement. Workshop on Scientific
Basis for Dealing with Chemical Toxic
Substances in the Great Lakes. Great
Lakes Basin Commission, Ann Arbor,
Mich. (1979). Submitted, J. Great
Lakes Res. (1980).

19. Di Toro, D.M., O'Connor, D.J., Thomann, R.V.,
St. John, J.P. Simplified Model of Partitioning
Chemicals in Lakes and Streams. Proc. of Workshop
"Modeling the Fate of Chemicals in Aquatic Environment". August 1981. Univ. of Mich. Biol. Sta.
Pellston, Mich. In press. Ann Arbor Science.

20. Mackay, D. Finding fugacity feasible. Env. Sci.
& Tech. 13(10), p. 1218. (1979).

21. Di Toro, D.M., Horzempa, L.M. Reversible and
Resistant Components of PCB Adsorption-Desorption: Isotherms. In press. Environ. Sci.
& Tech. (1982).

22. O'Connor, D.J., Connolly, J.P. The Effect of

Concentration of Adsorbing Solids on the Partition Coefficient". Water Research, 14, p. 1517. (1980).

23. Di Toro, D.M., Horzempa, L.M., Casey, M.M., Richardson, W. Reversible and Resistant Components of PCB Adsorption-Desorption: Adsorbent Concentration Effects. In press. J. of Great Lakes Res. (1982).

24. Horzempa, L.M., Di Toro, D.M. The Extent of Reversibility of Polychlorinated Biphenyl Adsorption. In press. Water Research. (1982).

CHAPTER 7

PCBs IN THE LAKE SUPERIOR
ATMOSPHERE 1978-1980

S. J. Eisenreich
  Environmental Engineering Program
  Department of Civil and Mineral Engineering
  University of Minnesota

B. B. Looney
  Environmental Engineering Program
  Department of Civil and Mineral Engineering
  University of Minnesota

G. J. Hollod
  E. I. du Pont de Nemours & Co.
  Environmental Transport Division
  Savannah River Laboratory
  Aiken, SC 29801

INTRODUCTION

Atmospheric concentrations of polychlorinated biphenyls (PCBs) have been measured over Lake Superior in the summers of 1978 to 1980. PCB concentrations are ~ 1.0 ng m^{-3}, exist almost totally in the vapor phase, and are primarily composed of the lower chlorinated isomers (Aroclor 1242, $\sim 70\%$). There is no substantive indication that the atmospheric burden of PCBs is decreasing following voluntary (1971) and legislated bans (1979) on production and usage. PCB concentrations of ~ 1 ng m^{-3} typify mid-continental background levels with higher concentrations resulting from local urban/industrial influences. With atmospheric PCB concentrations remaining elevated, reduction of inputs to Lake Superior is unlikely.

Polychlorinated biphenyls (PCBs) are ubiquitous in the global environment (1-3) with estimates of a total cumulative use of $\sim 6.1 \times 10^8$ Kg as of 1975 in the U.S. alone, of which 6.8×10^7 Kg occurs in a mobile environmental reservoir, 2.5 x 10^8 Kg have been degraded or incinerated, and 1.3×10^8 Kg exist in landfills or equipment dumps (3). This sum of 2.2×10^8 Kg of PCBs represents about 39% of total U.S. sales between 1930 and 1975, with the remainder still in service. PCBs have sufficiently high vapor pressures that volatilization from soil and water, and emissions from the incineration of municipal wastes and industrial products make airborne transport an important environmental pathway (4,5). Atmospheric transport has been implicated in recent studies as the most important pathway for the input of PCBs to the marine environment (6-12). A recent National Academy of Sciences report (3) concludes that the major sinks for PCBs in the U.S. and contiguous areas are freshwater sediments (primarily the Laurentian Great Lakes) and the water and sediments of the North Atlantic Ocean, representing 1.4 to 7.1×10^6 Kg and 6.7 to 69×10^6 Kg, respectively. PCBs have been measured at significant concentrations in the atmosphere of the North Atlantic (6), Bermuda (8), Grand Banks (6), British coastal areas (12), Gulf of Mexico (9), Newfoundland (11) and Enewetak Atoll in the North Pacific Ocean (10). Atlas and Giam (10) and Bidleman et al. (11) have conclusively demonstrated the presence of synthetic organic pollutants in the remote atmosphere, and established their long-range transport to the oceans. Atmospheric PCB concentrations in the marine atmosphere occur in the range of ~ 0.08 to 0.8 ng m^{-3}.

The Laurentian Great Lakes have been contaminated by PCBs since the mid-1950's, with commercial fish concentrations so elevated in one lake, Lake Michigan, that a ban was imposed on their interstate transport, and a warning issued to recreational consumers regarding food intake (13). Atmospheric transport of PCBs to the Great Lakes has been suggested as a significant if not the primary input pathway (14). Based on measurements of PCBs in air (13) and precipitation (16) in the Lake Michigan Basin, the total atmospheric input (wet + dry) has been estimated as ~ 6500 kg yr^{-1} (17), representing greater than 70% of inputs from all sources. Eisenreich et al. (14) have estimated a similar input to Lake Michigan (6900 kg yr^{-1}) using a simplified parameterization.

To obtain better estimates of the magnitude and importance of atmospheric deposition to the Great Lakes, and Lake Superior in particular, and to establish temporal concentration trends, we measured the airborne concentrations of PCBs in the summer atmosphere of Lake Superior in 1978 to 1980 from aboard ship, and in the urban atmosphere of Minneapolis, MN in 1978 and 1979 for comparison purposes. We summarize here these atmospheric PCB concentrations providing the only coherent data set of its kind against which future trends might be based and deposition estimates made. Lake Superior was selected for study because anomalously high levels of PCBs have been observed in fish, precipitation, water (18,19) and sediments (20-22) in several depositional areas and in the central region near Isle Royale. Lake Superior is particularly susceptible to atmospheric contaminant inputs because of its large surface area and surface to basin area ratio, long chemical and water residence times, significant annual hydrologic input derived from precipitation, (65%), and meteorological patterns accompanying eastward-moving low or high pressure systems that transport airborne pollutants from urban/industrial centers to sensitive areas.

The atmospheric sampler, described in detail elsewhere (23), consists of a Universal High Volume Air Sampler modified to include a back-up adsorbent. The air sampler held a 7.8 cm bed of XAD-2 macroreticular resin (600 mL) in an aluminum standpipe atop a motor housing, and following a glass fiber filter. Recoveries of Aroclor 1221, 1242 and 1254 isomeric standards at flow rates of 0.4 m^3 min^{-1} averaged 90-95% in laboratory studies. Atmospheric PCBs are thought to exist primarily as vapor (10,14,15), but operational distribution between filter residue and XAD-2 adsorbent is not sufficient to establish phase dominance, since PCB adsorbed to particles may volatilize during sampling. XAD-2 resins provide good recovery, collection efficiency and breakthrough behavior for airborne PCBs (23-25). The air sampler was affixed to the box of the ship and collected sample only when the ship was underway and the wind was within a vector of 60° off the bow. Air sampling times of 5 to 10 hours and collection volumes of 200 to 400 m^3 were typical. Flow rates of 0.4 to 0.5 m^3 min^{-1} were used. Urban samples were collected atop Shepherd Laboratories ∿ 30 m above ground on the University of Minnesota campus in Minneapolis. After sampling, the XAD-2 resin and filters were soxhlet extracted separately with petroleum ether for 12 to 18 hours, and the extract fractionated by liquid-solid chromatography on alumina (1978) or Florisil. The 1978 air samples were analyzed by packed-column gas chromatography with e^- capture detection on two columns and quantified by the method of Webb and McCall (26) and by comparison to Aroclor standards. The remaining samples were analyzed by glass capillary GC with e^- capture detection, and identified and quantified by a computer-generated reconstruction of Aroclor mixtures using a multiple linear regression program. At least 30 PCB peaks occurring in samples and standards were selected for direct comparison. PCB concentrations and isomeric distributions estimated by the Webb and McCall (26) and linear regression techniques on selected samples generally agreed within 10%.

Total PCB concentrations in the atmosphere over Lake Superior are reported in Table 1. Since the air sampling system does not differentiate between particle and vapor-phase PCB, reported concentrations are operationally defined as total values. However, none of the glass fiber filters contained detectable quantities of PCB. For a sampling volume of ∿ 300 m^3 and a total suspended particle concentration of ∿ 7 μg m^{-3} (27), the mean PCB concentration must have been

Table 1

PCB Concentrations in Air Over Lake Superior and Minneapolis, MN

Year	Mean	Total PCB Range ng m^{-3}	S.D.	n	% Aroclor 1242
		Lake Superior			
1978	1.5	0.9-3.5	0.8	13	79
1979	0.9	0.4-1.4	0.4	8	54
1980[a]	1.0	0.1-2.5	1.0	8	79
1978-1980	1.2	0.1-3.5	0.8	29	72
		Minneapolis, MN			
1978	7.0	1.3-20	5.0	26	75
1979	7.1	4.3-9.1	1.5	10	82
1978-1979	7.1	1.3-20	4.5	36	77

[a]Excludes value of 9.3 ng m^{-3}

less than 2.8 µg g^{-1}. This compares to the PCB concentrations measured in Lake Michigan aerosol of 4 µg g^{-1} (15) and 3 µg g^{-1} (16). Based on these data and those summarized by Eisenreich et al. (14,28,29), we conclude that greater than 90% of the airborne PCBs exist as vapor over Lake Superior. These results are consistent with airborne PCBs collected over Lake Michigan (15), North Pacific Ocean (10) and the North Atlantic Ocean (6), and calculated vapor-aerosol PCB distribution based on saturation vapor pressure and available surface area (30).

The three-year average for total PCBs in the summer air over Lake Superior is 1.2 ± 0.8 ng m^{-3} (n=29) with a range of 0.1 to 3.5 ng m^{-3}. This excludes a high value of 9.3 ng m^{-3} observed in early August, 1980 for a sample obtained along the eastern shore of the lake between Whitefish Bay and Michipicoten Island. The mean PCB concentration decreased by 33% from 1978 to 1979 and 1980 but no statistically valid decrease can be assigned. The PCB isomeric distribution as determined by matching at least 30 peaks of the sample and standard GC chromatograms established the predominance of the more volatile Aroclor 1242 mixture over Aroclor 1254. No indication of the presence of significant amounts of Aroclor 1248 and/or 1260 were evident. The lower chlorinated PCB isomers (mono, di) are not efficiently collected and retained by the XAD-2 resin, and therefore the atmospheric concentrations reported here underestimate the true atmospheric burden. Recent results for Lake Michigan (15) and North Pacific Ocean (10) also favor an Aroclor 1242 dominance. Atlas and Giam (10) believe that this behavior results from the shorter halflives of the heavier PCB isomers, but the greater volatility of the lighter isomers might also be argued as being important.

There are no significant correlations of atmospheric PCB concentrations with location, distance from shore, wind direction or speed, or wave height. However, the lowest concentrations (~ 0.1 ng m^{-3}) were observed following periods of mist or rain. This observation suggests that rain efficiently scavenges PCBs from the atmosphere. Using the average concentration of PCB in precipitation over Lake Superior as 30 ng L^{-1} (14,19) and an air concentration of 1.0 ng m^{-3}, the estimated mean scavenging coefficient is $\sim 3.6 \times 10^4$, which compares favorably with reported field values (14). If the atmospheric burden were in the vapor phase with an overall air-water partition coefficient (H') of 10^{-6} atm m^3 mole^{-1} (17), then the scavenging coefficient (RT/H') would be $\sim 2.5 \times 10^4$. Thus, PCB concentrations in rain/snow can be supported by vapor scavenging. If, however, H' is $\sim 10^{-4}$ atm m^3 mole^{-1} as has been suggested by some (4,14), then rain concentrations must be supported by efficient scavenging of particles.

The spatial distribution of PCBs for 1978 to 1980 is as follows: western basin, 1.7 ± 1.1 ng m^{-3}, central region; 0.8 ± 0.5 ng m^{-3}, eastern basin; 1.1 ± 0.5 ng m^{-3}. The somewhat higher values in the western basin may be due to the proximity to urban areas, but the highest concentration was observed in the eastern end of the lake.

Figure 1 depicts PCB concentrations over Lake Superior as compared to other "remote" environments and urban areas. Although the mean PCB concentration in air has not significantly changed from 1978 to 1980, the lower extreme of observed values decreased each year from 0.9 ng m^{-3} (1978), to 0.4 ng m^{-3} (1979), to 0.1 ng m^{-3} (1980). We are unconvinced this represents an overall decrease in emissions or atmospheric burden, but may more adequately reflect lake conditions of surface temperature and degree of stratification as well as meteorological factors. The values observed for

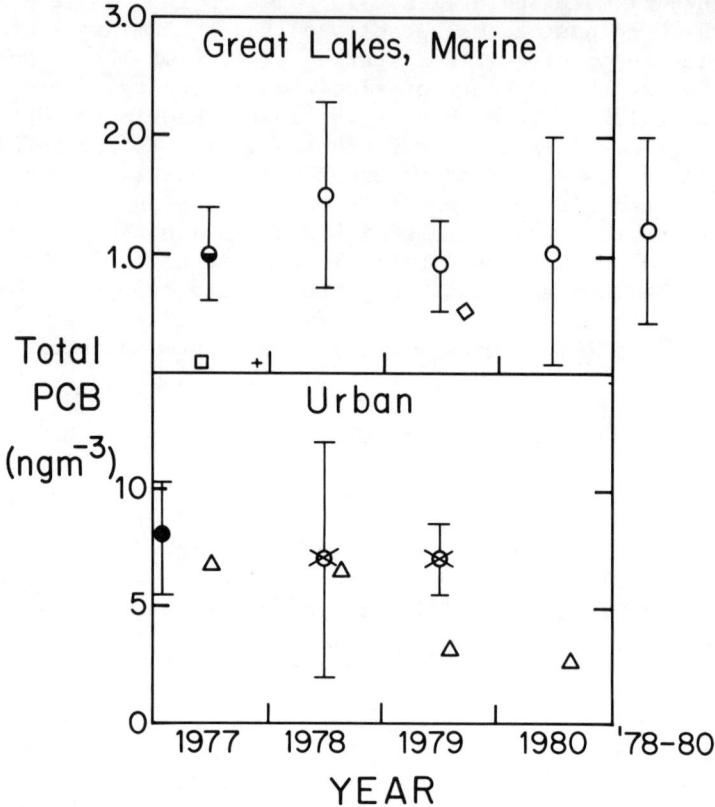

Figure 1. PCB concentrations in air over Lake Superior and Minneapolis, MN compared to other remote and urban areas.

O Lake Superior
● Lake Michigan (15,17)
◇ Enewetak Atoll (10)
□ Newfoundland (11)
⊠ Minneapolis, MN (27)
● Chicago, Ill. (16)
△ Columbia, S.C. (11,25,33)
+ Barbados (11)

Lake Superior are similar to those observed for southern Lake Michigan (15) which is much nearer the urban/industrial centers of Milwaukee, Wis., Chicago, Ill. and Gary, Ind. The over-lake value of ~ 1 ng m^{-3} total PCB probably represents general mid-continental transport from urban centers with higher concentrations due to more localized emissions and transport. In contrast, we observe from Figure 1 that mean PCB concentrations in urban air may vary from 1 to 30 ng m^{-3} with typical values of 5 to 10 ng m^{-3}, which is only 2 to 5 times over lake values. Based on these data and a compilation of the literature (28,29), we can assign PCB concentrations in air and precipitation to different geographical regions (Table 2). In this classification, rural represents non-urban, agricultural or forested areas not adjacent to marine systems, remote signifies continental background or pole sites, and Great Lakes represent specific values available from over-lake measurements. These values have been used to estimate atmospheric deposition to various regions in the U.S. (28).

Table 2. Atmospheric Concentrations of PCBs in Different Regions[a]

Location	Air Range ng m^{-3}	Mean	Precipitation Range ng L^{-1}	Mean
Urban	0.5-30	5-10	10-250	50
Rural	0.1-2	0.8	1-50	20
Great Lakes	0.1-5	1.0	10-150	10-30
Marine	0.05-2	0.1	0.5-10	.5-2
Remote	0.02-0.5	0.1	0.1-10	.2-2

[a]Modified from Eisenreich et al. (1981)

We have concluded that the atmospheric PCB concentration over Lake Superior during summer months is $\sim$ 1 ng m^{-3}, greater than 90% is transported in the vapor phase, $\sim$70% is composed of the Aroclor 1242 isomeric mixture, there is no apparent correlation with location or meteorological factors except rainfall, and are similar to those observed elsewhere in the Great Lakes. The poor correlations may be due in part to the lack of time resolution in the sampling program. Also, lower PCB concentrations in winter months may be due to decreased volatility (31). The atmospheric burden of PCBs is apparently decreasing only slowly, causing concern for the ecological welfare of Lake Superior. Recently, the flux of airborne PCBs across the air/water interface has been modeled, and limitations discussed in detail (17). Using the parameterizations listed in Doskey and Andren (17), a total atmospheric input to Lake Superior is calculated as 5000 to 7000 kg yr^{-1} (32). This represents more than 85% of the PCB inputs from all sources (27,32), and sources exceed sinks by 5000 to 7000 kg yr^{-1}. However, more recent information suggests that atmospheric inputs may be closer to 2000 to 3000 kg yr^{-1}.

The conclusions above could be altered if the air-water partition coefficient (H') for PCB transfer across the air/water interface selected is low by 10^1 to 10^3, and/or if the measured atmospheric PCB concentrations are in part due to aerosolization/volatilization from the lake surface. These possibilities are presently being examined. Based on a water residence time of $\sim$ 180 years, response of the lake to changes in PCB inputs ought to be slow. However, mass balance considerations suggest a chemical residence time for PCBs in Lake Superior of $\sim$ 3 to 5 years (32), with 95% achievement of a new steady-state concentration (if applicable) in 9 to 15 years. This is consistent with the chemical residence time calculated for DDT in Lake Superior of 3.5 years based on the rate of DDT decrease in fish (34).

Acknowledgments

This research was supported in part by the Large Lakes Research Station of the U.S. Environmental Protection Agency (CR806084), an institutional grant to the University of Minnesota from the Sea Grant Program of the National Oceanic and Atmospheric Administration (NOAA)-NA-79-AA-D-00134, NA-80-AA-D-0014, #127 and the Graduate School of the University of Minnesota. We thank the officers and crew of the R. V. Roger R. Simons and the R. V. Crockett for their cooperation and assistance. We also thank T. C. Johnson, A. W. Andren, M. R. Hoffmann, C. P. Rice and T. F. Bidleman for their helpful comments and assistance.

References

1. Ballschmiter, K., M. Zell. Baseline studies of the global pollution. I. Occurrence of organohalogens in pristine European and Antarctic aquatic environments. Intern. J. Environ. Anal. Chem., $\underline{8}$, 15-35 (1980).

2. Nisbet, I. C. T., A. F. Sarofim. Rates and routes of transport of PCBs in the environment. Environ. Health Perspec., $\underline{1}$, 21 (1972).

3. National Academy of Sciences (NAS), Polychlorinated Biphenyls, Washington, D.C., 1979, 182 pp.

4. Mackay, D., P. J. Leinonen. Rate of evaporation of low-solubility contaminants from water bodies to atmosphere. Environ. Sci. Tech., $\underline{9}$, 1178 (1975).

5. Mackay, D., A. W. Wolkoff. Rate of evaporation of low-solubility contaminants from water bodies to atmosphere. Environ. Sci. Tech., $\underline{7}$, 611 (1973).

6. Harvey, G. R., W. G. Steinhauer. Atmospheric transport of PCBs to the North Atlantic. Atmos. Environ., $\underline{8}$, 77 (1974).

7. Bidleman, T. F., C. E. Olney. Chlorinated hydrocarbons in the Sargasso Sea atmosphere and surface water. Science, $\underline{183}$, 516 (1974).

8. Bidleman, T. F., C. P. Rice, C. E. Olney, in Marine Pollutant Transfer, H. L. Windom and R. Duce (eds.), Lexington Books, Lexington, Mass., 1976, pp. 323-351.

9. Giam, C. S., E. L. Atlas, H. S. Chan, G. Neff. Phthalate esters, PCB, and DDT residues in the Gulf of Mexico atmosphere, Atmos. Environ., $\underline{14}$, 65 (1980).

10. Atlas, E. L., G. S. Giam. Global transport of organic pollutants: ambient concentrations in the remote marine atmosphere. Science, $\underline{211}$, 163 (1981).

11. Bidleman, T. F., E. J. Christensen, W. N. Billings, R. Leonard. Atmospheric transport of organochlorines in the North Atlantic gyre. J. Marine Res., $\underline{39}$ (in press).

12. Dawson, R., J. P. Riley. Chlorine-containing pesticides and PCB in British coastal water. Estuar. Coast. Mar. Sci., $\underline{4}$, 55 (1979).

13. Delfino, J. J. Toxic substances in the Great Lakes. Environ. Sci. Tech., 13, 1462 (1979).

14. Eisenreich, S. J., B. B. Looney, J. D. Thornton. Airborne organic contaminants in the Great Lakes ecosystem. Environ. Sci. Tech., 15, 30 (1981).

15. Doskey, P. V., A. W. Andren. Concentrations of airborne PCBs over Lake Michigan. J. Great Lakes Res., 7, 15 (1981).

16. Murphy, T. J., C. P. Rzeszutko. Precipitation inputs of PCBs to Lake Michigan. J. Great Lakes Res., 3, 305 (1977).

17. Doskey, P. V., A. W. Andren. Modeling the flux of atmospheric PCBs across the air/water interface. Environ. Sci. Tech., 15, 705 (1981).

18. Swain, W. R. J. Chlorinated organic residues in fish, water, and precipitation from the vicinity of Isle Royale, Lake Superior. J. Great Lakes Res., 4, 398-407 (1978).

19. Strachan, W. M. J., H. Huneault. PCBs and organochlorine pesticides in Great Lakes precipitation. J. Great Lakes Res., 5, 61-68 (1979).

20. Eisenreich, S. J., G. J. Hollod, T. C. Johnson. Accumulation of PCBs in surficial Lake Superior sediments. Environ. Sci. Tech., 13, 569 (1979).

21. Eisenreich, S. J., G. J. Hollod, T. C. Johnson, J. E. Evans. in Sediment-Contaminant Interactions, R. A. Baker (ed.). Ann Arbor Science Publishers, Ann Arbor, Mich., 1980. pp. 67-93.

22. Frank, R., R. L. Thomas, H. E. Braun, D. L. Gross, T. T. Davies. Organochlorine insecticides and PCB in surficial sediments of Lake Michigan (1975). J. Great Lakes Res., 7, 42-50 (1981).

23. Hollod, G. J., S. J. Eisenreich. Collection of atmospheric PCBs using XAD-2 resins. Anal. Chim. Acta, 124, 31 (1981).

24. Doskey, P. V., A. W. Andren. High volume sampling of airborne PCB with Amberlite XAD-2 resins. Anal. Chim. Acta, 110, 129 (1979).

25. Billings, W. N., T. F. Bidleman. Field comparison of polyurethane foam and Tenax GC resin for high volume air sampling of chlorinated hydrocarbons. Environ. Sci. Tech., 14, 679 (1980); also, T. F. Bidleman, Univ of South Carolina, Columbia; unpublished results.

26. Webb, R. O., A. C. McCall, Jr. Quantitative PCB standards for e capture gas chromatography. J. Chromatog. Sci., 11, 366 (1973).

27. Hollod, G. J., Ph.D. thesis, University of Minnesota, Minneapolis, 1979, 247 pp.

28. Eisenreich, S. J. et al., in Toxic Substances in Atmospheric Deposition: A Review and Assessment. J. N. Galloway, S. J. Eisenreich, B. C. Scott (eds.). NADP NC-141, 1980, pp. 83-113, and EPA Rept.

29. Eisenreich, S. J., B. B. Looney, J. D. Thornton, in Assessment of Airborne Organic Contaminants in the Great Lakes Ecosystem, Science Advisory Board and International Joint Commission, Windsor, Ontario, 1980.

30. Junge, C. E., in Fate of Pollutants in the Air and Water Environment, I. H. Suffet (ed.), Wiley-Interscience, New York, 1977, pp. 7-25.

31. Rice, C. P., B. J. Eadie, K. M. Erstfeld. Enrichment of PCBs in Lake Michigan surface films. J. Great Lakes Res., in press. (1982).

32. Hollod, G. J., S. J. Eisenreich, T. C. Johnson. Sources and sinks of PCBs in Lake Superior, in press, 1982.

33. Bidleman, T. F., E. J. Christensen, H. W. Harder, in Atmospheric Pollutants in Natural Water, Chapter 24, S. J. Eisenreich (ed.), Ann Arbor Science Publishers, Ann Arbor, Mich., 1981, pp. 481-508.

34. Bierman, V. D., W. R. J. Swain. Mass balance modeling of DDT and PCB dynamics in Lakes Michigan and Superior, in press 1982.

CHAPTER 8

PROCESSES DETERMINING THE FLUX
OF PCBS ACROSS AIR/WATER
INTERFACES

Anders W. Andren

University of Wisconsin
Madison, Wisconsin 53706

INTRODUCTION

In evaluating sources of PCBs to humans, it is now well recognized that fish consumption constitutes a primary route [1]. This is especially important in the Great Lakes region where the salmonid fishes regularly exceed FDA PCB concentration guidelines. In order to evaluate how PCBs accumulate in fish, it is instructive to ask various questions regarding the environmental behaviour of these compounds. Questions of importance are: What are the sources? What are the main mode of entry to a lake? How long will these compounds remain in the water column? How fast will they be buried in sediments? Are there any removal mechanisms such as revolatilization and degradation? Are PCB concentrations in various lake compartments (i.e. water column, bottom sediments, biota, air) increasing, decreasing, or remaining constant?

Methods for dealing with these questions are useful not only for PCBs but also for other contaminants. Measurements of concentrations in the various compartments, especially as a function of time, are, of course, necessary to answer many of these questions. However, problems in sampling and analysis, including cost and number of samples that must be collected, would be enormous, especially if long term programs for a multitude of chemicals must be considered. In addition, as in the case of PCBs, assessments usually are made after most of the emissions presumably occurred.

One assessment approach which presently is used is mass balance modeling. Eisenreich et al. [2] have for example, used a simple mass balance approach for Lake Superior to establish that atmospheric input of PCBs is important. More sophisticated mass balance models that include various transport equations have also been developed for Lake Michigan as described later in this book. While these approaches undoubtedly are useful, it is important at the same time to delineate the more detailed aspects of participating transfer processes. In the discussion below, several aspects of our ability to predict PCB transfer across air/water interfaces are presented. Problems associated with sampling PCBs in air have recently been discussed by Doskey and Andren [3] and are not elaborated upon here.

In this paper, models and problems involved in evaluating the flux of PCBs across lake surfaces are discussed. Fluxes to and from Lake Michigan are discussed in terms of: (1) wet deposition via vapor and particulate phase scavenging; and (2) dry deposition of vapor and particulate associated with particles, it is concluded that wet removal of this fraction accounts for 80 to 90% of total atmospheric input to Lake Michigan. Dry deposition of this fraction accounts for the remaining input. Henry's law constants, combined with concentration data, indicate that Lake Michigan should serve as a source of PCBs to the atmosphere. Arguments are presented which indicate that it is presently impossible to estimate emissions from the lake surface to the atmosphere.

AIR/WATER TRANSFER PROCESSES

A schematic of some of the most important processes that must be included in an air/water transfer model is presented in Figure 1. It should also be remembered that the mathematical formulations require, either explicitly or implicitly, data on basic physico-chemical properties for the compound of interest. For PCBs, which were sold as various Aroclor mixtures, such data has often been (and still is) incomplete. This lack of data continues to add uncertainty to flux predictions.

Each of the submodels are briefly discussed below although particle associated PCB deposition is emphasized in light of recent field experiments.

A. Vapor/particle partitioning

There is presently no uniform atmospheric sampling method which unequivocally distinguishes PCBs attached to aerosols from those in pure vapor form (see for example, Harvey and Steinhauer [4], Doskey and Andren [5]; Eisenreich et al. [2]). Using a model originally developed by Brunauer et al. [6] for multimolecular layer adsorption of gases on solids, with the assumption that the chemical nature of the adsorbent surface is invariant, Junge [7] made estimates on the partitioning of several chlorinated hydrocarbons in rural and urban atmospheres. The model expresses the partitioning, ϕ, as a function of particle surface area, θ,

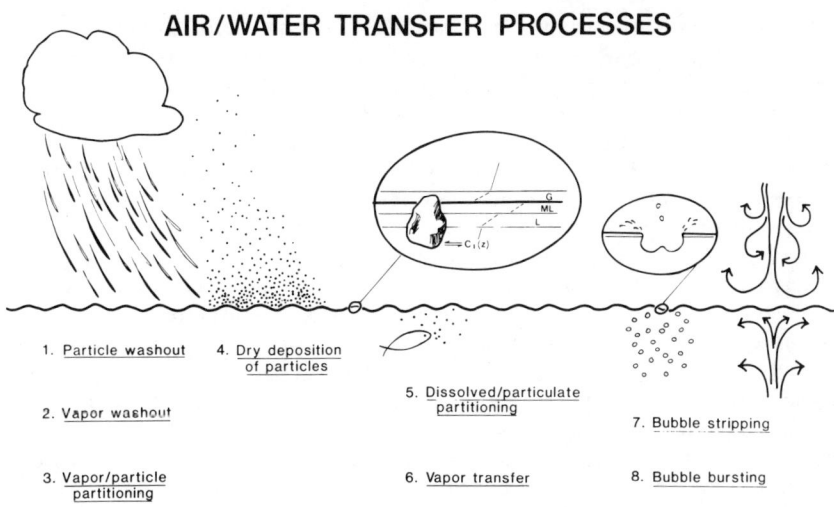

Figure 1. Processes involved in air/water transfer of PCBs.

and the saturation vapor pressure of the compound, P^s, i.e.:

$$\phi = \frac{c\theta}{P^s + c\theta} \qquad (1)$$

where c is a constant which includes the molecular weight of the adsorbate, temperature, gas constant, and an experimentally determined constant, inherent in the slope of the isotherm. Doskey and Andren [3] calculated the range of partitioning for several Aroclor mixtures in rural air (total aerosol concentration 25-50 µg/m^3), the results indicating that 10-45% of Aroclors 1254 and 1260 and less than 5% of Aroclors 1242 and 1248 should be associated with particulate matter. The model seems to give, at least qualitatively, reasonable predictions since measured values, reported in the literature [2,8], indicate that greater than 90% of atmospheric PCBs exist in the vapor.

B. Vapor Washout

Excellent reviews on the modeling of wet deposition have recently appeared in the literature [9,10] and need not be repeated here. The semi-empirical expression:

$$F = rJP = \frac{1}{H} JP \qquad (2)$$

is most often used. Here F = flux of PCB, r = washout coefficient, J = amount of precipitation, P = average partial presssure of PCBs at ground level, and H = Henry's law coefficient.

It is now becoming increasingly clear that H for individual PCB congeners seem to average around $10^{-4} - 10^{-5}$ atm-m^3-mole^{-1} (Murphy, T., this book). According to Mackay and Leinonen [11] this would make these compounds liquid-phase controlled. Doskey and Andren [3] calculated H indirectly, based on available vapor and precipitation data in air over Lake Michigan. Such a calculation is, however, inappropriate unless the amount of scavenged PCB vapor relative to particle associated PCB is known in the precipiation. A combination of recently measured H values - $1.65 \times 10^{-4} - 11.1 \times 10^{-4}$ atm-m^3-mole^{-1} (Murphy, T., this book) - and average atmospheric PCB vapor concentrations over Lake Michigan - about 1×10^{-9} g/m^3 [8] indicate that vapor phase PCB scavenging by rain should only amount to at most 5 kg/yr (precipitation rate = 0.75 m/yr). Thus, if it is correct to assume that PCB vapor rapidly equilibrates with a falling raindrop and that this partitioning obeys Henry's law, it must be concluded that PCB vapor scavenging by rain is not important over Lake Michigan.

C. Particle associated washout

A formulation similar to Eq. (2) is most often used to model particle washout. In this instance, r, the washout or scavenging coefficient, is a function of the particle-raindrop collision efficiency, rain-cloud height, and raindrop radius. Gatz [12] and NAS [1], have compiled r values for various elements and aerosol mass median diameters. If it is assumed that PCB particle size functions most closely resemble those of suspended particulate organic carbon [13], then a mass median diameter of 0.5 um produces $r \simeq 63,000$ [1]. The average particulate associated PCB concentration was found to be 0.13×10^{-9} g/m^3 [8]. This amount, together with J = 0.75 m/yr, produces an average PCB concentration in rain of about 10×10^{-9} g/L and a flux to Lake Michigan of about 650 kg/yr. Particulate PCB data presented by Doskey and Andren [8] indicate a range of 270 - 510 kg/yr could be expected. These calculations indicate that PCBs in precipitation are mainly due to the scavenging of particle associated PCBs.

D. Dry deposition of particle associated PCBs

A uniformly accepted technique for measuring particle fluxes to aqueous surfaces is presently not available [1,14]. Several investigators have used coated plates to measure the flux [15,16]. However, it is likely that these fluxes should only be used in a comparative sense since aerodynamic and hydrodynamic conditions of the collection plates and lake surfaces are quite different. Several modeling approaches now are available to estimate particle mass transfer of aerosols to water bodies [17,18]. Doskey and Andren [3] used a simplified version of a relationship originally proposed by Chamberlain [19], i.e.

$$F = V_D \bar{C}_z \qquad (3)$$

where F = flux, V_D = deposition velocity, and $\bar{C}_z$ = an average particle concentration at a reference height z. Particle associated PCB deposition velocities were chosen from Gatz [12] and Sehmel and Sutter [20] to be 0.005 m/sec. This formulation resulted in a calculated flux of particle associated PCBs of 1200 kg/yr for Lake Michigan. Available values for V_D represent time averaged measurements for a wide range of meteorological conditions and depositional surfaces. It is entirely possible that such calculations represent too much of an oversimplification and that more detailed analysis are required. A modified version of the model of Sehmel and Hodgson [18] has been evaluated for field conditions with the use of Pb-210 as a natural tracer [21]. Results are encouraging in that the calculated dry deposition of Pb-210, associated with particles, to Crystal Lake, Wisconsin agreed within ± 20% of that measured from sedimentation rates (the lake has no river input.)

This model was also used by Andren and Strand [13] to calculate atmospheric particulate deposition of organic carbon and polyaromatic hydrocarbons to Lake Michigan. Particle deposition is described by a one-dimensional steady-state continuity equation:

$$F = -(E+D)\partial C/\partial Z - V_t C \qquad (4)$$

where F = flux (g cm^{-2} sec^{-1})
E = particle eddy diffusivity (cm^2 sec^{-1})
D = Brownian diffusivity (cm^2 sec^{-1})
Z = height of measurement (cm)
V_t = terminal settling velocity of particle (cm/sec)

The particle flux is obtained from an integrated form of eq. (4):

$$F = \frac{V_t C_z \alpha}{1-\alpha} \qquad (5)$$

where: α = exp $(-V_t \text{Int}/U_*)$
U_* = friction velocity (cm sec^{-1})

Int. is the resistance integral which can be divided into two parts: Int_1 and Int_2, representing turbulent and surface resistances, respectively. It may be evaluated from the following expression:

$$\text{Int} = \text{Int}_1 + \text{Int}_2 = -U_* \int_{z_2}^{z_1} \frac{dz}{E} - \frac{U_*}{V_t} \ln \left(\frac{1}{1 - V_t/K} \right) \quad (6)$$

where K = deposition velocity at a specified height. Int_1 has been evaluated for different Obukhov's lengths (+500 to -500) for heights of 5m and 10m. Results are presented in Table 1. Int_2 has been evaluated by Sehmel and Hodgson [18] for z = 1cm via wind tunnel measurements, i.e.:

Table 1. Turbulent Resistances for Different Stabilities[a]

Obukhov's Length (cm)		Resistance Integral z = 5m	z = 10m
500	Stable	-33	-55
1000	Stable	-24	-35
2000	Stable	-20.5	-26
5000	Stable	-18.88	-22
	Neutral	-17.6	-20
-5000	Unstable	-17.0	-18.5
-2000	Unstable	-16.3	-17.5
-1000	Unstable	-15.5	-16.8
- 500	Unstable	-14.7	-15.5

[a] Calculated according to the method of Sehmel and Hodgson [18].

$$\text{Int}_2 = -\exp[-23.667 + 5.555 (\ln d/Z_o) - 0.07681 (\ln d/Z_o)^2 \\ + 0.9722 (\ln U_*/V_t) + 0.03799 (\ln U_*/V_t)^2 - 2.254 (\ln D/U_*Z_o) \\ -3.724 (\ln P_p U_* d^2/18 u Z_o) \quad (7)$$

where d = particle diameter (cm)
Z_o = roughness length (cm)
P_p = particle density (g cm^{-3})
u = kinematic viscosity of air (1.8 x 10^{-4} g cm^{-1} sec^{-1})

U_* has been found to be a function of the windspeed at 10m for oceanic conditions[22] so that:

$$U_* = 7.1 \times 10^{-3} \times U^{1.25} \quad (8)$$

Values for the roughness length, Z_o, were calculated according to eq. (9) which has been developed by Wesely et al. [23] for open water.

$$Z_o = 0.016 U_*/g + u/9.1 U_* \quad (9)$$

where g = acceleration due to gravity.

For the purposes of calculating flux of PCBs to Lake Michigan, a similar meteorological scheme to that used by Andren and Strand [13] was used. Calculated average U_* and Z_o for four different wind regimes are presented in Table II. Surface resistance integrals for the same wind-speed conditions are presented in Table III. Friction velocities and roughness lengths were taken from Table II. A mass median particle diameter, d, of 0.5 μm was used. Particle terminal settling velocity, V_t, was calculatd from Stoke's Law using the Cunningham slip correction and a particle density, P_p, of 1.5 g/cm³. Brownian diffusivity, D_f for a 0.5 um particle at 25°C is about 6×10^{-7} cm² sec⁻¹.

Table II. Calculated Average U_* and Z_o for four Wind-Speed Ranges Over Open Water.

Wind-Speed Range, U_{10} (m/sec)	% Occurrence in a Year[a]	Av. U_* (cm/sec)	Av. Z_o (cm)
5	40	7	0.001
7.5	40	28	0.013
12.5	15	53	0.046
20	5	95	0.147

[a] Cumulative frequency of wind-speed occurrence in Lake Michigan taken from Sievering et al. [24].

An average particulate associated PCB concentration of 0.13×10^{-9} g/m³ (range $0.10 - 0.19 \times 10^{-9}$ g/m³, N = 7) was used [8]. The average composition was 69% Aroclor 1242, 23% Aroclor 1254, and 8% Aroclor 1260. The expected annual average flux during the four windspeed conditions is presented in Table IV. The range was calculated from the range in measured particulate PCB only, i.e. not from other extreme ranges in meteorological conditions. The values thus represent estimates based from a total of about 3 weeks particulate PCB sampling over the lake and meteorological parameters taken from shore based stations [24].

Table III. Surface Resistance Integrals for Four Different Average Wind Speeds.

Wind Speed (m/sec)	Surface Integral (-)
5	7762
7.5	5639
12.5	3453
20	1815

A greater variation in atmospheric PCB concentrations undoubtedly occurs (Cliff Rice, personal communication). Available data indicate that the above estimate may be somewhat conservative although independent calculations from PCB accumulations in sediments (see Armstrong and Swackhamer, this book) suggest that these values are reasonable (assuming no wet input via vapor exchange-see below). The wet to dry input ratio of about 8:1 or 9:1 is also reasonable based on the Pb-210 data from Crystal Lake, Wis. [21].

Table IV. Calculated Average Particle Associated PCB Dry Deposition to Lake Michigan.

Wind Speed (m/sec)	% Time	Deposition (kg/yr)
5	40	2
7.5	40	5
12.5	15	6
20	5	20
	Total	33 (Range 25-50)

E. Vapor Phase Transfer

A detailed analysis of problems involved in modeling PCB vapor transfer across the air/water interface has recently been published [3]. They concluded that questions still remain as to direction and amount of transfer. With the aid of the stagnant film model, H calculated from solubility and vapor pressure data [25], and measured air and water concentrations from Lake Michigan, indications are that most of the resistance lie in the liquid phase. The direction of flux using such an analysis indicate that the lake serves as a source for the compounds to the atmosphere. Directly determined transfer coefficients, however, indicate that the direction of transfer may be in the opposite direction [23, 26].

As mentioned earlier, recent measurements of PCB solubilities and vapor pressures indicate that H for individual congeners range from about 10^{-4} to 10^{-5} atm-m^3-mol^{-1}. This magnitude of the H values indicate liquid phase control ranging from about 50% to 90%. A strict steady state analysis using either a stagnant film model [27,28], penetration theory model [29], surface renewal model [30], large eddy model [31], and eddy cell model [32], indicate that the direction of transfer should be from lake to atmosphere. However, several unresolved questions must be answered before the magnitude of this flux is determined.

The most important considerations are: (1) What is the concentration of PCBs in true solution? An operational definition of dissolved and particulate phases is always used. Henry's law may not be applied to PCBs that are associated with micelles, colloids, or particles; (2) What is the influence of surface organic microlayers(SOML)? It is not entirely clear whether it is appropriate to use a Henry's Law constant determined from pure air/water. Dissolved organic matter in the SOML may modify H; (3) Is it appropriate to assume steady state conditions? Rapid fluctuations in meteorological and hydrodynamic conditions during certain times may void this assumption; (4) What is the effect of adsorption/desorption processes (also should these processes be considered for the SOML or bulk water)? Recent data indicate that rate constants for these processes may have to be included in air/water transfer models (DiToro, this book). In addition, PCB removal from the air/water interface by sinking particles (due to irreversible adsorption) must also be included; (5) Is thorough mixing of the bulk liquid (or air) present? Indications are that this is not the case-at least for oxygen [33]. More sophisticated models, which incorporate additional hydrodynamic (and aerodynamic) parameters are desirable. Most present models, however, suffer from the basic limitation that they contain one or more arbitrary parameters which can not be specified a priori (stagnant film layer thickness, accommodation coefficient, surface renewal time constant, and fluid parcel contact time); (6) Are molecular diffusivities concentration independent? These constants are used in most models (f.ex. $k_1 = D/Z$ in stagnant film model, $k_1 = 2(D/\pi\theta)^{\frac{1}{2}}$ in penetration model, $k_1 = (Ds)^{\frac{1}{2}}$ in surface renewal model). Total Aroclor concentrations are on the order of 10^{-9} g/L (i.e. picomolar). Individual congeners thus have concentrations about 1/20-1/100 of the total. Most molecular diffusivites have been determined using higher concentrations [34] and Erdey-Gruz [35] argues that D is concentration dependent; (7) Does biological activity at the air/water interface exert any influence? Intense biological activity is often associated with this region [36]. Biological incorporation is thus a distinct possiblity; (8) How well do available models reflect transfer during storms? It is entirely possible that much of the transfer may be event related; and (9) Is it correct to model PCBs as Aroclor mixtures or should individual congeners be considered? Since Arolcors may contain about 60 congeners, future efforts should include the latter approach . The stagnant film model should thus be of the form:

$$\sum_{n=1}^{m} F_n = \sum_{n=1}^{m} K_{OL(n)} \ (Cn - Pn/Hn) \qquad (10)$$

where n = type of isomer and m = total number of isomers.

F. Bubble Stripping

Ample evidence that air bubbles are produced in natural water now exists [36,37,38]. Blanchard and Duncan [37] stated that bubbles in fresh waters tend to coalesce. These bubbles may strip hydrophobic substances from the water column and must be considered in any modelling effort. Matter-Muller et al. [39] have recently developed bubble stripping models for aeration systems. These models should be applicable to natural water systems. If upon bursting, PCB vapor in the bubble is in equilibrium with the liquid phase, the stripping process may be modeled using eq (11).

$$F = Q_g HC \left[1 - \exp\left(\frac{K_{ol} \cdot a \cdot V_L}{H \cdot Qg} \right) \right] \qquad (11)$$

where F = flux of PCB via bubble stripping
Q_g = amount of air carried by bubbles
a = bubble interfacial area per unit volume of liquid
V_L = total volume of liquid in bubble producing water layer
C = PCB concentration in water column
H = Henry's law constant
K_{OL} = overall mass transfer coefficient

This model must be evaluated using realistic values for bubble associated air transfer. Much remains to be learned about bubble size spectra and densities in natural waters.

G. Bubble bursting

Rising bubbles skim off the SOML and subsequently burst. Doskey and Andren [3] considered this transfer process for freshwater systems and concluded that an accurate estimate is presently not feasible. Much of the ejected material may redeposit onto the lake making it very difficult to evaluate the net input of particulate associated PCBs to the lake.

CONCLUSIONS

Problems involved in modeling the transfer of PCBs have been reviewed with particular emphasis on the atmospheric removal of the particle associated fraction. An analysis of the various processes involved in air/water transfer indicates that the following phenomena should be considered:(1) particle/vapor partitioning in the atmosphere; (2) particle washout; (3) vapor washout; (4) dry deposition of particles; (5) aqueous dissolved/particulate partitioning; (6) vapor transfer; (7) bubble stripping of vapor; and (8) bubble bursting. In addition, PCBs must be studied in the context of single congeners if a better understanding of environmental fate is desired.

Both theoretical calculations and field data support the notion that greater than 90% of atmospheric PCBs exist in the vapor phase. Sampling difficulties prevent the determination of the exact phase partitioning.

Based on Henry's law constants, derived from individual congener solubility and vapor pressure data, it is concluded that rain (and snow) scavenging of vapor phase PCBs from the atmosphere is negligible.The estimated input to Lake Michigan is at most a few kilograms per year.

Washout of PCBs associated with particles appears to be the most important atmospheric removal process. Model calculations indicate that observed PCB concentrations in rain may be due to this process. Annual input of PCBs via particulate washout is estimated to range from about 300 kg to 500 kg. This range is considerably lower than previous estimates.

The dry deposition of particle associated PCBs was analyzed with the aid of a model that previously had been tested with Pb-210 tracer for a recharge lake in northern Wisconsin. Results indicate a deposition of 25 kg/yr to 50 kg/yr. It is emphasized that limited temporal and spatial data sets on particulate associated PCBs over Lake Michigan are available. It is likely that the above estimate will change as future measurements are made. However, previous estimates seem too high and the present estimate should be considerably better than order of magnitude.

Several processes were discussed which explain difficulties involved in modeling PCB vapor transfer. An analysis based strictly on physico-chemical properties of individual congeners indicate that PCBs are mainly liquid-phase controlled. This fact, combined with available measurements of air and water concentrations, indicate that Lake Michigan should serve as a net source of PCBs to the atmosphere. The magnitude of this is presently difficult to ascertain. Additional unknown lake sources include bubble stripping and bubble bursting.

The gross input of PCBs to Lake Michigan is thus calculated to range from about 300 to 550 kg/yr (wet/dry ratio 8:1). This amounts to a deposition of 0.54×10^{-3} to 1.0×10^{-3} ug-cm^{-2}-yr^{-1}. From mass sedimentation data discussed by Armstrong and Swackhamner (this book) this would result in Lake Michigan bottom sediment PCB concentrations ranging from 0.08 - 0.14 µg/g. The above authors present surface sediment concentrations in the range of 0.2 - 0.09 ug/g. Similar ranges have been observed by Eisenreich et al. [23] for Lake Superior. This suggests that, at least for Lake Superior where no known large river input PCB sources have been documented, a large part of the sediment burden may be accounted for via atmospheric particle associated input. Also, although PCBs are liquid-phase controlled, a net accumulation in lakes seems to occur. Experiments which include studies on the effects of biota, aqueous particle dynamics, and hydrodynamics on PCB air/water transfer must be undertaken to explain these observations.

ACKNOWLEDGEMENTS

Paul Doskey sampled and analyzed for PCB over Lake Michigan. Helen Grogan and Jean Schneider patiently typed the manuscript. Most of the research discussed was supported by grant from NOAA-SEA GRANT to the University of Wisconsin.

REFERENCES

1. National Academy of Sciences "Polychlorinated Biphenyls". Washington, D.C. (1979).

2. Eisenreich, S.J., Hollod, G.J., and Johnson, T.C. In "Atmospheric Pollutants in Natural Waters". Eisenreich, S.J. [Ed.]. Chapter 21, Ann Arbor, Michigan (1981).

3. Doskey, P.V., and Andren, A.W. Envir. Sci. Technol., 15, 705 (1981).

4. Harvey, G.R., and Steinhauer, W.G. Atm. Environ., 8, 777 (1974).

5. Doskey, P.V., and Andren, A.W., Anal. Chim. Acta, 110, 129 (1979).

6. Brunauer, S., Emmett, P.H., and Teller, E. J. Am. Chem. Soc., 60, 309 (1938).

7. Junge, C.E. In "Fate of Pollutants in the Air and Water Environments" Part I. Suffet, I.H. [Ed.]. Wiley-Interscience, New York. (1977).

8. Doskey, P.V., and Andren, A.W. J. Great Lakes Research 7, 15 (1981).

9. National Academy of Sciences "The Tropospheric Transport of Pollutants and Other Substances to the Oceans". Washington, D.C. (1978).

10. Scott, B.C. In "Atmospheric Pollutants in Natural Waters". Eisenreich, S.J. [Ed.]. Chapter 1. Ann Arbor, Michigan (1981).

11. Mackay, D., and Leinonen, P.J. Env. Sci. Technol., 9, 1178 (1975).

12. Gatz, D.F. Water, Air, and Soil Pollution, 5, 239 (1975).

13. Andren, A.W., and Strand J.W. In "Atmospheric Pollutants In Natural Waters". Eisenreich, S.J. [Ed.]. Chapter 23. Ann Arbor, Michigan (1981).

14. Sehmel, G.A. In "Atmospheric Sulfur Deposition", Shriner, D.S., Richmond, C.R., and Lindberg, S.E. [Eds.] Chapter 25. Ann Arbor (1980).

15. Sodergren, A. Nature (London), 236, 395 (1972).

16. Young, D.R., McDermott-Erhlich, D., and Heesen, T.C. "PCB Inputs to the Southern California Bight". Southern California Coastal Research Project, El Segundo, Calif. (1975).

17. Slinn, W.G.N., and Slinn, S.A. In "Atmospheric Pollutants In Natural Waters". Eisenreich, S.J. [Ed.]. Chapter 2. Ann Arbor, Michigan (1981).

18. Sehmel, G.A., and Hodgson, W.H. In "Atmosphere, Surface Exchange of Particulate and Gaseous Pollutants" ERDA Symposium Series 38. (1974).

19. Chamberlain, A.C. Proc. Roy. Soc. London, Ser. A., 236, 45 (1966).

20. Sehmel, G.A., and Sutter, S.L. J. Rech. Atmos. 3, 911 (1974).

21. Talbot, R.W. Ph.D. Thesis, Water Chemistry Department, University of Wisconsin, Madison (1981).

22. Wu, J.J. J. Geophys. Res., 74, 444 (1969).

23. Wesely, M.L., Cook, D.R., and Williams, R.M. Boundary Layer Met. (In Press).

24. Sievering, H., Dolske, D., Dave, M., Hughes, R.L., and McCoy, P. EPA Report EPA-905/4-79-016 (1979).

25. Westcott, J.W., Simon, C.G., and Bidleman, T.F. Env. Sci. Technol., 15, 1375 (1981).

26. Hetling, L., Horn, E., and Tofflemire, J. New York State Department of Environmental Conservation, Technical Report No. 51 (1978).

27. Whitman, W.G. Chem. Metall. Eng. 29, 146 (1923).

28. Liss, P.S., and Slater, P.G. Nature (London, 247, 181 (1974).

29. Higbie, R. Trans. AICHE, 31, 365 (1935).

30. Danckwerts, P.V. Ind. Eng. Chem., 43, 1460 (1951).

31. Fortescue, G.E., and Pearson, J.R.A. Chem. Eng. Sci., 22, 1163 (1967).

32. Lamont, J.C., and Scott, D.S. AICHE, 16, 513 (1970).

33. Lee, Y.H., Tsau, G.T., Wankat, P.C. AICHE, 26, 1008 (1980).

34. Reid, R.C., and Sherwood, T.K. "The Properties of Gases and Liquids", 2nd ed.; McGraw-Hill, New York (1966).

35. Erdey-Gruz, T. "Transport Phenomena in Aqueous Solutions". John Wiley and Sons, New York (1974).

36. MacIntyre F. In "The Sea" Vol. 5. Goldber, E.D. [Ed.]. Chapter 8. Wiley-Interscience, New York.

37. Blanchard, D. C., and Woodcock, A.H. Tellus, 9, 145 (1957).

38. Johnson, B.D., and Cooke, R.C. Science, 213, 209 (1981).

39. Matter-Muller, C., Gujer, W., and Griger, W. Wat. Res. 15, 1271 (1981).

CHAPTER 9

EVIDENCE FOR THE ATMOSPHERIC
FLUX OF POLYCHLORINATED BI-
PHENYLS TO LAKE SUPERIOR

S. J. Eisenreich
  Environmental Engineering Program
  Department of Civil and Mineral Engineering
  University of Minnesota

B. B. Looney
  Environmental Engineering Program
  Department of Civil and Mineral Engineering
  University of Minnesota

INTRODUCTION

Polychlorinated biphenyls (PCBs) are present in marine (1-6), the Great Lakes (7-11) and urban atmospheres (7,8,10,12,13) at concentrations such that atmospheric transport has been implicated as a primary input pathway to freshwater and marine ecosystems. For example, the North Atlantic Ocean (water plus sediments) is the principal sink for PCBs produced and emitted in North America from 1930 to 1976 (14). A second major sink is freshwater sediment, primarily in the Laurentian Great Lakes, where atmospheric inputs may be partially responsible for sedimentary accumulation of PCBs. Eisenreich et al. (15) concluded that atmospheric input of PCBs to Lake Superior can support their accumulation in surficial sediments. Also, PCB concentrations in the Lake Michigan (7) and Lake Superior atmospheres (9,10) of ~ 1 ng m^{-3} may support inputs of $\sim$ 6500 (16) and 7000 Kg yr^{-1} (17), respectively. Accurate estimates of PCB loading to the Great Lakes from all sources are especially important because the 1-10 ng L^{-1} PCBs in the water column are concentrated in the biota supporting elevated toxic concentrations in larger fish (18).

Recent theoretical (19) and operational field studies in marine and freshwater atmospheres (4,5,7,10,20) show that PCB transport is dominated by the vapor phase. Doskey and Andren (7) found that an average of 87% of the 1.0 ng m^{-3} airborne PCBs over southern Lake Michigan were collected on an

XAD-2 adsorbent trap following a glass fiber filter. Eisenreich et al. (10) concluded that 90 to 100% of the PCBs in the Lake Superior atmosphere also existed in the vapor phase. Thus, calculation of PCB vapor flux across the air/water interface is critically important in quantifying atmospheric inputs. However, Henry's Law constants (H) or air/water partition coefficients (H') reported in the literature describing the equilibrium distribution between vapor and solution phases differ by 10^{-3} (Table 1). The values listed in Table 1 result from laboratory experiments or theoretical calculations. We report here the first field determination of H' by measuring the surface flux necessary to support a PCB concentration profile observed in the epilimnetic waters of Lake Superior in July, 1978, with comparison to atmospheric fluxes based on various H' values. Estimation of a true thermodynamic Henry's Law constant from environmental parameters is obscured by surface microlayers or association of aqueous PCB with small particles. Thus, while the estimated air/water partition coefficient (H') may be incorrect in a strict thermodynamic sense, it is an important parameter needed to estimate the "real world" flux of PCBs to lakes.

In this paper, the flux of PCBs across the air/water interface has been determined using reported air/water partition coefficients (H') and comparing the results to a surface flux estimated from PCB and temperature profiles observed in Lake Superior, July 1978. The surface influx of PCB from the atmosphere was estimated as $\sim$ .02 to .2 $\mu g\ m^{-2}\ day^{-1}$ as compared to the gas-phase controlled influx calculated as $\sim$ .17 $\mu g\ m^{-2}\ day^{-1}$. Air/water partition coefficients of 10^{-6} to $10^{-7}\ atm\ m^3\ mol^{-1}$ are justified by the observed profiles. Conditional constants of this magnitude do not necessarily indicate thermodynamic "gas phase control" due to physicochemical factors such as adsorption or the presence of a microlayer.

EXPERIMENTAL

Atmospheric PCB concentrations were measured in the summers of 1978 to 1980 from aboard ship on Lake Superior. The atmospheric sampler, described in detail elsewhere (21), consisted of a Universal High Volume Air Sampler (BGI, Inc.) modified to include a back-up adsorbent. The air sampler held a 7.8 cm bed of XAD-2 macroreticular resin (600 mL) in an aluminum standpipe atop a motor housing, and following a glass fiber filter. Recoveries of Aroclor 1221, 1242 and 1254 isomeric standards at flow rates of 0.4 $m^3\ min^{-1}$ averaged 90-95% in laboratory studies. XAD-2 resins provide good recovery, collection efficiency and breakthrough behavior for

Table I. Air/Water Partition Coefficients (H') for PCB Aroclor Mixtures

	H' atm m^3 mol^{-1} Aroclor				
Reference	1221	1016	1242	1254	Gas (G) or Liquid (L) Phase Control
MacKay and Leinonen (29)			5.7×10^{-4}	2.8×10^{-3}	L
Paris et al. (35)		$1.4 \pm 7 \times 10^{-2}$	$7.6 \pm 4.5 \times 10^{-3}$		L
Atlas et al. (36)					
DW			7.8×10^{-4}		L
SW			2.9×10^{-3}		
Hetling et al. (34)	$7.4 \pm 10 \times 10^{-6}$	$2.7 \pm 5 \times 10^{-5}$			G
Murphy and Rzeszutko (12)			4×10^{-6}	9.8×10^{-7}	G
Doskey and Andren (16)			$2.8 \pm 1.8 \times 10^{-7}$	$1.4 \pm 9.7 \times 10^{-7}$	G

airborne PCBs (21,22). The air sampler was affixed to the bow of the ship and collected sample only when the ship was underway and the wind was within a vector of 60° off the bow. Air sampling volumes typically were 200 to 400 m^3 at flow rates 0.4-0.5 m^3 min^{-1} (10). Water samples were obtained by hydrocast using a 30 L Oceanics Go-Flo water sampling bottle specially designed for collection of trace organics in water. A resin column packed with $\sim$ 70 mL of slurried XAD-2 was used to isolate dissolved plus particulate PCBs from replicate 20L volumes of sample. A flow rate of 100-200 mL min^{-1} was used.

XAD-2 resins used for air collection were rinsed for 15 minutes with distilled H_2O and air dried at 60°C. The resins were soxhlet extracted for 48 hours with petroleum ether and the solvent changed at 24 hours. The resins were subsequently air dryed at 60° C for 24 hours and stored dry in glass containers. The resins used in water sampling were cleaned by sequentially extracting with methanol, hexane/acetone and petroleum ether, re-equilibrating with acetone and water, and stored in water.

After the air sampling event, the XAD-2 resins were soxhlet extracted for 24 hours using petroleum ether, fractionated on an alumina/florisil column (2:1 w/w) and eluted with hexane. Recoveries of applied Aroclor standards were $\sim$ 98%. PCB analysis was performed on an HP 5840 gas chromatograph (GC) equipped with a ^{63}Ni e$^-$ capture detector using packed columns, 2.4 m x 2mm i.d. (1.5% OV-17 + 1.95 QF-1; 4% SE-30 + 6% OV-210). Chromatographic conditions were: column temperature, 190° C; injector temperature, 225° C; detector temperature, 350° C; nitrogen flow, 30 mL min^{-1}.

XAD-2 resins used in the water sampling were soxhlet extracted with hexane/acetone (1:1) with the acetone removed by back-extraction with H_2O. The extract was concentrated using a Kudurna-Danish evaporator to $\sim$ 0.3 mL and fractionated by high performance liquid chromatography (HPLC) in the reverse-phase mode to remove interfering substances. The clean-up was performed on a Waters Model 202 HPLC equipped with two Model 6000A pumps, a Model 660 solvent programmer, a Model UK6 injector, a Model 440 UV detector and two 30 cm µ-Bondapak C-18 columns connected in series. The entire 0.3 mL of sample extract (hexane plus rinse) was injected and isocratically chromatographed for 43 minutes with an acetonitrile/H_2O mixture (50:50) followed by a linear gradient to 100% acetonitrile over 5 minutes at 1 mL min. The fraction containing PCBs and chlorinated pesticides quantitatively eluted from 56 to 68 minutes. The PCB fraction was back-extracted into hexane, dried over anhydrous sodium sulfate, concentrated and chromatographed on an HP 5840 GC as above on a glass capillary column (SP-2100, 25 m x .22 cm i.d.) at 190° C and 1 mL min^{-1} N$_2$ flow. PCBs from both the packed and glass capillary column output were identified and quantified by comparison of samples to Aroclor standards and applying a modified Webb and McCall procedure (23).

RESULTS AND DISCUSSION

Atmospheric concentrations of PCBs over Lake Superior average ~ 1 ng m^{-3} (8-10) although concentration ranges may span an order of magnitude. Doskey and Andren (7) measured similar total PCB concentrations over southern Lake Michigan in 1977, ranging from 0.5 to 1.5 ng m^{-3} (x = 1.0 ng m^{-3}) of which 87% was operationally in the vapor phase. Eisenreich et al. (10), studying airborne PCBs over Lake Superior in the summers of 1978 to 1980, measured concentrations of 0.1 to 3.5 ng m^{-3} (x = 1.2 ng m^{-3}) of which 90 to 100% collected on the back-up XAD-2 adsorbent.

The calculation of PCB inputs from the atmosphere necessarily must include aspects of the vapor-phase flux of PCBs across the air/water interface. Of the models developed to predict gas transfer across the air/water interface (24-26), the two-film resistance model of Whatman (24) has been most popular and will be used here, although the "penetration" (25) and "surface renewal" models (26) have also been used. These models have recently been expanded upon by Liss and Slater (27), Tsivoglou et al. (28), Mackay and Leinonen (29) and Smith et al. (30).

In the two-film resistance model, the air and water reservoirs are assumed to be well-mixed except for thin, stagnant films of air and water at the interface. The rate of transfer is governed by molecular diffusion across these interfacial layers and is driven by the concentration gradients between equilibrium concentrations at the interface and the concentrations at the interface in the bulk air and water reservoirs. For steady-state transfer, the flux (F) is given by (31):

$$F = K_{OL} (C - P/H) \qquad (1)$$

$$\frac{1}{K_{OL}} = \frac{1}{K_L} + \frac{RT}{HK_g} \qquad (2)$$

where F = flux (mol m^{-2} hr^{-1}); K_L, K_g = liquid and gas-phase mass transfer coefficients (m hr^{-1}); K_{OL} = overall mass transfer coefficient (m hr^{-1}); H = Henry's law constant (atm m^3

mol^{-1}); C = solute concentration (mole m^{-3}); P = gas partial pressure (atm); T = absolute temperature (°K); R = gas constant.

Resistance to gas transfer occurs in both the liquid and gas films. Mackay et al. (31) have shown that for H > 10^{-3} resistance lies almost totally in the liquid phase and the flux is:

$$F = K_L (C-P/H) \quad (3)$$

If H ≤ 5 x 10^{-5}, resistance lies almost totally in the gas phase and the flux across the air/water interface is given by:

$$F = K_g \left(\frac{CH-P}{RT}\right)$$

For H values between these extremes, resistance to mass transfer occurs in both liquid and gas phases. Knowledge of Henry's Law constant (H) is essential to predict air/water exchange because the organic concentration in equilibrium with the vapor phase must be calculated. If the water is undersaturated with respect to the atmosphere, there would be a net transfer to the water. The rate and amount of mass transferred depends on whether the compound is gas or liquid-phase controlled.

At the present time, considerable confusion exists as to which H or H' value should be used to calculate atmospheric inputs of PCBs to water via dry vapor deposition. The range of reported values (Table 1) is such that either gas or liquid-phase control, and deposition versus volatilation may be argued. The variability of literature values of H or H' may be due to many parameters. Mackay and Leionen's values (29) are calculated thermodynamically, while the others were experimentally determined in the laboratory or inferred from precipitation measurements (12). Part of the confusion arises in that H constants are for individual compounds, whereas the values reported in Table 1 refer to mixtures. In addition, PCB isomers are solids at room temperature while Aroclors are complex mixtures of PCB isomers and occur as viscous liquids. Thus, H values may be too high if calculated as the ratio of the saturation vapor pressure and aqueous solubility for mixtures. H' values apply to specific environmental conditions. Also, vapor-phase organics of low water solubility entering the water column may remain dissolved or unassociated, bind with dissolved or colloidal organics, adsorb onto a particle surface, or be absorbed into organic detritus. Only the dissolved and unassociated species in solution may be in equilibrium with the vapor phase. The net effect of solute association in solution is to decrease the apparent air/water partition coefficient (H').

It appears from the literature, that thermodynamic H lies in the range of 10^{-3} to 10^{-4} atm m^3 mol^{-1} for all isomers. Those studies which have estimated H' from environmental parameters, and those which simulate environmental conditions generally measure H' values which range from 10^{-6} to 10^{-7} atm m^3 mol^{-1}. Doskey and Andren (16) recently described in detail problems of modeling the flux of PCBs across the air/water interface.

We estimated the flux of PCBs across the air/water interface using several values of H' and comparing the results to a surface flux estimated from PCB and temperature profiles observed in Lake Superior in July 1978. In this manner, H' may be construed. The parameters for calculation of PCB flux are given in Table 2. Total PCB concentrations in Lake Superior surface water (< 10 m) in 1978 were 1.9 ± 2.0 ng L^{-1} (n = 10) and in 1979 were 3.8 ± 2.1 ng L^{-1} (n=16) of which 45 to 50% consisted of Aroclor 1242. The surface concentration (1 m) at Site 5 to the SW of Isle Royale (47° 34.8' N, 89° 39.0' W) was 1.1 ng L^{-1} of which 50% was Aroclor 1242. Based on a partition coefficient of 5×10^4 (32) a suspended solids concentration of 0.15 mg L^{-1}, and an average PCB concentration in surficial sediments of 130 ng g^{-1} (15), we calculate that 99% of the aqueous concentration is in solution.

Atmospheric concentrations of PCBs over Lake Superior averaged 1.2 ± 0.8 ng m^{-3} (n = 29) in the summers of 1978 to 1980 (10). The atmospheric PCB concentration at the site of interest was 1.1 ng m^{-3} and was composed of $\sim$ 80% Aroclor 1242 isomeric mixture. The air sampling time was $\sim$ 10 hrs., and therefore greater time resolution was not possible. However, the average concentration of airborne PCBs from 1978 to 1980 in the same area of the lake was $\sim$ 1 ng m^{-3}.

Using the PCB concentrations in solution and air given in Table 2, and K_g from Doskey and Andren (16), the dry flux of gaseous PCBs across the air/water interface was calculated for three different sets of reported H or H' values. Case 1 employs H calculated from vapor pressure and solubility measurements (29) and indicates apparent liquid-phase control while Cases 2 and 3 represent determination of H' from precipitation (12) or laboratory experiments under simulated environmental conditions (16). Cases 2 and 3 indicate apparent gas-phase control for the transfer of gaseous PCB across the interface.

Case 1 predicts a volatilization of total PCBs out of the water at a rate of 1.6 µg m^{-2} day^{-1} while Cases 2 and 3 predict a flux to the water of $\sim$ .17 µg m^{-2} day^{-1}. If maintained on an annual basis, Case 1 predicts the loss of PCB to the air from the water of $\sim$ 48 x 10^3 Kg yr^{-1}, while Cases 2 and 3 (gas-phase control) predict inputs to the lake of $\sim$ 5.1 x 10^3 Kg yr^{-1} (Table 3). Similar calculations and results have been reported for Lake Michigan (16).

In July 1978 a unique set of environmental conditions permitted the build-up of a PCB concentration gradient in the surface waters of Lake Superior at a sampling location to the SW of Isle Royale. The period preceding the water sampling event was unusually calm ($\leq$ 0-1 m sec^{-1} wind speed) and warm (16-27° C) for three days. Figure 1 shows the temperature profile ranging from 8.5° C at 1 m to 4.3° C at depth. The PCB concentration decreased from 1.1 ng L^{-1} at the surface to 0.4 ng L^{-1} at a depth of 25 m, and stabilized at $\sim$ 0.6 ng L^{-1} to a depth of 180 m. The PCB profile indicated a net flux from the air to the water.

At a steady state condition in which the concentration gradient (dc/d$_z$) must be maintained for the flux to be non-zero, the surface flux (F) supporting the gradient is

$$F = - D_z (dc/dz) = \text{constant} \qquad (4)$$

where D_z = vertical eddy diffusivity (cm^2 sec^{-1})

dc/d_z = concentration gradient (µg cm^{-3} cm^{-1})

For this equation to apply, we assume that PCBs are being "removed" by dilution in the bulk water of the lake at a rate equal to the atmospheric flux. D_z can be related to the buoyancy frequency (N) which is

$$N^2 = g \, \alpha \frac{\delta T}{\delta z} \, (\sec^{-2}) \qquad (5)$$

where N = buoyancy frequency (sec^{-2})

g = gravitational acceleration (cm sec^{-2})

$\alpha = (-1/\rho) \frac{\delta \rho}{\delta t}$ (deg^{-1}) = coefficient of thermal expansion

ρ = density of water (g cm^{-3})

$\frac{\delta T}{\delta z}$ = temperature gradient (deg cm^{-1})

Figure 1.
Site 5. R/V R.R. Simons - Lake Superior - 1978.
Temperature, turbidity, suspended solids, total organic carbon (TOC), total phosphorus (TP), and PCB water profiles taken in Lake Superior (47°34.8'N, 89° 39.0'W) in July, 1978.

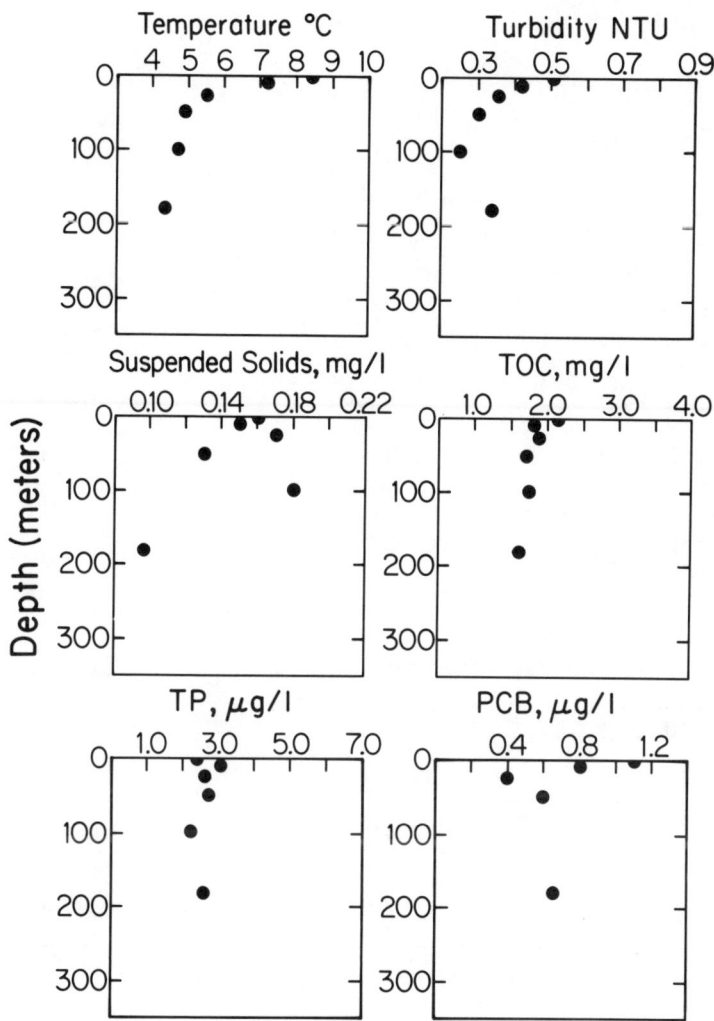

Table II. Parameters for PCB Flux Calculation

	Aroclor 1242	Aroclor 1254
Conc. in Air (P) (atm)	7.4×10^{-14}	1.5×10^{-14}
(ng m^{-3})	1.1	
Conc. in Water (C) (mols m^{-3})	2.2×10^{-9}	1.7×10^{-9}
(ng L^{-1})	1.1	
Air/Water Partition Coefficient (H') (atm m^3 mol^{-1})		
Case 1 (29)	5.7×10^{-4}	2.8×10^{-3}
Case 2 (12)	4×10^{-6}	9.8×10^{-7}
Case 3 (16)	2.8×10^{-7}	1.4×10^{-7}
Overall Gas-Phase Mass Transfer Coefficient (K_{OG}) (m hr^{-1})	7.9	7.0
Overall Liquid-Phase Mass Transfer Coefficient (K_{OL}) (m hr^{-1})	5.7×10^{-2}	6.7×10^{-2}
RT (25° C) (atm m^3 mol^{-1})	2.45×10^{-2}	2.45×10^{-2}
MW	256	324

Table III. Comparison of Calculated and Measured Surface Fluxes of PCBs

	Flux[a]		
	Aroclor 1242	Aroclor 1254	Total PCB
		$\mu g\ m^{-2}\ day^{-1}$	
Measured Surface Flux			−.02 to .2
Case 1 Liquid-Phase Control	+.74	+.86	1.6
Case 2 Gas-Phase Control	−.14	−.03	−.17
Case 3 Gas-Phase Control	−.15	−.03	−.18

[a] Negative flux means transfer from the air to the water.

The higher the value of N^2 or stronger the density gradient, the more stable is the water column against vertical mixing, and lower the value of D_z. Also, a negative temperature gradient gives a positive value of the stability frequency. For this particular case, $N^2 = 7 \times 10^{-5}$ sec^{-2} and D_z is 0.1 to 1.0 cm^2 sec^{-1}, typical of surface waters of large stratified lakes and oceans under some conditions (33). For a PCB gradient over the surface 25 m of .02 to .03 µg m^{-4}, a vertical flux of 0.02 to 0.2 µg m^{-2} day^{-1} into the water is calculated. This is interpreted as the flux from the air to the water necessary to maintain the concentration gradient under steady state conditions. The measured surface flux of .02 to .2 µg m^{-2} day^{-1} encompasses the PCB flux across the air/water interface under gas-phase control (Cases 2 and 3) of $\sim$ .17 µg m^{-2} day^{-1}. Under the conditions observed in this specific environmental situation, air/water partition coefficients of $\sim$ 10^{-6} to 10^{-7} atm m^3 mol^{-1} for Aroclor 1242 and 1254 are justified. We cannot conclusively state that gaseous PCBs will be transferred to the water from the air under all environmental conditions involving different concentrations, wind speeds, temperatures, etc. It is important to note the downward flux of PCBs in this situation - a flux which is not predicted by thermodynamic considerations of H and which supports the fact that we find these compounds in the water and bottom sediments of Lake Superior. It is quite possible that PCBs may be lost from the water to the air, or that the magnitude of the downward flux would be altered by different environmental conditions or by aerosolization. These considerations are supported by the water burden. For a lake-wide mean concentration of 1 ng L^{-1}, the PCB burden in Lake Superior water (11,920 Km^3) is $\sim$ 12,000 Kg with about an additional 20,000 Kg buried in the sediments (8) for a total of 32,000 Kg.

CONCLUSIONS

The flux of PCBs across the air/water interface has been determined using reported air/water partition coefficients and comparing the result to a surface flux estimated from PCB and temperature profiles observed in Lake Superior, July 1978. The surface influx of PCBs was estimated as $\sim$ .02 to .2 µg m^{-2} day^{-1} as compared to the gas-phase controlled influx of $\sim$ .17 µg m^{-2} day^{-1}. Air/water partition coefficients of 10^{-6} -10^{-7} atm m^3 mol^{-1} are justified under the conditions described.

REFERENCES

1. Harvey, G. R., Steinhauer, W. G., "Atmospheric Transport of Polychlorinated Biphenyls to the North Atlantic," Atmos. Environ. 8:777 (1974).

2. Bidleman, T. F., Olney, C. E., "Chlorinated Hydrocarbons in the Sargasso Sea Atmosphere and Surface Water," Science, 183:516 (1974).

3. Bidleman, T. F., Rice, C. P., Olney, C. A., "High Molecular Weight Hydrocarbons in the Air and Sea," In Marine Pollutant Transfer, H. L. Windom and R. Duce (eds.), Lexington Books, Lexington, Mass., pp. 323-351 (1976).

4. Giam, C. S., Atlas, E. L., Chan, H. S., Neff, G., "Phthalate Esters, PCB and DDT Residues in the Gulf of Mexico Atmosphere," Atmos. Environ., 14:65 (1980).

5. Atlas, E. L., Giam, C. S., "Global Transport of Organic Pollutants - Ambient Concentrations in the Remote Marine Atmosphere," Science, 211:163 (1981).

6. Dawson, R., Riley, J. P., "Chlorine Containing Pesticides and Polychlorinated Biphenyls in British Coastal Waters," Estuar. Coast. Mar. Sci., 4:55 (1977).

7. Doskey, P. V., Andren, A. W., "Concentrations of airborne PCBs over Lake Michigan. J. Great Lakes Res.. 7:15 (1981).

8. Hollod, G. J., "Polychlorinated Biphenyls (PCBs) in the Lake Superior Ecosystem: Atmospheric Deposition and Accumulation in the Bottom Sediments." Ph.D. Thesis, University of Minnesota, Minneapolis, 247 pp.

9. Eisenreich, S. J., Hollod, G. J., Johnson, T. C., "Atmospheric Concentrations and Deposition of Polychlorinated Biphenyls to Lake Superior." In Atmospheric Pollutants in Natural Waters, S. J. Eisenreich (ed.), Ann Arbor Science Publishers, Ann Arbor, Mich., pp. 425-444 (1981).

10. Eisenreich, S. J., Looney, B. B., Hollod, G. J., "PCBs in Lake Superior Atmosphere: 1978-1981," submitted for publication (1982).

11. Rice, C. P. Personal communication GLRD, University of Michigan, Ann Arbor (1981).

12. Murphy, T. J., Rezszutko, C. P., "Precipitation Inputs of PCBs to Lake Michigan," J. Great Lakes Res., 3:305 (1977).

13. Bidleman, T. F., Christensen, E. J., Harder, H. W., "Aerial Deposition of Organochlorines in Urban and Coastal South Carolina." In Atmospheric Pollutants in Natural Waters, S. J. Eisenreich (ed.), Ann Arbor Science Publishers, Ann Arbor, Mich., pp. 481-508 (1981).

14. National Academy of Sciences (NAS), Polychlorinated Biphenyls, Washington, D.C., 182 p. (1979).

15. Eisenreich, S. J., Hollod, G. J., Johnson, T. C., "Accumulation of Polychlorinated Biphenyls (PCBs) in Surficial Lake Superior Sediments," Environ. Sci. Tech., 13:569 (1979).

16. Doskey, P. V., Andren, A. W., "Modeling the Flux of Atmospheric Polychlorinated Biphenyls Across the Air/Water Interface," Environ. Sci. Tech., 15:705 (1981).

17. Hollod, G. J., Eisenreich, S. J., Johnson, T. C., "Sources and Sinks of PCBs in Lake Superior," submitted for publication (1981).

18. Delfino, J. J., "Toxic Substances in the Great Lakes," Environ. Sci. Tech., 13:1462 (1979).

19. Junge, C. E., "Basic Considerations about Trace Constituents in the Atmosphere as Related to the Fate of Global Contaminants." In Fate of Pollutants in the Air and Water Environment, I. H. Suffet (ed.), Wiley-Interscience, New York, pp. 7-25 (1977).

20. Eisenreich, S. J., Looney, B. B., Thornton, J. D., "Airborne Organic Contaminants in the Great Lakes Ecosystem," Environ. Sci. Tech., 15:30 (1981).

21. Hollod, G. J., Eisenreich, S. J., "Collection of Atmospheric Polychlorinated Biphenyls on Amberlite XAD-2 Resins," Anal. Chim. Acta, 110:129 (1979).

22. Doskey, P. V., Andren, A. W., "High Volume Sampling of Airborne Polychlorinated Biphenyls with Amberlite XAD-2 Resins," Anal. Chim. Acta, 110:129 (1979).

23. Webb, R. G., McCall, A. C., "Quantitative PCB Standards for Electron Captu-e Gas Chromatography," J. Chromatogr. Sci., 11:366 (1973).

24. Whitman, W. G., "Preliminary Experimental Confirmation of the Two Film Theory of Gas Adsorption," Chem. Metall. Eng., 29:146 (1923).

25. Higbie, R., "Rate of Adsorption of a Pure Gas into a Still Liquid During Short Periods of Exposure," Trans. AIChE, 31:365 (1935).

26. Danckwerts, P. V., "Gas Absorption Followed by Chemical Reaction," AIChE Journal, 1:456 (1955).

27. Liss, P. S., Slater, P. G., "Flux of Gases Across the Air/Sea Interface," Nature, 247:181.

28. Tsivoglou, E. C., O'Connell, R. L., Walter, C. M., Godsil, P. J., Logsdon, G. S., "Tracer Studies of Atmospheric Reaeration-1: Laboratory Studies," J. Wat. Poll. Contr. Fed., 37:1343 (1965).

29. MacKay, D., Leinonen, P. J., "Rate of Evaporation of Low Solubility Contaminants from Water Bodies to Atmosphere," Environ. Sci. Tech., 9:1178 (1975).

30. Smith, J. H., Bombarger, D. C., Haynes, D. L., "Prediction of the Volatilization Rates of High Volatility Chemicals from Natural Waters," Environ. Sci. Tech., 14:1332 (1980).

31. MacKay, D., Yuen, A. K. T., "Transfer Rates of Gaseous Pollutants Between the Atmosphere and Natural Waters," in Atmospheric Pollutants in Natural Waters, S. J. Eisenreich (ed.), Chapter 3, Ann Arbor Science Publishers, Ann Arbor, Michigan, pp. 55-65.

32. Pavlou, S. P., Dexter, R. N., "Distribution of Polychlorinated Biphenyls in Estuarine Ecosystems - Testing the Concept of Partitioning in the Marine Environment, Environ. Sci. Tech., 13:65-71 (1979).

33. Lerman, A., Geochemical Processes: Water and Sediment Environments, Wiley-Interscience, New York, pp. 69-72 (1979).

34. Hetling, L., Horn, E., Toftlemire, J., Albany, N.Y. Summary of Hudson River PCB Study Results, New York State Dept. of Environmental Conservation, Technical Rept. No. 51 (1978).

35. Paris, D. F., Steen, W. C., Baughman, G. L., "Role of Physicochemical Properties of Aroclors 1016 and 1242 in

　　　　Determining their Fate and Transport in Aquatic Environments," 7:319 (1978).

36.　Atlas, E., Foster, R., Giam, C. S., "Air-Sea Exchange of High-Molecular Weight Organic Pollutants: Laboratory Studies," Environ. Sci. Tech., 16:382-286 (1982).

CHAPTER 10

ROLE OF SURFACE MICROLAYERS
IN THE AIR-WATER EXCHANGE OF
PCBS

C.P. Rice
Great Lakes Research Division
University of Michigan

P.A. Meyers
Dept. of Atmospheric and Oceanic Science
University of Michigan

G.S. Brown
Great Lakes Research Division
University of Michigan

INTRODUCTION

Compartmentalization is frequently used in modeling as a means of describing, simplifying and defining complex systems [1]. Cycling and transport of environmental contaminants are complex processes which are conveniently studied in terms of compartments. The major environmental compartments used by the National Research Council [2] for describing PCB movement through the environment are the atmosphere, hydrosphere, and lithosphere. At our present level of understanding of environmental contaminants, the importance of compartments such as these is generally regarded in terms of their holding capacity for various contaminants or in terms of the reservoir strength of a given contaminant which then leads to an environmental exposure assessment for that material.

The air, water, and soil-sediment concept has historically been adequate for most of the materials of concern in terms of reservoirs and contaminant exposure to biota and man. Recently, however, interest has been growing in trying to explain cycling between the major reservoirs and in describing exchange processes, and hence new compartments are being defined. One of these is the surface microlayer. Because of its extreme thinness (0.001 to 200 μm [3]) this compartment is hardly important as a major reservoir for contaminants. However, its role as a biologically active zone for neustonic pollutant cycling and as a collector and recycler of both natural and anthropogenic, hydrophobic compounds have only recently been appreciated.

Two measurements of critical importance to determining the importance of the microlayer to PCB transfer are precise measurements of PCB concentrations in the surface microlayer and determination of residence times for PCBs in this region. A number of measurements of PCBs and related toxicants in microlayers from both lakes and oceans can be found in the literature. Seba and Corcoran [4] published one of the first reports on organochlorine enrichment in microlayers, and Bidleman et al. [5] summarized much of the marine literature on PCBs and DDT. Baumann-Ofstad and co-workers [6] recently measured up to 13 times more PCB in surface microlayers relative to subsurface water in marine locations around Norway.

Information on microlayer enrichment of organochlorine in freshwater systems is rather sparse. Andren et al. [7] and Eisenreich et al. [8] reported DDT enrichments in the microlayer of Lake Mendota, Wisconsin, and Elzerman [9] and Rice et al. [10] reported significant enrichments of PCB in surface microlayers of Lake Michigan.

Despite the many measurements of PCBs in surface microlayers, there have been no direct studies to determine residence times at the air-water interface. Modeling considerations of air-water transfer of PCBs have generally considered the microlayer as unimportant. Slinn et al. [11] stated that, without the water surface being constantly covered by a surface film, it is difficult to see how it could be important to transport. MacKay [12] and Liss [3] have stated that from a theoretical standpoint the microlayer would seem to offer at most only a mildly increased resistance to transfer at the air-water interface. Both of these suppositions, however, are based more on conjectural rather than on empirical foundations. The extent of coverage of the water surface by microlayer is not well known, and theoretical estimates are far from precise in terms of modeling air-water exchange of PCBs. Air-water exchange coefficients presently in use have been based on laboratory experiments that only partially reproduce the real system. For example, the particulate-to-dissolved associations of PCBs in the microlayer have not been considered, biological involvements in this transfer process have been largely ignored, and the actual chemical composition of the lipophilic materials in this region is only poorly understood.

PCBs, because of their high affinity to surfaces [13], would seem to be readily transported by particle and bubble transport. Recent research has disclosed that bubbles and particles may function as important vehicles of transport of materials contained in microlayers. Södergren [14] presented evidence from a laboratory study which shows that bubbles discharging through the microlayer surface can remove significant portions of lipid components of the microlayer. Johnson and Cook [15] studied the rate of bubble dissolution in the formation of aggregates and particles. Many of the processes described by Johnson and Cook as "below-surface phenomena" may actually be taking place at the sea surface as this surface film can be loosely construed as a large bubble surface separating the air from the lake. Wheeler [16] demonstrated that particles could be produced by compressing surface films. Thus particle generation by and in surface films may be an important mechanism leading to transfer of substances from this zone. A net result of this process would be a downward flux of sedimenting particles with their associated organic contaminants.

In this paper we present data on the concentration of PCBs in the microlayer and subsurface waters of Lake Michigan. These data show that the portion of PCBs associated with the particulate phase is related to their cycling at this interface.

MATERIAL AND METHODS

Samples for this study were collected from Lake Michigan during R/V Laurentian cruises in 1979 and 1980. The stations occupied are shown in Figure 1. Some of the samples were collected near shore in the vicinity of the Grand River (inset in Figure 1). The details of sampling are described in Rice et al. [10]. Surface film collection was done primarily with a screen sampler of the type first described by Garrett [17] in order to allow better comparison with the many other studies which have used screen samplers. Film thicknesses collected by this method were calculated to be about 385 μm ± 55. A plate sampler [18] was also tested for comparative purposes. This sampled an average film thickness of 174 ± 103 μm. The plate sampler consisted of a rectangular plexiglass sheet (50 x

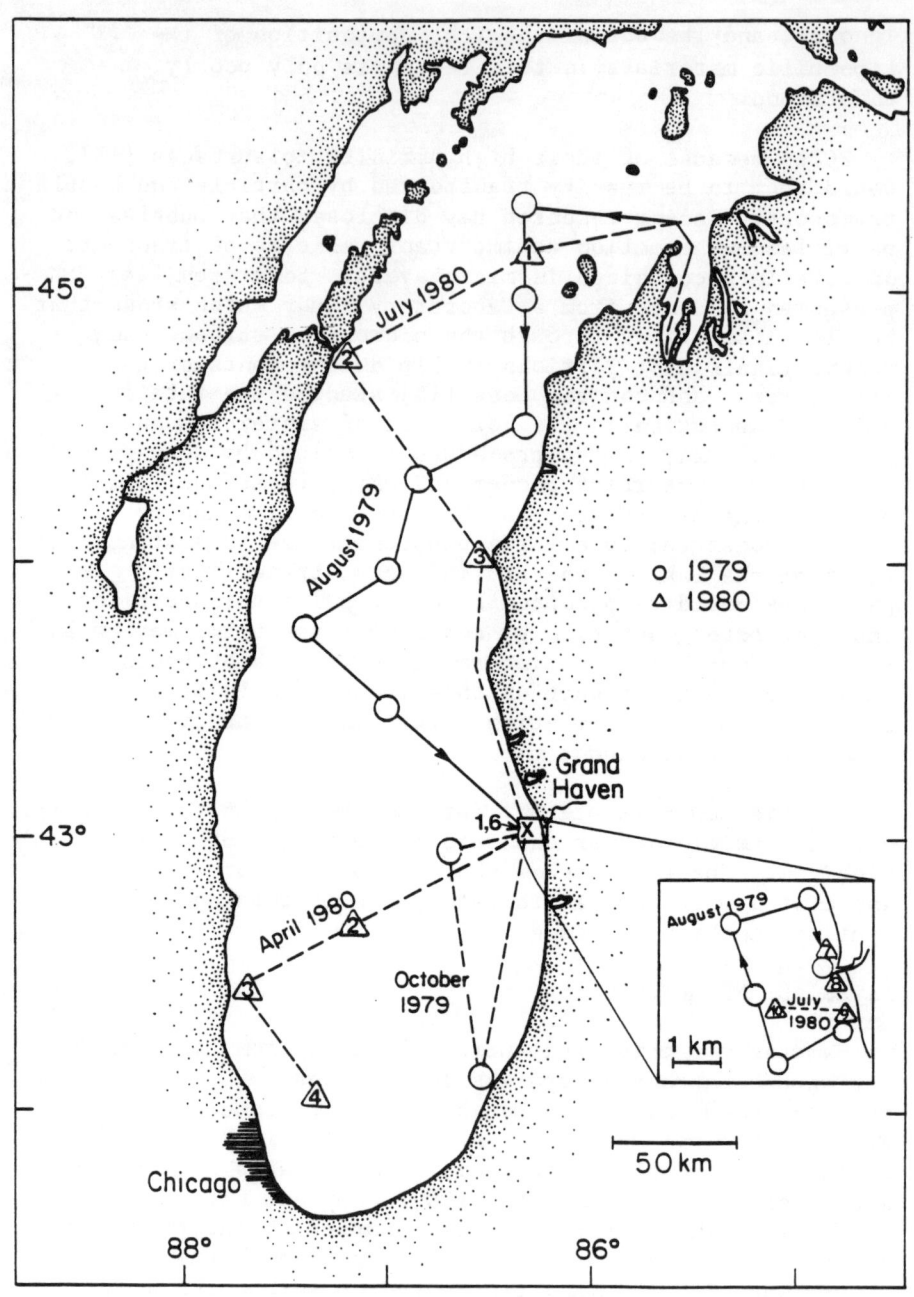

Figure 1. Cruise tracks and sampling stations occupied in 1979 and 1980 for collection of microlayer and subsurface water from Lake Michigan. (The solid arrows connecting the circles indicate that air sampling results are available for this cruise.) The numbered stations refer to Figures 2-4.

46 x 0.25 cm) attached to a nylon line. It was lowered in a vertical position through the air-water interface. When withdrawn from the water, the plate carried with it a thin film of surface microlayer material held onto its sides by surface tension. The values for both types of samplers include a correction for the collection efficiency determined from a visual estimate of the percentage of the total surface of the collector which was coated by film. Owen and Meyers [19] gave details of estimating collection efficiencies. The efficiencies in our study varied from 90 to 50%. Even though a thinner film was collected by the plate method, there was no significant difference in the qualitative results obtained by the two methods.

The material collected by the microlayer samplers was drained into glass bottles. For subsurface water collection, a 10-L stoppered bottle was submerged to a depth of approximately one meter before being opened. After collection, each of the samples was immediately filtered through precombusted Gelman glass fiber filters which were stored frozen for subsequent later extraction. The dissolved fractions were extracted by mixing with pesticide grade methylene chloride (10 water:1 methylene chloride). Some of the dissolved fractions of the microlayer samples were split into two equal portions for duplicate analyses. Replicate samples of microlayer and subsurface water were also collected at some stations. The replicate and split samples were used to monitor the analytical and sampling precision in this study. The precision calculated from the split samples of microlayer gave a value of 0.58 relative standard deviation. This was slightly less than the relative standard deviation of the duplicate station collections, i.e. 0.69. The implication from this is that the higher variability of the duplicate samples reflects the patchiness which seems to be a common characteristic of microlayers [20]. Spike and recovery experiments of the methylene chloride extraction procedure were carried out using C^{14}-labelled trichlorobiphenyl and C^{14}-labelled hexachlorobiphenyl. Recovery of the two PCB isomers averaged 82% for this extraction method. Procedural blanks were analyzed for each of the collection and extraction procedures. The values quantitated from these blanks were used to correct the results for each of the sample types. In most cases, these values were less than 1/2 the sample result. The coefficient for variation of the blanks was less than 0.20.

To address the question of possible shipboard contamination, some microlayer samples were collected at a distance from the ship with a small fiberglass boat. Care was taken to position the boat upwind of the ship and at a distance no closer than 30 m. There was no significant difference in samples collected from the ship versus those collected from the small boat.

Organic carbon concentrations of the dissolved and particulate phases were determined as described by Owen and Meyers [19]. The procedure is based upon one developed by Menzel and Vaccaro [21] and modified by Strickland and Parsons [22]. Total organic carbon is measured in water samples before and after filtration by oxidizing the organic matter and measuring the amount of resulting carbon dioxide with an Oceanography International Total Carbon Analyzer. The post-filtration values yield dissolved organic carbon concentrations while the differences between unfiltered and filtered values give particulate organic carbon concentrations. In addition, particulate POC on the filters was measured directly by combusting the capsulated filters in a Hewlett-Packard 185-B CHN Analyzer.

Each of the sample fractions - microlayer particulate and dissolved, and subsurface particulate and dissolved - was analyzed for PCBs. The PCB methods have been described by Rice et al. [10]. They involve solvent extraction of the dissolved material with methylene chloride and Soxhlet extraction of the filters with a 1:1 mixture of acetone:hexane. For preparation of the samples for electron capture gas chromatographic (EC/GC) analysis, each of the separate particulate and dissolved fractions was cleaned of contaminants by alumina chromatography [23] and then these extracts were separated by silicic acid column chromatography [5] into their various organochlorine classes. These classes were analysed by injection into an EC/GC chromatograph equipped with two 1.8-m packed gas chromatographic columns, the columns were 2-mm i.d. glass columns packed with 1.5% SP-2250/1.95% SP-2401 on 100/120 mesh Supelcoport (Supelco Inc., Bellefonte, PA) and 5% SP-2100 on 100/120 mesh supelcoport. The GC conditions are described in Rice et al. [10].

For this paper, only the PCB results are reported; a subsequent report will present information about ppDDT, ppDDD, ppDDE, dieldrin, alpha- and gamma-chlordane, and other organochlorines identified in the samples. PCB distributions in the samples were matched against a number of Aroclor mixtures likely to appear in these samples (i.e. Aroclor 1221, 1016, 1242, 1248, 1254, and 1260). Aroclor 1242 and 1254 accounted for most of the peaks in the sample chromatograms. Identification was accomplished by matching retention times of sample peaks to those of standards, with exclusion of those peaks in the samples that appeared to be out of proportion to the expected ratios of those peaks in the standards. For statistical processing, the University of Michigan's statistical program package (MIDAS) was used. Suspended solids determinations were carried out for the 1980 collections. These were obtained by suction filtering a 1- to 3-L portion of each sample through a filter supported on a Millipore sintered glass filter holder. Forty-two-mm diameter Gelman® type AE glass fiber filters were used.

RESULTS AND DISCUSSION

Because of the separation of each sample into particulate and dissolved components, and because PCBs are a mixture of numerous homologues and isomers of chlorobiphenyl compounds, there are a number of ways the data could be presented. Each of these samples is a subset of either a microlayer or a subsurface water sample, and each of these can be subdivided into an Aroclor 1242 and an Aroclor 1254 concentration component. To compile a total PCB concentration, each particulate phase concentration has been combined with its dissolved complement and the Aroclor levels added together. The total PCBs in each region - surface microlayer or subsurface water - for each sampling period in 1979 and 1980 are summarized in Table I. A test of variance on the totals for each of these four periods indicated that there was no significant difference between them. For the subsurface values in August 1979, it was difficult to do any statistical treatment as there are apparent outliers in the data set. Using a d^2 outlier test [24], a value of 4.79 ng/L was determined for the subsurface concentration of PCBs in this data set. However, one of the more suspect values that the d^2 rule would not exclude was measured at a nearshore sampling location (see inset in Figure 1). By discarding this result, a value of 2.88 ng/L, was calculated. Using the qualified value of 2.88 ng/L, it appears that there might have been a higher concentration of PCB in the 1980 subsurface waters than in 1979.

Table I. Average Concentration of Total PCB in Lake Michigan Surface Microlayer and Subsurface Waters (1979-1980)

	Number of Samples	Mean Values (ng/L)	s.d. (ng/L.)
Microlayer			
August 1979	12	15.70	12.28
October 1979	5	21.38	17.02
April 1980	11	22.03	8.91
July 1980	10	17.63	3.71
Subsurface Water			
August 1979	9	4.79	6.56
August 1979 (qualified)*	8	2.88	3.37
October 1979	------- no samples taken -------		
April 1980	4	5.66	1.12
July 1980	7	6.36	1.30

* see text.

    The August 1979 values have the highest variance of the three subsurface groupings. Although we have no satisfactory explanation of this high variation, problems with analytical precision seem to be the likeliest possibility. We have attempted to correct this aspect of the study by requiring, subsequent to the August analyses, that only one person do all the analyses. Because of the problem with the August 1979 samples, they are not included in any of the statistical presentations in this paper. The results of this cruise, however, are presented as an additional indicator of the general concentration of PCBs in the upper waters of Lake Michigan.

The April and July collections from 1980 offer the most complete data sets for statistical processing. Figure 2 illustrates the particle and dissolved distribution of total PCBs in the microlayer and in the subsurface water in these samples. It can be seen in this figure that the proportion of PCBs carried on particulates is always higher in microlayer than in subsurface samples. Also, the relative amount of PCBs on the particles was higher in the microlayer and subsurface samples in July than in April. This observation is expressed numerically in the test of variances of this factor (percentage of total PCB on the particles) listed in Tables II and III. An explanation for this preferential partitioning onto particles in July might be related to a probable higher abundance of organic-containing particles in July as compared to April [25]. In addition, elevated temperatures at this time result in higher rates of turnover [26]. Although this explanation is reasonable, organic carbon measurements were not carried out in April, and so we were unable to test this supposition. Another reasonable possibility is the existence of a greater amount of suspended particulate material in Lake Michigan waters in July than in April. However, our measurements show that for July the suspended solids averaged 1.66 mg/L in the microlayer and 1.57 mg/L in subsurface samples, whereas for April these values were respectively 2.95 and 2.01 mg/L. Therefore a greater particle density did not correlate with the higher particulate partitioning, suggesting that the quality of the particles seems to be more important than their quantity.

Careful examination of the peaks produced by the packed column analysis of the samples showed that various blends of Aroclor 1242 and Aroclor 1254 could best account for most of the G.C. peaks resolved in our analyses. Hence, each sample subset is reported in terms of an individual concentration of Aroclor 1242 and Aroclor 1254. In Tables II and III this parameter is given as a relative Aroclor 1242 concentration and has been examined statistically by a test of variance for the amount of Aroclor 1242 in the particulate phase alone. This was analyzed separately for each cruise. For 1980 microlayer samples, it appears that the relative amount of Aroclor 1242 on the particles is significantly correlated with the percentage of particulate-associated PCBs for both April and July sampling times (Figure 3). In contrast, there appears to be a negative correlation of the percentage of particulate-associated PCBs with the proportion of Aroclor

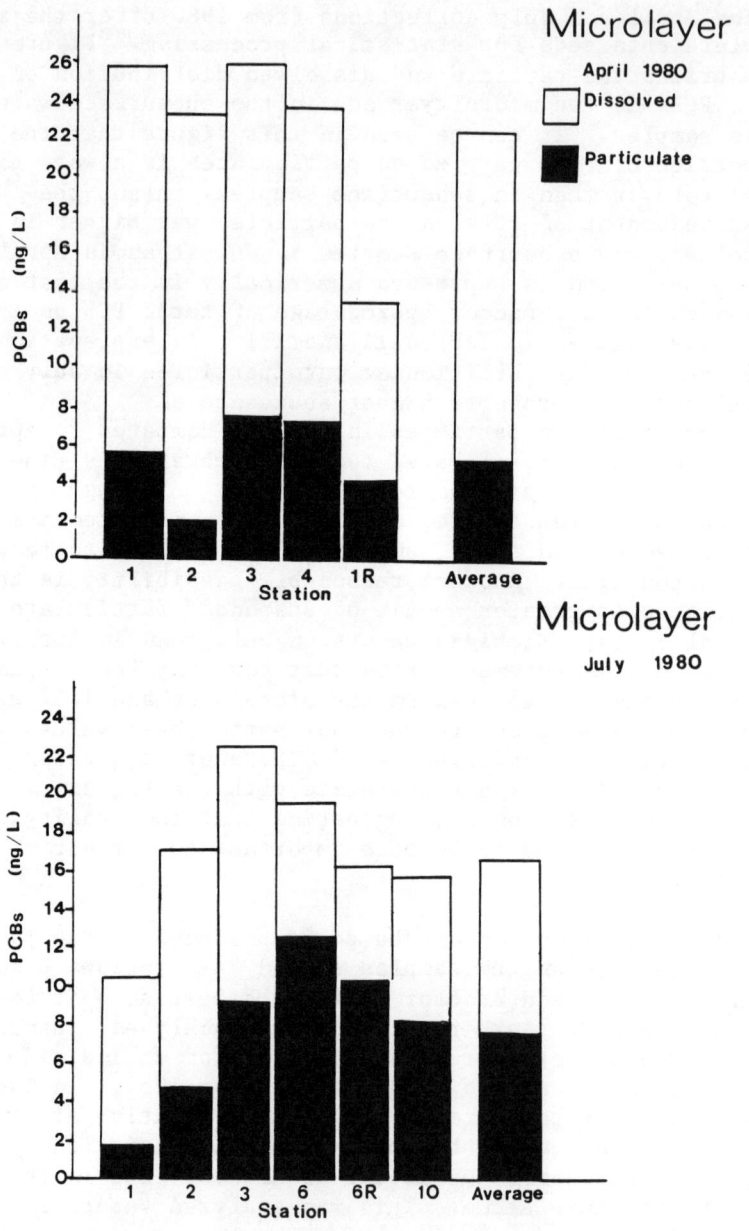

Figure 2A. The particle and dissolved distributions of total PCB in samples of surface microlayer from Lake Michigan. The station numbers refer to the stations identified in Figure 1.

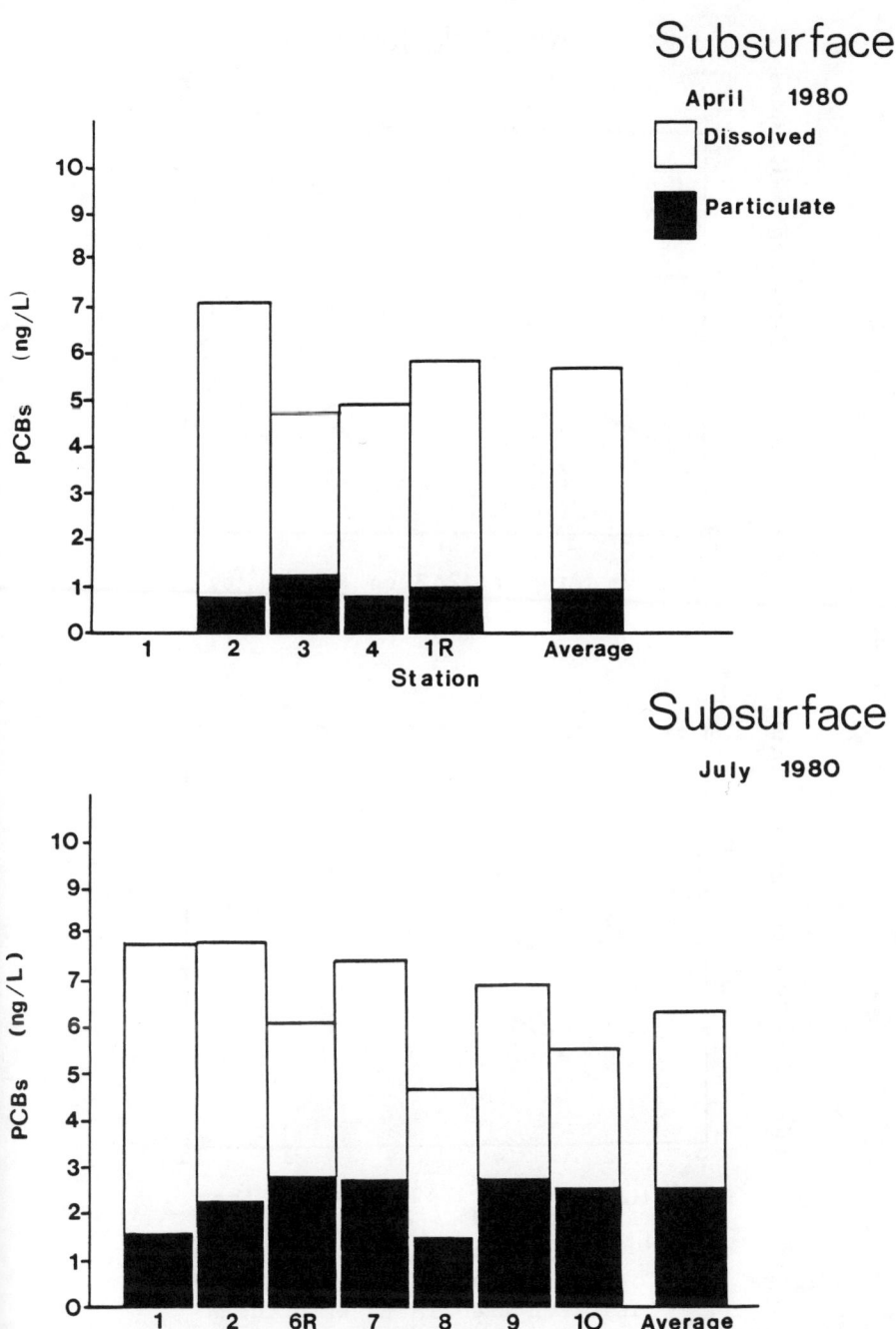

Figure 2B. The particle and dissolved distributions of total PCB in samples of subsurface water from Lake Michigan. The station numbers refer to the stations identified in Figure 1.

Microlayer

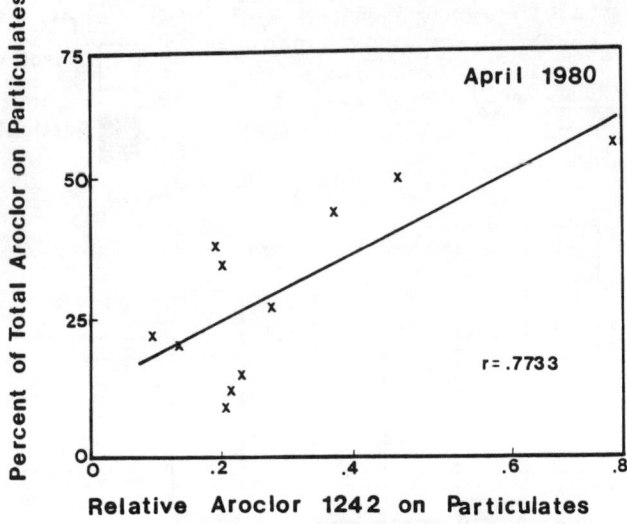

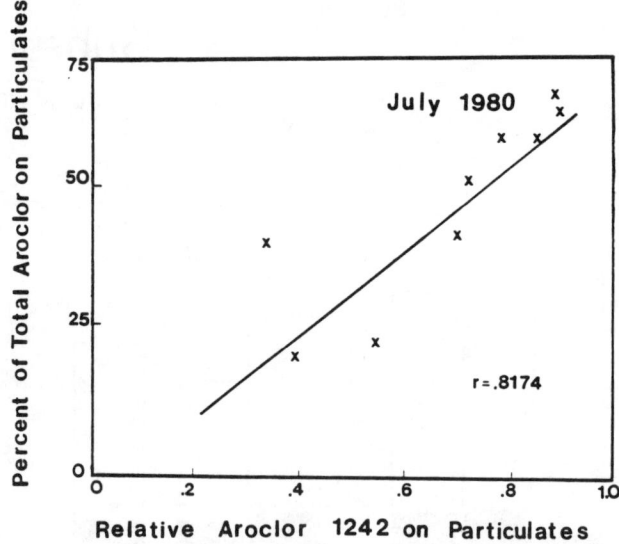

Figure 3. Correlation between the amount of total PCB associated with the particulates in the microlayer samples from April 1980 and July 1980 versus the percentage of the total amount of the PCB on the particles which was measured as Aroclor 1242.

Table II. Statistical test of Variance on PCBs associated with particulates for the microlayer data on (a) the percentage of the total PCB (particulate plus dissolved) which is on the particulates in the microlayer samples from each of the cruises, August 1979, October 1979, April 1980, and July 1980, and (b) the relative amount of Aroclor 1242 with the sampled particulates (expressed as percentage of the total Aroclor, Aroclor 1242, and Aroclor 1254) for each of the four cruises.

Analysis of Variance
Microlayer

(a)
Percentage of Total PCB on Particulates

Cruise	n	mean	variance	std dev
August 1979	12	65.997	455.38	21.340
October 1979	5	51.555	189.46	13.765
April 1980	11	29.961	269.29	16.410
July 1980	10	44.057	393.66	19.841

n = 38 of 53   significance = .0007

(b)
Relative Aroclor 1242 on Particulates

Cruise	n	mean	variance	std dev
August 1979	12	.4186	.0312	.1767
October 1979	5	.4115	.0185	.1361
April 1980	13	.2733	.0329	.1815
July 1980	19	.7523	.0315	.1775

n = 49 of 53   significance = .0000

Table III. Statistical test of variance on PCBs associated with particulates for the subsurface data on (a) the percentage of total PCB (particulte plus dissolved) which is on the particulates in the subsurface samples from each of the four cruises, August 1979, April 1980, and July 1980; and (b) the relative amount of Aroclor 1242 with the sampled particulates (expressed as percentage of the total Aroclor, Aroclor 1242 plus Aroclor 1254) for each of the four cruises.

Analysis of Variance
Subsurface

(a)
Precentage of Total PCB on Particulates

Cruise	n	mean	variance	std dev
August 1979	10	21.674	107.17	10.353
April 1980	4	17.439	50.806	7.1228
July 1980	7	39.725	56.106	7.490

n = 21 of 27　　　　　　　　significance = .0007

(b)
Relative Aroclor 1242 on Particulates

Cruise	n	mean	variance	std dev
August 1979	11	.3418	.0361	.1901
April 1980	8	.3306	.0109	.1044
July 1980	7	.8161	.015	.1467

n = 26 of 27　　　　　　　　significance = .0000

Subsurface

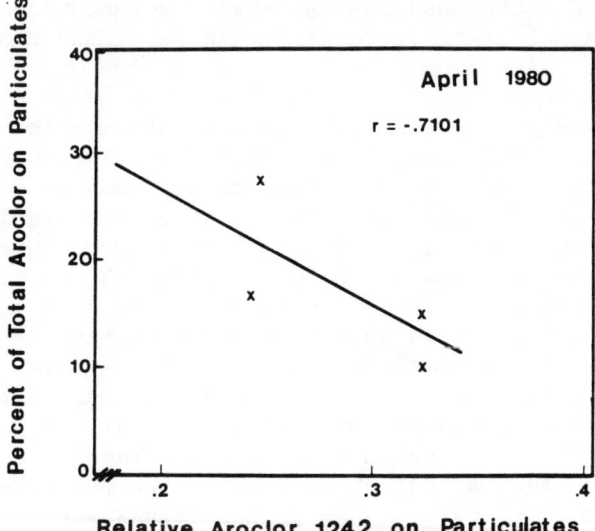

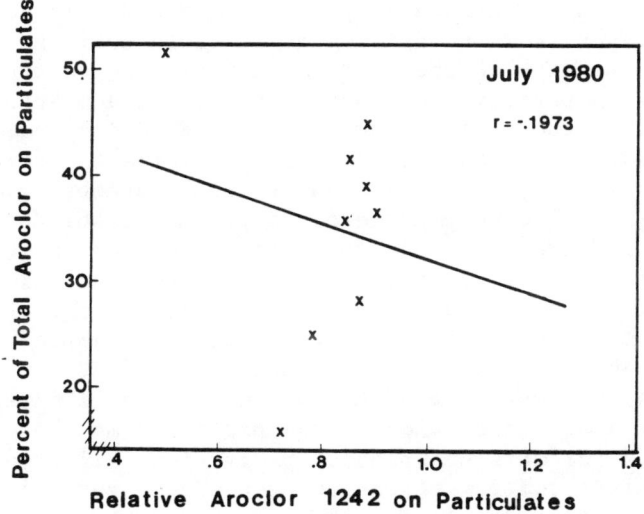

Figure 4.   Correlation between the amount of total PCB associated with the particulates in the subsurface samples from April 1980 and July 1980 versus the percentage of the total amount of PCB on the particles which was measured as Aroclor 1242.

1242 on the subsurface particles (Figure 4). For the subsurface region this agrees with predicted partitioning behavior of PCBs, because the less-water-soluble Aroclor 1254 should associate more readily with particles than would Aroclor 1242.

A useful way to express partitioning behavior is to calculate a partition coefficient (Kp) for the PCBs in each of the phases. Partition coefficients are used to describe the relative partitioning of substances between liquids and solids. For calculation of Kp's, the usual procedure is to divide the concentration of the material on the particles by the concentration of the substance dissolved in the solution; therefore, this number adjusts for the concentration of suspended solids. In samples from July 1980 we were also able to normalize these values to organic carbon content as suggested by Karichoff et al. [27]. Partition coefficients calculated for 1980 samples are shown in Table IV. A test of difference was performed using a Student's t test. None of the values was statistically different. After applying the Karichoff organic carbon correction to the July values, the mean Kp's were 4.5×10^6 and 1.5×10^6, respectively, for the microlayer and subsurface samples. The greater Kp value in the microlayer seems surprising as it would seem that organic components of the microlayer would reduce the difference in partitioning such that the surface Kp's would be lower than those in the subsurface. For instance, the greater level of dissolved lipophilic materials generally present in microlayers [20] might keep hydrophobic materials in solution and therefore produce lower Kp's in the surface microlayer. Furthermore, there was no difference in particulate organic carbon concentrations in these surface and subsurface samples. The average organic carbon content of particulates in microlayer samples collected in July was 25.2% as compared to 25.3% for subsurface samples. Although some unmeasured difference in the biochemical composition of the particles may be responsible for this unexpected observation, differences in organic matter content of the surface microlayer and of subsurface waters do not seem to be responsible.

Table IV. Average Partition Coefficients (Kp) for Total PCBs in Microlayer and Subsurface Water Collected from Lake Michigan (April and July 1980)

	Microlayer		
	No. of Samples	Suspended Solids (mg/L)	Kp
April	11	2.95	2.5×10^5
July	9	1.66	9.4×10^5

	Subsurface		
	No. of Samples	Suspended Solids (mg/L)	Kp
April	4	1.57	1.4×10^5
July	9	2.01	8.5×10^5

Partition coefficients for the subsurface PCBs were calculated by Aroclor type and the relative values compared (Table V). Solubility differences would predict a stronger tendency for Aroclor 1254 to be associated with the particles than Aroclor 1242 [13]. However, in general there was little difference in Kp's for the two Aroclor mixtures. Eisenreich [28] also observed this same lack of difference for the Aroclor mixtures occurring on particles in Lake Superior water. The implication from these results is that an equilibrium based solely on solubility consideration is not the only thing operative in these samples.

Table V. Average Partition Coefficient Values by Aroclor Type for April and July Cruises, 1980

Microlayer		
	Aroclor 1242	Aroclor 1254
April 1980	0.389×10^6	0.297×10^6
July 1980	2.50×10^6	0.823×10^6

Subsurface		
	Aroclor 1242	Aroclor 1254
April 1980	0.124×10^6	0.184×10^6
July 1980	1.20×10^6	0.620×10^6

The correlation of percent particulate sorption of total PCBs with the percent Aroclor 1242 content of the particulates has some interesting implications (Figures 3 and 4). It suggests that when the amount of PCB in the microlayers is more highly associated with particulates, the relative concentration of Aroclor 1242 is higher. This is contrary to what solubility rules would dictate as Kp's are reported to decrease with increasing solid concentrations [29] or at least to stay constant. However, as the amount of Aroclor 1242 increases disproportionately to the Aroclor 1254, this correlation appears to relate to disequilibrium in the microlayer, perhaps from a recent input phenomenon. The likeliest source of higher than typical ratios of Aroclor 1242 to Aroclor 1254 is the atmosphere. Both the microlayer and subsurface Kp's could be influenced by a direct source input of particles from the atmosphere. In this case partitioning has occurred through vapor phase partitioning on a solid surface (i.e. the airborne particle). Measurements have shown Aroclor

1242 to constitute 60% of air sampled over Lake Michigan [10]. Doskey and Andren [30] also found Aroclor 1242 to be a major constituent of air over Lake Michigan. Therefore, it appears that the highly correlated particle association of Aroclor 1242 to microlayer content of particulate PCBs has a direct link to an atmospheric input loading of this material. Murphy and Rzeszutko [31] reported particles in rain to contain more Aroclor 1254 and 1260 than Aroclor 1242; therefore rainout of particles is not a likely source of the Lake Michigan PCBs. A more plausible source of microlayer particulates would be from dryfall of particulates. A further implication of these results is that the particles once in the microlayer may play an important role in downward PCB transfer. Processes have been reported which might create particulate flux out of the microlayer [16] and lead to downward transport of PCBs and similar hydrophobic organochlorine materials into deeper portions of the water column.

SUMMARY AND CONCLUSIONS

The results of this two-year period of study of PCBs in the air-water interface of Lake Michigan provide new information about the probable significance of surface microlayers to the cycling and transport of contaminants. This information can be summarized as:

1. Concentrations of total PCBs are three times greater in the surface microlayer than in waters 1 m below the surface. Although considerable spatial and temporal variability exists in these concentrations, subsurface values appear to be greater in samples collected in 1980 than in 1979 samples.

2. A greater proportion of total PCBs is in the particulate phase in the surface microlayer than in subsurface waters. This behavior does not correlate with organic matter nor with particulate matter concentrations.

3. Aroclor 1242 and Aroclor 1254 are the major PCB types present in Lake Michigan waters. The amount of Aroclor 1242 with the particulates of the microlayer correlates positively with the proportion of total PCBs associated with particles. In subsurface samples, there is a negative correlation.

4. Particulate-dissolved partitioning coefficients (Kp's) of total PCBs are greater in the microlayer than in subsurface water, although organic matter contents of particulates in both regions are the same. Kp's of Aroclor 1242 and of Aroclor 1254 are not significantly different despite the possible effect of solubility differences.

These observations indicate that there is a significant enhancement of particulate-phase PCBs in the surface microlayer of Lake Michigan. The most probable origin of this phenomenon is dryfall input of airborne particles which contain PCBs and their transitory accumulation in the air-water surface. These particles have the potential of becoming sufficiently aggregated to sink into deeper parts of this lake.

ACKNOWLEDGMENTS

We wish to thank Orest E. Kawka, Cynthia C. Carl, and William A. Frez for their able technical assistance on this study and the National Oceanic and Atmospheric Administration, Michigan Sea Grant and Office of Marine Pollution Assessment - Great Lakes Environmental Research Laboratory, for financial support under NOAA Grants NA-79 RA-H-00001 and NA-80 AA-D-00072.

REFERENCES

1. Smith, J.H., Mabey, W.R., Bohonos, N., Holt, B.R., Lee, S.S., Chom, T.W., Bomberger, D.C., and Mill, T. "Environmental Pathways of Selected Chemicals in Freshwater Systems." Environ. Res. Lab., EPA. EPA-600/17-77-113 (1977).

2. National Research Council. "Polychlorinated Biphenyls." Washington, D.C. p. 182 (1979).

3. Liss, P.S. "Chemistry of the Sea Surface Microlayer." In: Chemical Oceanography, Vol. 2, (ed. J. P. Riley and G. Shirrow), Academic Press, London, pp. 193-243 (1975).

4. Seba, D.B. and Corcoran, E.F. "Surface Slicks as Concentrators of Pesticides in the Marine Environment." Pestic. Monit. J. $\underline{3}$, 190 (1969).

5. Bidleman, T.F., Matthews, J.R., Olney, C.E., and Rice, C.P. "Separation of Polychlorinated Biphenyls, Chlordane, and p,p'-DDT from Toxaphene by Silicic Acid Column Chromatography." J. Assoc. Offic. Anal. Chem. $\underline{61}$, 820 (1978).

6. Baumann-Ofstad, E., Gulbrand, L., and Drangsholt, H. "Chlorinated Organic Compounds in the Fatty Surface Film on Water." Intern. J. Environ. Anal. Chem. $\underline{6}$, 119 (1979).

7. Andren, A.W., Elzerman, A.W., and Armstrong, D.E. "Chemical and Physical Aspects of Surface Organic Microlayers in Freshwater Lakes." J. Great Lakes Res. $\underline{2}$, Supplement $\underline{1}$, 101 (1976).

8. Eisenreich, S.J., Elzerman, A.W., and Armstrong, D.E. "Enrichment of Micronutrients, Heavy Metals, and Chlorinated Hydrocarbons in Wind-generated Lake Foam." Environ. Sci. Technol. $\underline{12}$, 413 (1978).

9. Elzerman, A.W. "Surface Microlayer-microcontaminant Interactions in Freshwater Lakes." Ph.D. Thesis, Water Chemistry Program, University of Wisconsin-Madison (1976).

10. Rice, C.P., Eadie, B.J., and Erstfeld, K. M. "Enrichment of PCBs in Lake Michigan Surface Films." J. Great Lakes Res. 8(2), 265 (1982).

11. Slinn, W.G.N., Hasse, L., Hicks, B.B., Hogan, A.W., Lal, D., Liss, P.S., Munnich, K.O., Sehmel, G.A., and Vittori, O. "Some Aspects of the Transfer of Atmospheric Trace Constituents Past the Air-Sea Interface." Atmospheric Environment 12, 2055 (1978).

12. MacKay, D. "Volatilization of Pollutants from Water." In: Aquatic Pollutants: Transformations and Biological Effects. I.H. Lelyreld and B.C.J. Zoeteman. pp. 175-185. Proceedings of the Second International Symposium on Aquatic Pollutants, Amsterdam, The Netherlands, Sept. 26-28, 1977.

13. Haque, R. and Schmedding, D.W. "Studies on the Adsorption of Selected Polychlorinated Biphenyl Isomers on Several Surfaces." J. Environ. Sci. Health B 11(2), 129 (1976).

14. Södergren, A. "Origin of ^{14}C and ^{32}P Labelled Lipids Moving To and From Freshwater Surface Microlayers." Oikos 33, 278 (1979).

15. Johnson, B.D. and Cooke, R.C. "Organic Particle and Aggregate Formation Resulting from the Dissolution of Bubbles in Seawater." Limnol. Oceanogr. 25(4), 653 (1980).

16. Wheeler, J.R. "Formation and Collapse of Surface Films." Limnol. Oceanogr. 20(3), 338 (1975).

17. Garrett, W.D. "Collection of Slick-forming Materials from the Sea Surface." Limnol. Oceanogr. 10, 602 (1965).

18. Harvey, G.W. and Burzell, L.A. "A Simple Microlayer Method for Small Samplers." Limnol. Oceanogr. 17, 156 (1972).

19. Owen, R.M. and Meyers, P.A. "Petroleum Hydrocarbons and Heavy Metals in Great Lakes Surface Films." Mich. Sea Grant. Tech. Rep. 60 (1978).

20. Meyers, P.A. and Kawka, O.E. "Fractionation of Hydrophobic Organic Materials in Surface Microlayers." J. Great Lakes Res. 8, 288 (1982).

21. Menzel, D.W. and Vaccaro, R.F. "The Measurement of Dissolved Organic and Particulate Carbon in Seawater." Limnol. Oceanogr. 9, 138 (1974).

22. Strickland, J.D.H. and Parsons, T.R. "A Practical Handbook of Seawater Analysis," second edition. Ottawa: Alger Press Ltd. (1972).

23. Christensen, E.J., Olney, C.E., and Bidleman, T.F. "Comparison of Dry and Wet Surfaces for Collecting Organochlorine Dry Desposition." Bull. Environ. Contam. Toxicol. 14, 679 (1979).

24. Afifi, A.A. and Azen, S.P., "Statistical Analysis: a Computer Oriented Approach." Academic Press, Inc., London. 366 pp. (1973).

25. Wetzel, R.G. "Limnology." W.B. Saunders Company, Philadelphia, London, Toronto. p. 743 (1975).

26. Paerl, H.W. "Microbial Organic Carbon Recovery in Aquatic Ecosystems." Limnol. Oceanogr. 23, 927 (1978).

27. Karickhoff, W.W., Brown, D.S. and Scott, T.A. "Sorption of Hydrophobic Pollutants on Natural Sediments." Water Research 13, 241 (1979).

28. Eisenreich, S.J., Capel, P., and Looney, B. "PCB dynamics in Lake Superior Water." In Proceedings: Conference on the Physical Behavior of PCB in the Great Lakes. University of Toronto, Toronto, Canada, December 1981.

29. O'Conner, D. and Connolly, J.P. "The effect of concentration of adsorbing solids on the partition coefficient." Water Research 14, 1517 (1980).

30. Doskey, P.V. and Andren, A.W. "Modeling the flux of atmospheric polychlorinated biphenyls across the air/water interace." Environ. Sci. Technol. 15, 705 (1981).

31. Murphy, T.J. and Rzeszutko, C.P. "Precipitation Inputs of PCBs to Lake Michigan." J. Great Lakes Res. 3, 305 (1977).

CHAPTER 11

PCB DYNAMICS IN LAKE SUPERIOR
WATER

S. J. Eisenreich
  Environmental Engineering Program
  Department of Civil and Mineral Engineering
  University of Minnesota

P. D. Capel
  University of Minnesota

B. B. Looney
  University of Minnesota

INTRODUCTION

Polychlorinated biphenyls (PCBs), as a class of anthropogenic organic chemicals, are ubiquitous in the marine and freshwater environments of our planet (1,2). The global distribution of PCBs is demonstrated by their occurrence in fish products from as dissimilar environments as the Antarctic Ocean, the North Atlantic Ocean, North Sea and Alpine Lakes (3-5), the Great Lakes (6-9) and various rivers and streams of the terrestrial regime (10,11). Since 1929 $\sim$ 2 x 10^9 kg of PCBs have been commercially produced, the majority by Monsanto, and $\sim$ 2 x 10^8 kg remain in mobile environmental reservoirs. The voluntary and legislative bans instituted since 1971 on the production, sale and use of PCBs in open systems suggests that the tropospheric and hydrospheric burdens should be decreasing. Decreasing concentrations in Great Lakes' fishes (12), sediment (13), and air (14) support this hypothesis. However, PCBs are refractory compounds resisting chemical and biological degradation and will remain active for decades to come (15).

Since 1977, we have conducted investigations into the sources and sinks of PCBs in Lake Superior, and the dynamics responsible for their distribution in various environmental compartments. The objectives of these studies were to better understand the cycling of PCBs in large lakes, and to use

PCBs as a tracer of similar hydrophobic, low-water soluble organics. Our studies have shown that atmospheric transport and deposition in the sparsely-populated Lake Superior basin is the dominant input pathway, and that loss to the sediments is the dominant sink (16). Tributary and municipal/industrial discharges represent the remaining inputs, and outflow through the St. Mary's River the remaining loss term. Preliminary mass budgets for PCBs in Lakes Superior (16) and Michigan (17) indicate that inputs exceed outputs by a factor of 2 to 3. This means that 1) atmospheric inputs have been overestimated; 2) important loss terms have been omitted (e.g., volatilization, aerosolization); and/or 3) the water burden is increasing, although recent information suggests that reported concentrations are high by a factor of 5 to 100.

This paper represents a summary of our findings regarding PCBs in the waters of Lake Superior following investigations from 1978 to 1980. PCB concentrations averaged 1.3 ± 1.3 ng L^{-1} in 1978 (n = 30), 3.6 ± 1.9 ng L^{-1} in 1979 (n = 35) and 0.9 ± 0.4 ng L^{-1} in 1980 (n = 62). Aroclor 1242 and 1254 commercial mixtures each accounted for $\sim$ 50% of the total PCBs. Concentrations were generally higher in the surface waters than deeper waters (> 50 m), and were highest in the central region of the lake between Thunder Bay, Ontario and the Keweenaw Peninsula. Depth profiles at selected sites in 1978 and 1980 suggested that air/water and sediment/water exchanges were contributing to higher PCB concentrations in surface and bottom waters, respectively. Individual isomer profiles (n = 3,4,5) suggested the atmosphere as a primary source of n = 3 and 4 species in surface waters, whereas all isomers exhibited high concentrations above the sediments. PCBs partition between suspended solids (SS) and water to yield partition coefficients (K_p) of 5×10^4 to 10^7. The relationship of K_p to SS is expressed by the equation: $\log K_p = 1.2 \log SS + 5.4$. That is, K_p is inversely proportional to the SS concentration. PCBs exist $\sim$ 75% in the "dissolved" phase, and $\sim$ 25% in the "particulate" phase. Consideration of PCB concentrations and fluxes leads us to believe that the hydrospheric burden of PCBs in Lake Superior is decreasing in response to diminished inputs. The estimated residence times of PCBs is $\sim$ 2 to 6 years.

METHODS

During the summers of 1978, 1979, and 1980 three extensive sampling trips were conducted on Lake Superior. The cruise of July 1978 was limited to the western arm and central basin of Lake Superior, whereas the 1979 (June) and 1980 (August) cruises covered the entire lake, with the most complete data set generated in 1980. In many instances, water samples were taken from nearly identical sites in different years.

Most water samples were obtained by hydrocast, using 30 L General Oceanics Go-Flo sampling bottles specially coated on the inner surfaces to minimize sorption of trace organics from water. The sampler was thoroughly cleaned prior to use with laboratory detergent and water, and equilibrated in lake water. The water sampler was passed through the air/water interface closed, opened at $\sim$ 3 m by triggering a hydrostatic valve and adjusted to the desired depth. Surface water samples (1 m, 10 m) for isolation of particulate PCBs were obtained by pumping through thick-walled Tygon tubing directly through a stainless-steel filtration unit (Millipore, 292 mm diameter) containing a pre-combusted glass filter containing no bonding agent. The bulk water samples and filtrate were transferred to 20 L pre-cleaned glass carboys, and 20 to 40 L were collected for each sample. Total volumes of lake water filtered ranged from 60 to 120 L. The filters were transferred wet to cleaned-glass jars with Al-lined caps, and $\sim$ 50 mL acetone was added to initiate extraction.

The water samples (bulk, filtrate) were pumped through short columns of cleaned XAD-2 macroreticular resin at 100-200 mL min^{-1}. The columns were made of glass with the XAD-2 dimensions being $\sim$ 2 x 15 cm holding $\sim$ 70-100 mL of slurry. Glass wool at each end of the column held the XAD-2 resin in place. XAD-2 resin is effective at isolating hydrophobic organics from water at ambient concentrations.

Table 1 describes the analytical procedure for isolating PCBs from XAD-2 resins and particulate-laden filters. For 1979 and 1980, the XAD-2 resin and filters were extracted using a 1:1, hexane: acetone mixture in a soxhlet apparatus for 24 hours, then extracted three times with Milli-Q water to remove the acetone. The extracts were dried with anhydrous Na_2SO_4, concentrated to $\sim$ 1 mL using a micro Kudurna-Danish unit, and fractionated on Florisil eluting with hexane. Following volume reduction to $\sim$ 0.5 mL with high-purity N_2, the extracts were analyzed for PCBs by glass capillary gas chromatography (GC) and e$^-$ capture detection. The columns were 25 m in length and either 0.22 mm (SP-2100, 1979) or .32 mm i.d. (SE-54, 1980). The quantification and identification of Aroclor mixtures and individual isomers was performed with a computer-generated least square analysis. Overall recovery of spiked water samples averaged $\sim$ 70%. No correction of generated data for experimental losses were made. Additional details of this procedure are available (18).

The filter extracts were fractionated using a larger Florisil column prepared by heating Florisil to 550°C for 4 hours, and then at 130°C overnight. The Florisil was deactivated by addition of 1.25% water which permitted improved separation of PCBs, chlorinated pesticides and toxaphene. The PCB fraction was eluted with 60 mL hexane, concentrated as before to ∿ 0.5 mL and analyzed. PCB analyses were performed on a HP GC Model 5840A equipped with a ^{63}Ne̅ capture detector, splitless injection and autosampler. Column conditions were as follows:

	1978, 1979	1980
Column	SP 2100	SE 54
Length	25 m	25 m
Temp. 1	150°C	150°C
Time 1	4 min.	4 min.
Temp. 2	200°C	200°C
Rate	1°C min^{-1}	1°C min^{-1}
Run Time	75 min.	65 min.
Inject. Temp.	230°C	230°C
Detector Temp.	340°C	340°C
Carrier	N_2	N_2
Flow	1 mL min^{-1}	1 mL min^{-1}

The 1978 XAD-2 resin extracts, after removal of acetone, were concentrated to ∿ 0.3 mL and fractionated by high performance liquid chromatography (HPLC) in the reverse-phase mode to remove interfering substances. The entire 0.3 mL of sample extract was injected into a series of two, 30 cm μ-Bondapak C-18 columns, and isocratically chromatographed for 43 minutes with an 1:1 acetonitrile:water mixture, followed by a linear gradient to 100% acetonitrile over 5 minutes at 1 mL min^{-1}. The fraction containing PCBs and chlorinated pesticides quantitatively elute at 56 to 68 minutes. The PCBs were quantified and identified as for the 1979 samples above.

Analysis of PCBs in water samples by high resolution glass capillary GC provides a chromatogram with more than 60 individual peaks representing PCB isomers. A computer program was written to estimate the contribution of Aroclor 1242 and 1254 to the total PCB composition. These commercial mixtures best approximated the PCB composition in air, water and sediments of Lake Superior (13,16). The computer program was based on an "n-observables in two-parameters least square analysis) (19) and generated total and mixture concentrations. For the 1979 samples, 50 (n=50) diagnostic peaks were used, whereas 35 peaks (n=35) were used for 1980 samples.

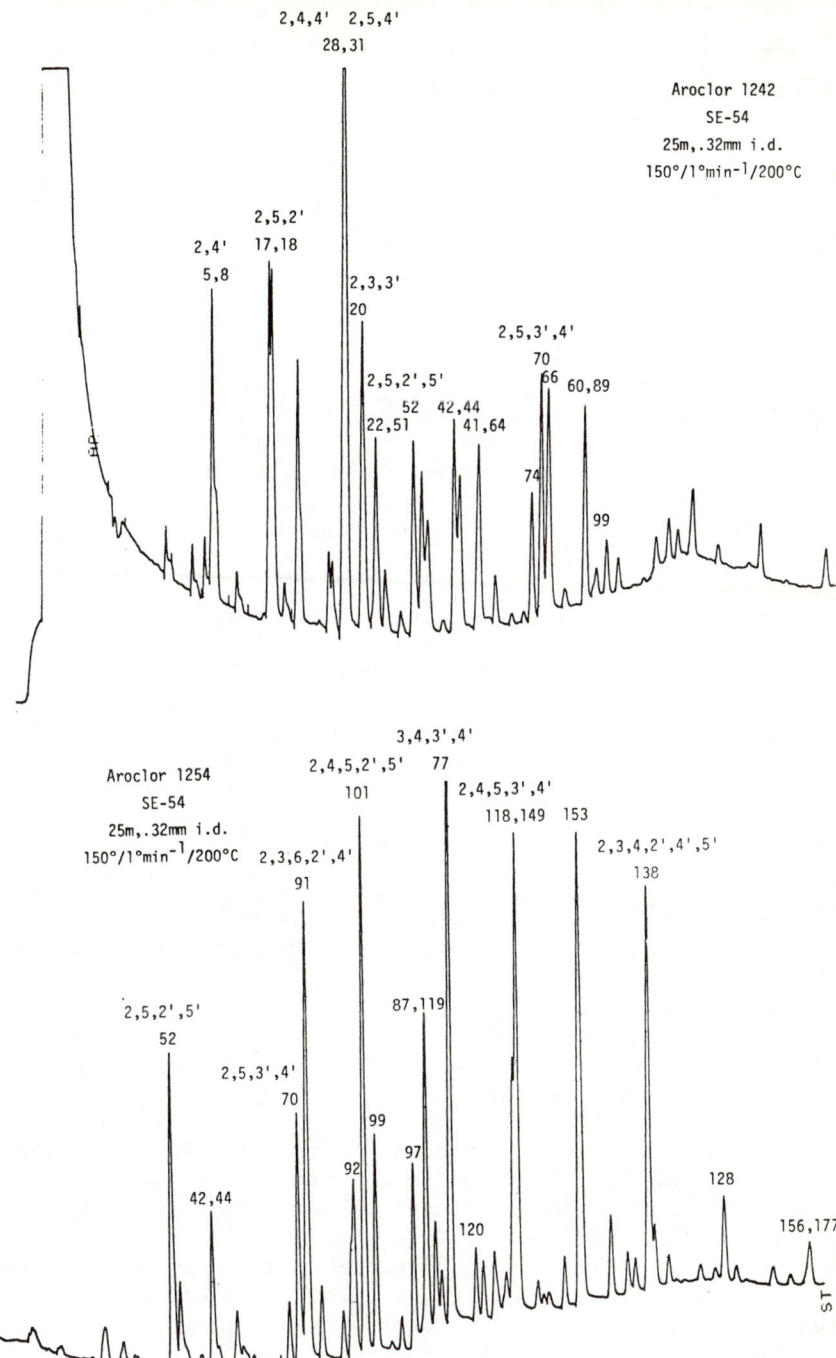

Figure 1a and b   Gas Chromatograms of Aroclor 1242 and 1254

Table 1. Analytical Procedure

ISOLATION:	40-L; XAD-2 100-200 mL min^{-1}
EXTRACTION:	Soxhlet 1:1 hexane:acetone, 24 hrs. Back-extraction with water Hexane dried with anhd. Na_2SO_4
CONCENTRATION:	Kuderna-Danish to 8 mL Micro K-D to 1 mL
FRACTIONATION/ CLEAN-UP:	Florisil, 0.3 g Hexane, 4 mL Hexane: diethyl ether, 9:1, 4 mL
CONCENTRATION:	Gentle evaporation to $\sim$ 0.5 mL with N_2
ANALYSIS:	Glass-capillary GC e^- capture detection SP-2100 or SE-54; 25 m
QUANTIFICATION:	Computer-generated least square analysis Individual isomer identification
RECOVERY:	$\sim$ 70%

PCB isomers in Aroclor mixtures and samples were identified by comparison of the Aroclor standards to single isomer standards and the characterization of PCB isomers provided by Ballschmiter and Zell (4). Twelver PCB isomer standards were used to check the accuracy of literature identifications and the identity of the other isomers established by comparison to relative retention times.

Concentrations of individual PCB isomers were determined by quantifying the PCB isomer concentrations in Aroclor standards. This was accomplished by taking the relative molar percentages of individual PCB isomers in Aroclor 1242 (20) and Aroclor 1254 (21) and generating relative molar responses of individual isomers on the e^- captive detector. The net result is the capability to identify and quantify individual PCB isomers in environmental samples. Figures 1a, b show the glass capillary gas chromatograms of Aroclor 1242 and 1254 run on a SE-54 column (25 m x .32 mm i.d.). The numbering sequence for PCB isomers is identical to that used by

Ballschmiter and Zell (3,46) and prominent PCB isomers are labelled for reference. Table 2 gives the systematic number, structural identification, weight % of each isomer in Aroclor 1242 and 1254 and the relative GC retention time under the chromatographic conditions described above.

RESULTS AND DISCUSSION

The concentrations of PCBs in Lake Superior have been reported over the last ten years to range from 0.8 to $\sim$ 110 ng L^{-1} in the open waters, and perhaps up to 200 ng L^{-1} in the nearshore waters and harbors (6-9,22). By comparison, the concentrations of PCBs in Lake Michigan have been reported to range from 1 ng L^{-1} to as high as 200 ng L^{-1} in open waters, and higher in harbors and embayments. PCB concentrations in the contaminated Waukegan Harbor of southern Lake Michigan ranged during 1979 from 40 to 100 ng L^{-1}, while 0.5 km offshore, concentrations of 1 to 10 ng L^{-1} were typical (23). Recent information provided by D. Armstrong (Univ. of Wisc., pers. comm.) and C. Rice (Univ. of Mich., pers. comm.) suggest that the PCB concentration is probably 1 to 5 ng L^{-1} in the open waters of Lake Michigan. Reported PCB concentrations apparently decrease over time which probably results more from improved analytical capabilities than an actual decrease in water burden. Thus, the initial objective of our research was to establish the PCB concentrations in the aqueous compartment. In reviewing the literature, Capel (18) has estimated for comparison purposes the PCB concentrations in the Great Lakes, the marine environment, and typical rivers of the continental U.S. The range of concentrations in the individual Great Lakes suggest that internal processes of adsorption and sedimentation maintain PCBs at low and relatively constant concentrations independent of loading.

Temporal and Spatial Variations

The concentrations of total (dissolved plus particulate) PCB in the waters of Lake Superior were measured in the summers of 1978 to 1980. The PCBs were isolated by pumping the water from glass carboys through glass columns containing XAD-2 resin with glass wool at each end. Columns connected in series showed that under the sampling conditions employed (100-200 mL min^{-1}; 20-40 L water), no PCBs were detected in test back-up columns. PCBs were isolated by a combination of adsorption of dissolved PCBs onto XAD-2, and filtration of particulate PCBs.

PCB concentrations averaged 1.3 ± 1.3 ng L^{-1} (n=30), 3.6 ± 1.9 ng L^{-1} (n=35) and 0.9 ± 0.4 ng L^{-1} (n=62) in 1978,

Table II. PCB Isomer Identification[1]

Systematic Number[2]	Structure	Wgt. % 1242[3]	1254[4]	Ret. Time
4	2,2'	3.4	--	10.06
5	2,3	---	--	12.80
6	2,3'	1.06	.05	12.35
7	2,4	.89	--	11.63
8	2,4'	7.66	--	12.80
9	2,5	.26	--	11.63
10	2,6	.11	--	10.06
14	3,5	.30	--	14.37
15	4,4'	.84	--	16.47
16	2,2',3	3.21	--	18.11
17	2,2',4	2.88	--	16.47
18	2,2',5	9.23	.06	16.47
19	2,2',6	.96	--	14.37
20	2,3,3'	3.59	--	23.05
22	2,3',4'	2.60	--	22.88
24	2,3,6	---	--	17.29
26	2,3'5	.54	--	20.22
28	2',4,4'	13.12	--	20.97
29	2,4',5	---	--	20.02
31	2,4',5	4.29	.57	20.97
32	2,4',6	2.12	--	18.11
34	2',3,5	---	--	20.22
36	3,3',5	---	--	24.37
37	3,4,4'	1.59	.16	27.96
40	2,2',3,3'	.17	.23	30.12
41	2,2',3,4	1.87	---	29.11
42	2,2',3,4'	---	1.95	27.59
44	2,2',3,5'	1.19	--	27.59
47	2,2',4,4'	1.85	.46	26.00
49	2,2',4,5'	3.67	1.47	25.62

Table II. PCB Isomer Identification[1] (continued)

Systematic Number[2]	Structure	Wgt. %1242[3]	1254[4]	Ret. Time
51	2,2',4,6'	---	--	22.88
52	2,2',5,5'	4.56	3.89	25.13
60	2,3,4,4'	.23	--	35.66
62	2,3,4,6	---	--	26.00
64	2,3,4',6	---	--	29.11
66	2,3',4,4'	.91	2.00	33.47
67	2,3',4,5	---	--	31.09
69	2,3',4,6	---	--	25.62
70	2,3',4',5	1.24	4.24	32.92
74	2,4,4',5	2.26	.27	32.34
77	3,3',4,4'	.43	.11	42.25
79	3,3',4,5'	.26	.21	36.97
83	2,2',3,3',6	.48	1.72	39.21
84	2,2',3,3',6	.48	1.72	36.33
85	2,2',3,4,4'	.50	2.15	41.32
87	2,2',4,4,5'	.11	3.81	40.75
89	2,2',3,4,6'	---	--	35.66
91	2,2',3,4',6	---	5.00	33.47
92	2,2',3,5,5'	.15	.63	36.33
95	2,2',3,5',6	.66	--	34.37
97	2,2',3',4,5	---	2.59	39.98
99	2,2',4,4',5	.69	6.10	37.67
101	2,2',4',5,5'	.34	6.98	36.97
105	2,3,3',4,4'	.31	--	49.22
108	2,3',4,5'	.58	.55	47.36
118	2,3',4,4',5	---	8.09	46.36
119	2,3',4,4',6	---	--	40.75
120	2,3',4,5,5'	.39	.15	43.79
128	2,2',3',4,4'	---	1.45	58.98
129	2,2',3,3',4,5	---	--	55.55

Table II. PCB Isomer Identification[1] (continued)

Systematic Number	Structure[2]	Wgt. % 1242[3]	1254[4]	Ret. Time
130	2,2',3,3',4,5'	---	--	53.52
134	2,2',3,3',5,5'	---	.42	47.56
135	2,2',3,3',5,6'	---	.22	44.93
137	2,2',3,4,4',5	---	--	53.08
138	2,2',3,4,4',5	.11	4.60	54.36
141	2,2',3,4,5,5'	---	--	52.05
144	2,2',3,4,5',6	---	--	46.36
147	2,2',3,4',5,6	---	--	44.93
149	2,2',3,4',5',6	---	3.96	46.36
151	2,2',3,5,5',6	---	.36	44.24
153	2,2',4,4',5,5'	.03	3.67	50.18
159	2,3,3',4,4',5	---	.20	64.47
177	2,2',3,3',4',5,6	---	---	64.47
179	2,2',3,3',5,6,6'	---	.68	52.05
187	2,2',3,4',5,5',6	---	.58	60.00

[1] SE 54 column (25 m, .932 mm i.d.), 150°C/4 min/1°C min^{-1}/ 200°C
[2] Ballschmiter and Zell (1980) - ref. 46
[3] Albro and Parker (1979) - ref. 20
[4] Albro et al. (1981) - ref. 21

Table III. Spatial Distribution of PCBs in Lake Superior

	Western Arm					
	1978		1979		1980	
	S	B	S	B	S	B
C_1 ng L^{-1}	1.2	0.7	3.2	3.4	0.8	0.9
R	.7-2.0	.5-1.2	---	---	.5-1.0	.4-2.1
SD	0.5	0.3	---	---	0.2	0.6
% 1242	45	47	71	45	60	57

Table III. (continued)

Central Region

C, ng L^{-1}	2.3	1.4	5.3	2.9	1.1	0.8
R	.4-7.6	.6-1.8	3.5-8.4	.9-6.0	.5-1.9	.4-1.6
SD	2.6	0.5	1.9	2.0	0.4	0.4
% 1242	62	48	35	41	57	52

Eastern Region

C, ng L^{-1}	---	---	3.1	3.9	1.1	1.0
R	---	---	.3-6.0	3.3-5.2	.5-1.5	.3-1.8
SD	---	---	1.9	0.7	.3	.5
% 1242	---	---	37	38	51	56

1979 and 1980, respectively (Figure 2). In general, bulk water PCB concentrations were distributed evenly between Aroclor 1242 and 1254. The range in observed concentrations in 1978 (0.4-7.4 ng L^{-1}) and 1979 (0.8-8.4 ng L^{-1}) differed considerably from the range observed in samples collected in 1980 (0.3-2.1 ng L^{-1}). Thus, both the lakewide mean and range in observed concentrations decreased over the three years in which detailed measurements were made. At present there is no justification for concluding that water concentrations are decreasing based on these measurements alone. However, PCB profiles in sediment cores collected in 1977 and 1978 suggest that recent inputs are somewhat less than prior to 1970 (13,24). More importantly, the airborne PCB concentrations over Lake Superior have definitely decreased in the period 1978 to 1981 (25). In this period, mean airborne PCBs decreased from $\sim$ 1.5 ng m^{-3} to $\sim$ 0.3 ng m^{-3}.

The higher concentrations observed lakewide in 1979 were at first alarming. However, the annual variation in PCB levels are greatly affected by seasonal variations in quantity and composition of particulate matter and in the hydrographic regime (26). The severity of the previous winter, and the time of year in which sampling occurred are important considerations. The sampling trip in 1979 occurred in late June following a long, snowy and cold winter in which Lake Superior totally froze over. The sampling trips in 1978 and 1980 occurred later in the summer. Also, no open-lake stratification had been established at the time of sampling with the result that surface water temperatures were ∿ 3 to 4°C in 1979, and ∿ 6 to 20°C in 1978 and 1980. We suspect that the combination of ice cover during a severe winter and an early sampling cruise led to the elevated concentrations. We have no reason to suspect that the PCB concentrations measured in 1979 were inaccurate. The higher PCB concentrations likely result from increased sediment resuspension and snow inputs in conjunction with decreased PCB volatility under ice cover and colder water temperatures. Seasonal variations in hydrophobic pollutant concentrations in the water column need to be further examined.

The PCB burden in Lake Superior is not uniformly distributed with respect to vertical and horizontal space (Figure 2, Table 3). In 1978 to 1980, the concentrations were highest in the central region between Thunder Bay and the Upper Peninsula of Michigan, and were always highest in the 1 m samples than at depths of 5 m above the bottom. This behavior suggests regional surface loading of PCBs from the atmosphere or horizontal transport from nearshore areas. Sediment (27) and fish (6) PCB concentrations in the central region of Lake Superior also appear to be higher than the rest of the lake. Swain (6) has found elevated concentrations of PCBs and other CHs in fish and water in Siskiwet Lake on Isle Royale, and in nearby Lake Superior water. There is little doubt that loading/transport processes enhance PCB concentrations in the central region.

Figures 3a, b show the PCB concentrations in surface (1 m) and bottom waters (5 m above bottom) in 1980. The surface water concentrations were lowest in the western region with a mean of .8 ng L^{-1}, and about 1.1 ng L^{-1} in the central and eastern areas. The mean concentration in bottom waters was 0.9, 0.8 and 1.0 ng L^{-1} total PCB in the western, central and eastern regions, respectively. In general, the lowest PCB concentrations occurred in the surface waters of the western arm where the suspended solids (SS) load was highest. The SS concentration in the western, central and eastern regions averaged .66 ± .23, .45 ± .25 and .20 ± .06 ng L^{-1}, respectively. The higher SS load in the western region of Lake Superior, derived from erosion along Wisconsin's north shore and the sediment load of the St. Louis River (28-30) in conjunction with the lower PCB concentrations in the water column suggests that particles adsorb hydrophobic organics and remove them to the sediment. Higher fluxes of sediment and PCBs to the lake bottom in this area support this hypothesis (13,31). In summary, the concentration of total PCBs in the waters of Lake Superior is approximately 1 ng L^{-1}, about equally distributed between Aroclor 1242 and Aroclor

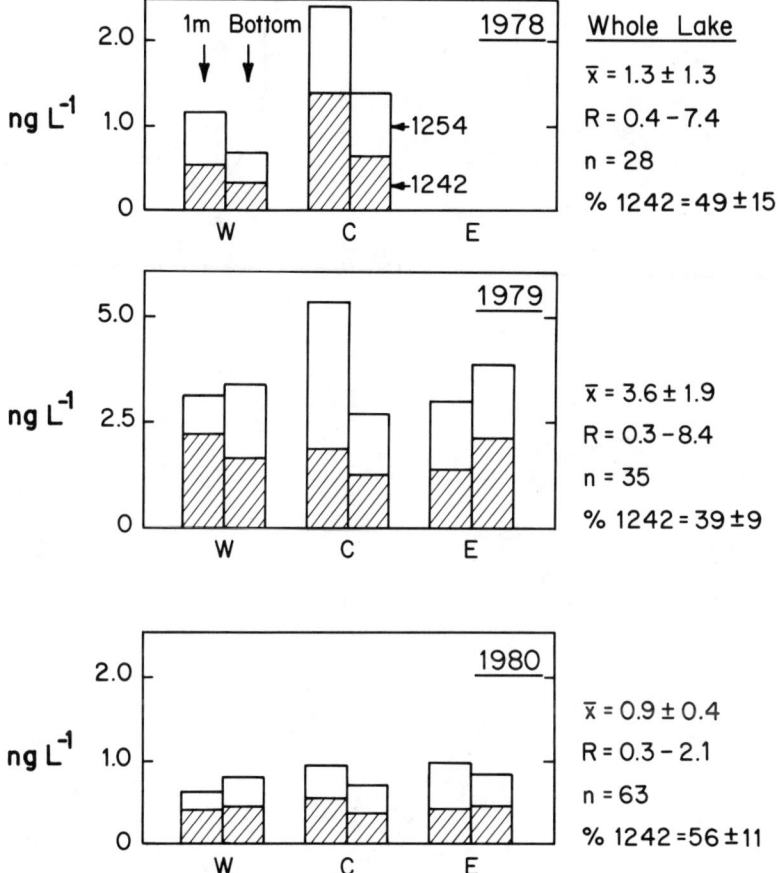

Figure 2 Total PCB Concentrations in Water of Lake Superior for the Years 1978 to 1980

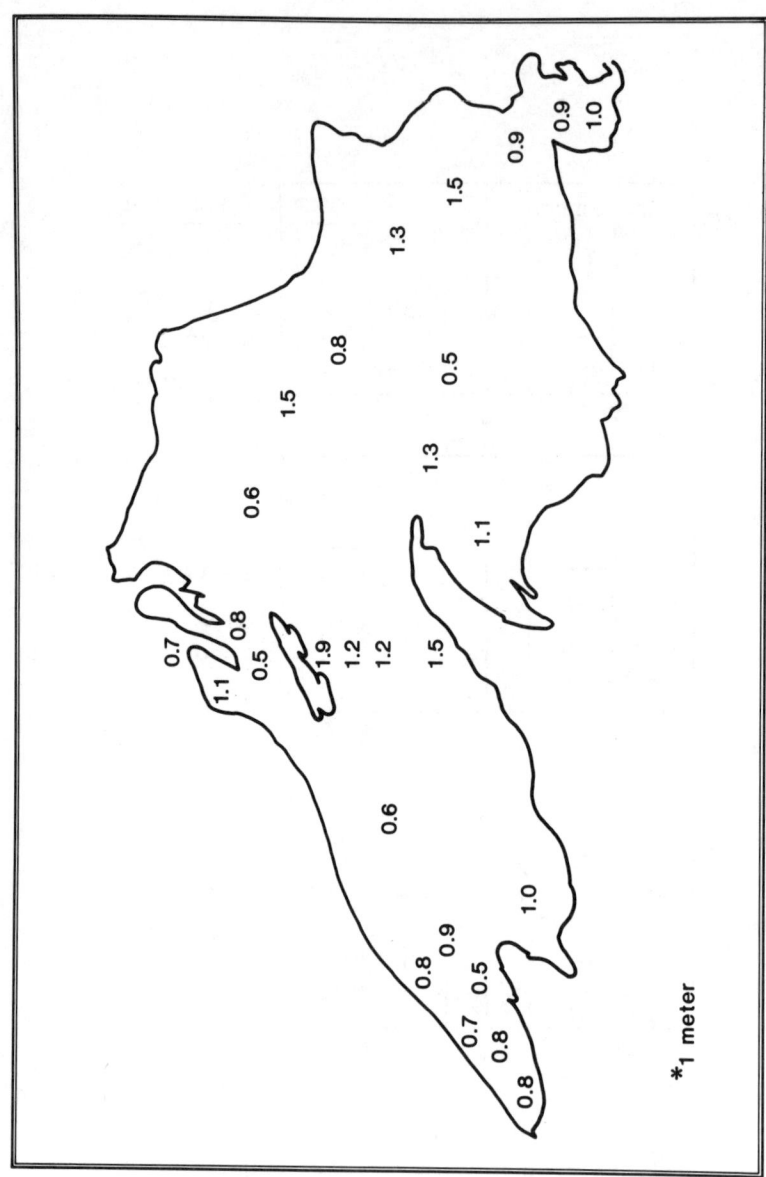

Figure 3(a) Spatial Distribution of Total PCBs in the Surface (1m) Water of Lake Superior in August, 1980. Units are ngL^{-1}.

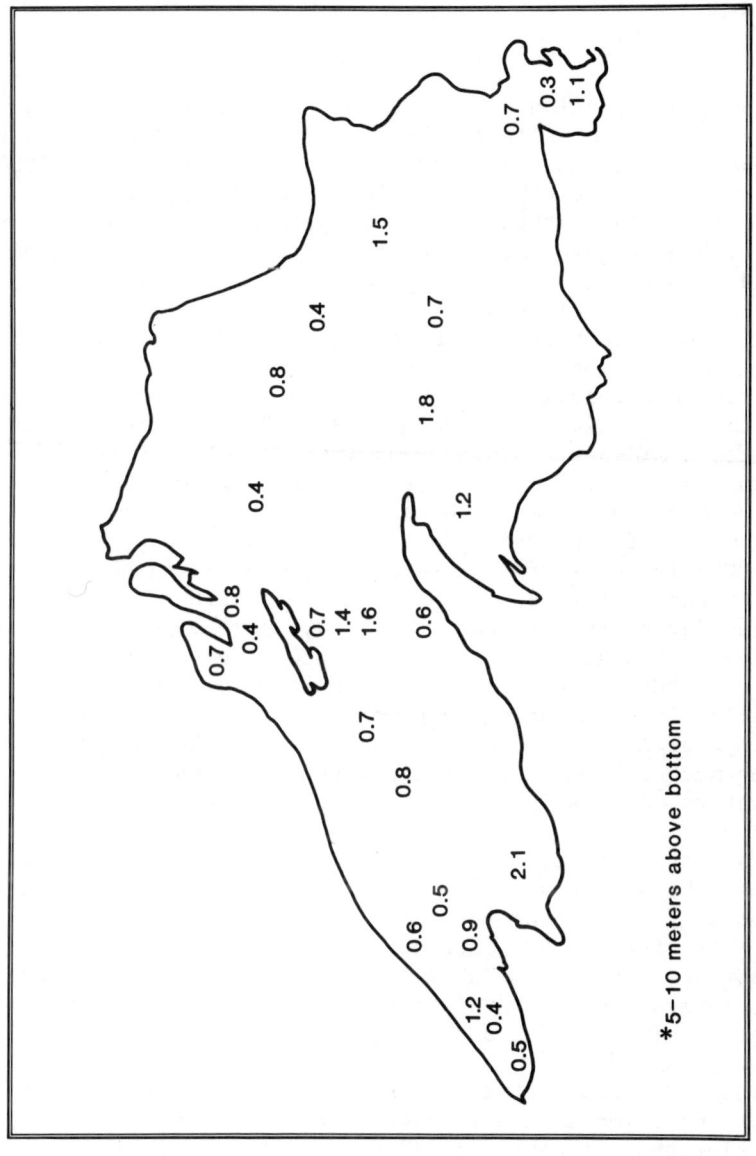

Figure 3(b) Spatial Distribution of Total PCBs in deep (5-10 m off bottom) Water of Lake Superior in August, 1980. Units are ngL^{-1}.

1254, with concentrations somewhat higher in the central region of the lake between Thunder Bay, Ontario and the Keweenaw Peninsula of Upper Michigan. At a PCB_4 concentration of 1 ng L^{-1} and a whole-lake volume of 1.2 x 10^4 km^3, the PCB burden of Lake Superior's water is ∿ 12,000 kg.

PCB Depth Profiles

The concentration of PCBs in surface waters (1 m) was usually equal to or slightly less than concentrations at 5 to 10 m above the bottom. On several occasions in 1978 to 1980, depth profiles were constructed for PCBs, SS and various water quality parameters. Two such profiles will be presented for 1980. Figure 4a,b shows the depth profiles at sites 14 (midway between Isle Royale and the Keweenaw Peninsula), and site 24 (central area of western region) observed in August, 1980. In both instances, surface concentrations were higher than those at 5 or 10 meters. At site 24, concentrations remained relatively constant with depth, even though SS and turbidity varied by factors of 2 to 3. At site 14, PCBs exhibited a mid-depth minimum and increased toward the bottom. The temperature profile exhibited a thermocline depth of ∿ 30 m, matched by thermocline maxima in SS and turbidity. The distribution of Aroclor 1242 on a percentage basis followed total PCBs in the surface waters, while remaining constant in the bottom waters at ∿ 50%. Since the airborne PCBs strongly favor Aroclor 1242 over 1254, there appears to be a coupling of the atmospheric and surface water regimes with respect to PCBs. That is, the % Aroclor 1242 in surface waters is directly influence by air/water exchange. The distribution of Aroclor 1242 with depth at site 24 is similar, with as much as 70% in the lighter PCB fraction in the surface waters, decreasing to ∿ 49% in the bottom waters.

With the lake stratified, the SS and turbidity profiles at these sites ought to show maxima near the bottom of the thermocline, reflecting accumulation of planktonic populations. At site 14, SS, turbidity and total PCB profiles show surface maxima at ∿ 30 m near the bottom of the thermocline, suggesting an association. The profile at site 24 has only a hint of such a relationship.

The PCB depth profile for site 14 (Figure 4a) shows concentration maxima in the surface water and near the sediment, suggesting that atmospheric and sedimentary exchange may be important input pathways. Since atmospheric PCBs are more heavily dominated by the lighter Aroclor 1242 (16,25,32) than are the sedimentary PCBs (13), individual isomer profiles may provide some insight as to the source of PCBs in the water. Figure 5 and Table 4 give the concentrations of

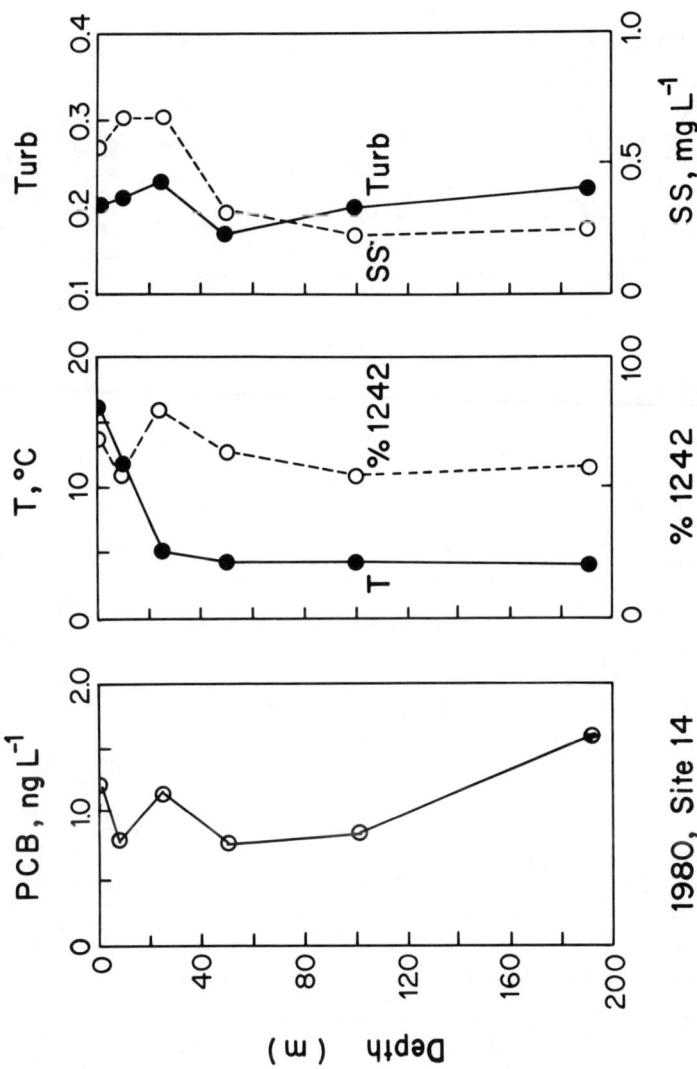

Figure 4(a) Depth Profiles of Total PCBs % Aroclor 1242, Temperature (°C), Turbidity and Suspended Solids at Two Sites in Lake Superior in August, 1980. Site 14 is Midway Between Isle Royale and the Keweenaw Peninsula.

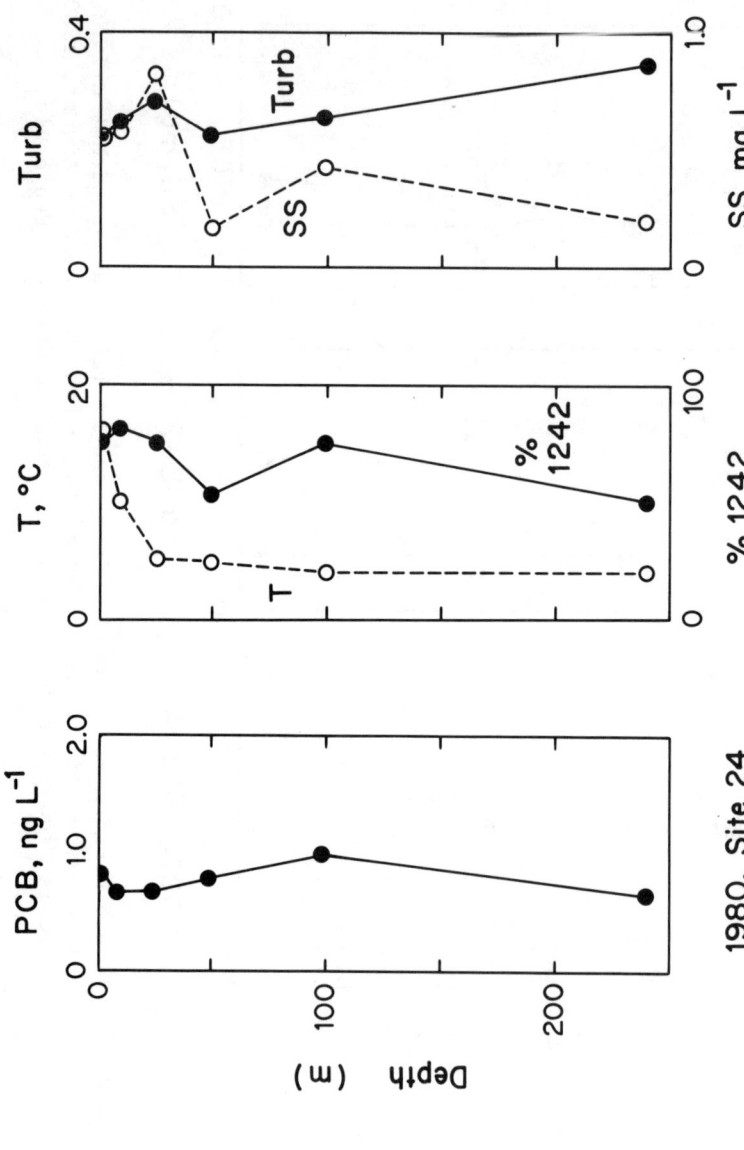

Figure 4(b) Depth Profiles of Total PCBs, % Aroclor 1242, Temperature (°C), Turbidity and Suspended Solids at Two Sites in Lake Superior in August, 1980. Site 24 is in the Central Part of the Western Half of Lake Superior.

typical 3, 4 and 5-chlorinated PCB isomers at depth at site 14. The 3 and 4 chlorinated isomers typical of the less chlorinated Aroclor 1242 follow in general the concentration profile for total PCBs. In contrast, the penta-chlorinated isomers have relatively constant concentrations with increasing depth until increasing near the bottom. We tentatively conclude that the PCBs in the water column at any given open lake station are a mix of atmospherically-derived lighter isomers and sediment-derived heavier isomers. Under the influence of stratification, the 2, 3 and 4-Cl PCB isomers under the influence of air/water exchange increase in concentration, while the the near bottom waters may be influenced by sediment resuspension and/or pore water advection.

Table IV. PCB Isomer Concentrations in Depth Profiles - Site 14, 1980 -

Isomer	Depth (m)					
	1	10	25	50	100	190
			($pg\ L^{-1}$)			
2,4,4'	74	36	67	33	47	82
2,3,3'	30	19	28	18	18	39
2,2',5,5'	35	28	29	14	28	53
2,2',4,5'	18	15	21	10	16	34
2,2',4,5,5'	38	33	34	31	36	95
2,2',4,4',5	21	20	27	12	19	45
Total PCB	1200	800	1100	700	900	1600

PCB Partitioning to Suspended Sediment

The availability of PCBs to the biota and the removal of PCBs from the lake is largely dependent on the extent of partitioning between the dissolved and particulate phases. Biota can take up PCBs from the dissolved phase by direct water uptake or from the particulate phase via ingestion. The pathways are not totally independent in that the dissolved phase supports the concentrations in both compartments. Weininger (33) and Weininger and Armstrong (17) have clearly shown that Lake Michigan lake trout, a predator fish, accumulates > 90% of its PCB burden via the dietary route, and little by direct uptake from water. Non-predator fish and plankton (especially algae) derive most of their PCB burden from the dissolved phase. Clayton et al. (34) suggest that PCB accumulation in zooplankton is predominantly controlled by equilibrium partitioning of the chemical between internal lipid pools and ambient water. Harding and Phillips (35) have found the phytoplankton can accumulate PCBs by the same process plus utilize a portion of the particulate PCBs by inducing desorption. Since sedimentation is the major loss process for PCBs in Lake Superior, the percentage in the particulate phase coupled to suspended solid flux controls water-column concentrations.

The factors controlling adsorption of PCBs and other hydrophobic pollutants have been described (36-37) and will only be summarized here. The partitioning of PCBs between the dissolved and particulate phase is governed by physical/chemical properties of the sorbent and sorbate. For the sorbent, organic adsorption increases with decreasing particle size and increasing organic content. Thus, the partition coefficient-K_p ($(ng/g)_s/(ng/g)_w$) was modified by the fractional organic content of the solid to yield: $K_{oc} = K_p/oc$. Recently, O'Connor and Connolly (38) and Di Toro et al. (39) provided evidence that K_p (and K_{oc}) decrease with increasing SS concentration. The latter group suggests that this anamolous behavior may be explained by the kinetics of PCB adsorption/desorption of exchangeable and non-exchangeable species.

The extent to which hydrophobic pollutants partition to SS can be predicted from solubility (S) and octanol-water partition coefficients (36):

$$\log K_{oc} = -.54 \log S + .44$$

$$\log K_{oc} = \log K_{ow} - .21$$

Although Karickhoff et al. (38) found no evidence to suggest that the partitioning process was not reversible, others suggest that the rate of desorption is less than the rate of adsorption, with the disparity increasing with increasing adsorption times.

In our 1980 studies, 24 water samples with volumes between 60 and 120 L were filtered through 0.6 μm glass-fiber filters (23 at 1m; 1 at 10 m) and the filtrate passed through XAD-2 columns. Measurement of SS concentration was performed by filtering 1 to 2 L water through similar glas fiber filters and determining the dry residue. Our objective was to determine the fraction of total PCB in the dissolved and particulate phases, with the corresponding K_p values.

For SS concentrations ranging from .13 to 4.95 mg L^{-1} ($\bar{x}_s$ = .64 ± .96 mg L^{-1}), K_p values ranged from 0.5 to 74 x 10^5 (log K_p = 4.65 to 6.87) with a mean value of 16.6 ± 23 x 10^5 (log K_p = 6.22). By comparison, the Saginaw Bay study

(Lake Huron) yielded a mean log $K_p \sim$ 4.1 to 4.6 for SS levels of 4 to 18 mg L^{-1} (39). In the estuarine waters of Puget Sound, Pavlou and Dexter (40) found log $K_p \sim$ 4 to 5 for SS concentrations of 4 to 20 mg L^{-1}. A log-log plot of K_p and SS (Figure 6) clearly shows that K_p is inversely related to the SS concentration through the best-fit linear equation:

$$\log K_p = -1.21 \log (SS) + 5.41$$
$$r^2 = .75$$

Thus, the apparent log K_p decreases from 6.6 at .1 mg L^{-1} to 5.4 at 1 mg L^{-1} and 4.2 at 10 mg L^{-1} SS. Also, the log K_p values observed in Saginaw Bay and Puget Sound at higher SS concentrations may be accurately estimated from the empirical equation developed for Lake Superior. This phenomenon is supported by the observations of Connolly and O'Connor (38) in many diverse systems. Although no firm explanation yet offers a plausible mechanism at the molecular level to explain the observations, its relative importance is not diminished. It is possible that the relationship described above is an artifact of the way in which the K_p is calculated, or indeed measured.

The percent of total PCB associated with particles ranged from 9 to 67%, and is a function of the quantity and composition of the SS (biotic, mineral), and the total PCB concentration. Figure 7 shows that 27 ± 12% (n = 24) of the total PCB is in the particulate phase. Although there is significant scatter in the data, the percent dissolved is not strongly affected by SS concentration in the range of 0.1 to 1 mg L^{-1}, suggesting that particle composition is important. Unfortunately, no measurement of particulate organic carbon was performed on these filters.

The solubility and vapor pressure of individual PCB isomers decreases with increasing numbers of chlorine (41), whereas K_{ow} increases. We would expect that the more highly-chlorinated PCB isomers would preferentially adsorb to particles. As a test of this hypothesis, log K was plotted against the GC retention time of individual PCB isomers identified and quantified in both particulate and dissolved phases of the 1 meter sample taken at site 14 (Figure 8). The plot should show higher K_p values with increasing RT (i.e., increasing Cl number). However, there is no discernible trend in K_p values. Murphy (this book) has shown that Henry's Law constants vary little with increasing chlorine number (i.e., ratio of vapor pressure to aqueous solubility is constant). Perhaps a similar relationship occurs between the activities in the particulate and aqueous phases for PCB isomers of in-

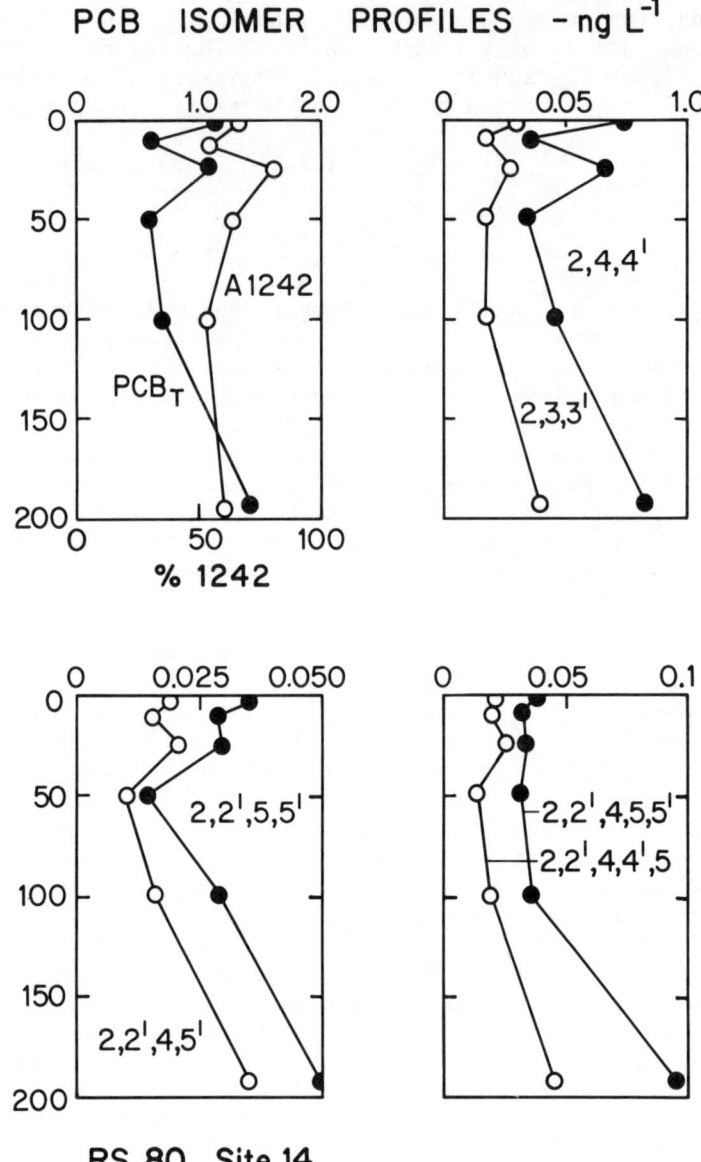

Figure 5 Depth Profiles of Total PCBs, % Aroclor 1242 and Two 3-CBs, 4-CBs and 5-CBs at Site 14 During 1980.

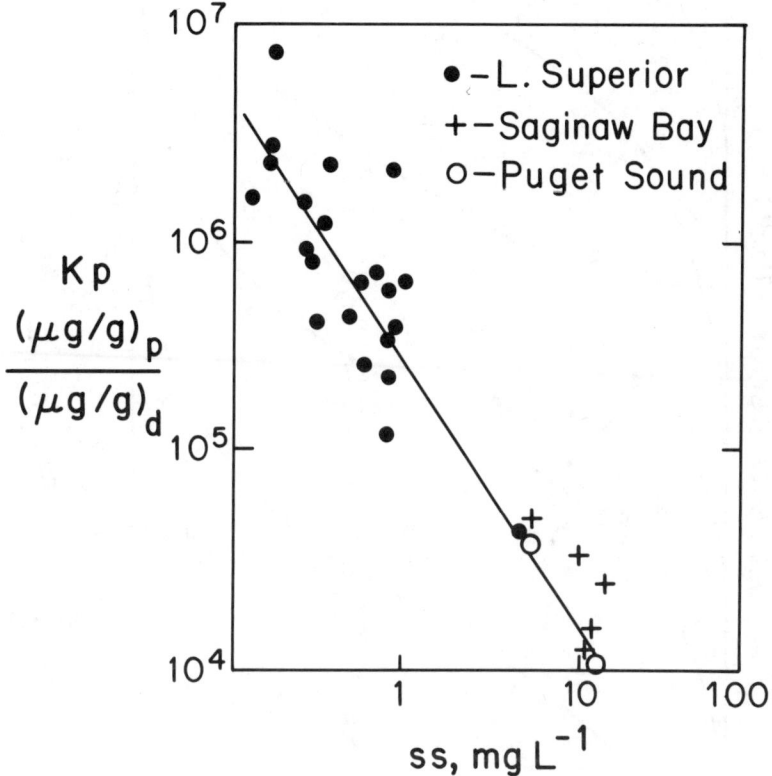

Figure 6 Relationship of the Partition Coefficient and SS for PCBs in Lake Superior During August, 1980. Saginaw Bay Data: ref. 39 Puget Sound Data: ref. 40.

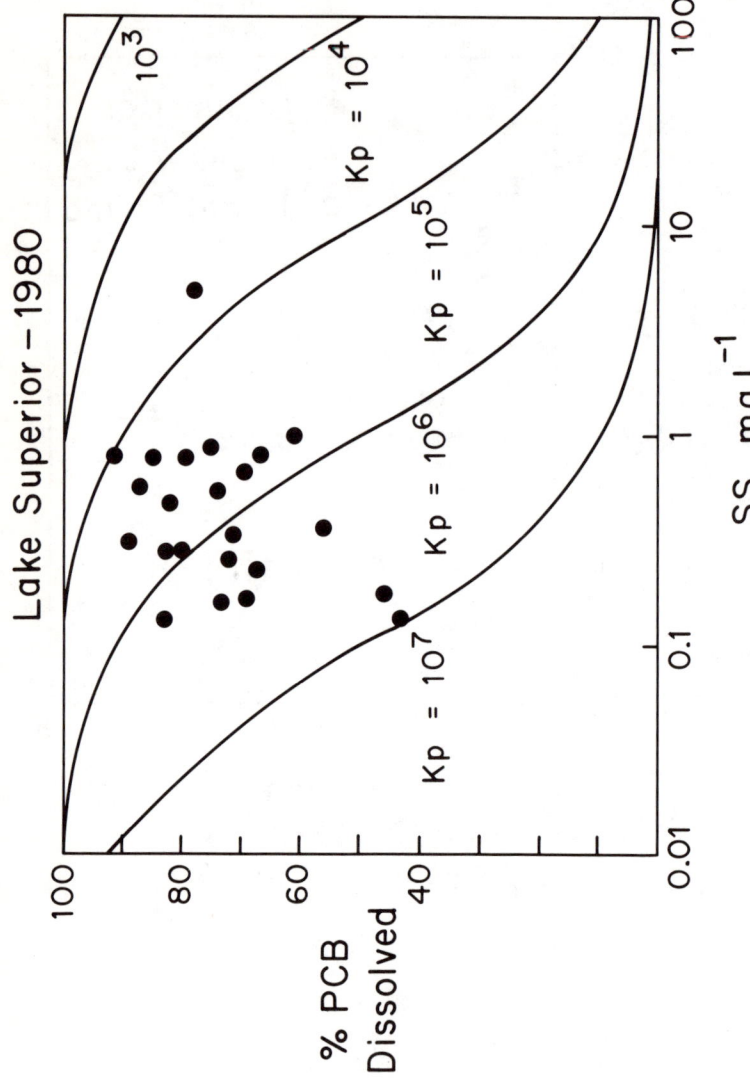

Figure 7 Relationship of % Dissolved PCB to SS Concentration in Lake Superior During August 1980.

creasing Cl number.

Comments on Sources and Sinks of Water-Column PCBs

The major sources of PCBs to Lake Superior are atmospheric deposition and tributary inputs. The magnitude of the inputs relative to major outputs (sedimentation, volatilization) is very much open to question and deserves comment.

Sources vs. Sinks: Doskey and Andren (42) and Hollod et al. (16) have attempted to model the atmospheric flux of PCBs to Lakes Michigan and Superior, respectively. Since more than 85% of airborne PCBs probably occur in the vapor phase (32,44), the importance of Henry's Law Constant (H) in

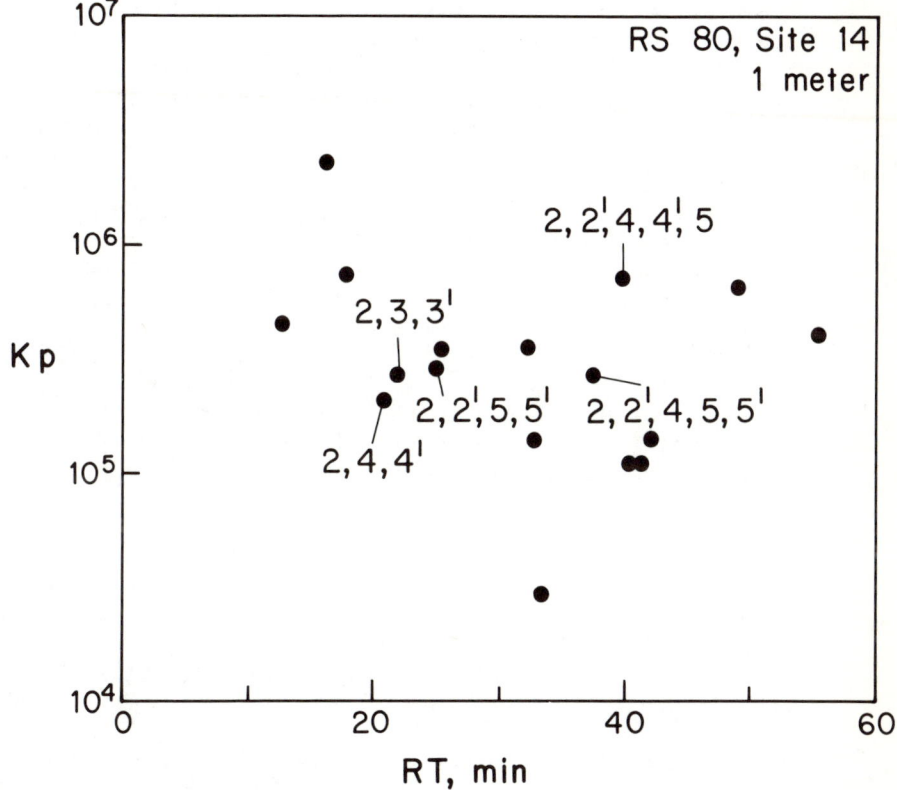

Figure 8 Relationship Between K_p for Selected PCB isomers (n = 2-6) in a 1 m Sample at Site 14, 1980 to GC Relative Rentention Time on a SE-54 Glass Capillary Column (25 m x .32 mm i.d.)

estimating precipitation scavenging and vapor partitioning across the air-water interface is emphasized. Formerly, there were two schools: one promoting H values equal to 10^{-3} to 10^{-4} atm m^3 mol^{-1} (liquid-phase control) and H equal to 10^{-5} to 10^{-6} atm m^3 mol^{-1} (gas-phase control). Given surface water and air concentrations of PCBs observed in Lake Superior and Lake Michigan, the difference in H corresponded to PCB flux from the water to the air (liquid-phase control) or flux from air to water (gas-phase control). What evidence do we have that would assist us in evaluating the magnitude of sources and sinks?

a. Data reported in this book clearly indicates that H values, determined in laboratory experiments or from thermodynamic properties, for PCB isomers are in the range of 10^{-3} to 10^{-4} atm m^3 mol^{-1}.

b. PCB profiles in surface waters for both totals and the lighter isomers show a decrease in concentration with increasing depth. The concentration and fugacity gradients in the surface waters suggest air-to-water transfer. However, concentration and fugacity gradients between surface water and airborne PCBs suggests a tendency to volatilize. That is, the water concentration is higher than it should be in equilibrium with the atmospheric vapor phase. Recent data obtained by Capel (18) show that the air-water partition coefficients for 17 PCB isomers (n = 2 to 6) in Lake Superior range only from 0.14 to 0.50 ng m^{-3}/ng L^{-1} or from 6.9 to 11 x 10^{-3} atm m^3 mol^{-1}. Henry's Law constants are approximately 2.6 to 4.0 x 10^{-4} atm m^3 mol^{-1} (41). However, some values reported for H (41) are in the range of 2 to 8 x 10^{-3} atm m^3 mol^{-1}. It appears that airborne PCBs are in near equilibrium with the PCB in the surface waters of Lake Superior.

c. The PCB burden of Lakes Superior ($\sim$ 39,000 Kg) and Michigan ($\sim$ 70,000) (see Eisenreich, 1982) are such that atmospheric inputs must have been important, and volatilization relatively unimportant until recently. This observation is particularly true for Lake Superior where other sources are improbable.

d. PCB profiles in Lake Superior sediments collected with a box corer in 1978 show 20 to 100% lower concentrations in the uppermost 0.25 cm segment than in successively-deeper sediments (13). This behavior strongly suggests that the flux of PCBs to the sediments is decreasing in response to decreased inputs.

e. Bierman and Swain (45) have documented a decrease in the DDT concentrations in certain fish species in Lakes Superior and Michigan which suggest a lake residence time of $\sim$ 1 to 3 years. The similarity between DDT and PCBs implies an analogous rapid response of the lakes to a decrease in PCB inputs.

Acknowledgements

This research was mostly supported by the U.S. Environmental Protection Agency under grants R805172-01 and R806084-01, and to a lesser extent by the National Science Foundation under grants OCE76-02255, OCE76-09432 and OCE78-08166 awarded to my close colleague, Thomas C. Johnson of the University of Minnesota, whose cooperation and intellectual assistance was crucial. Special thanks is extended to the captain(s) and crew(s) of the R.V. Roger R. Simons and the R.V. Crockett (now Rachael Carson). Additional support was provided under Minnesota Sea Grant NA-79-AA-D-00134, NA-80-AA-D-00114, and NA-81-AA-D-00114, Res. #122.

REFERENCES

1. National Academy of Sciences (NAS), Polychlorinated Biphenyls, Washington, D.C., 182 pp. (1979).

2. Nisbet, I. C. T., Sarofim, A. F., "Rates and routes of transport of PCBs in the environment," Environ. Health Perspective, $\underline{1}$, 21-38 (1972).

3. Zell, M., Ballschmiter, K., "Baseline studies of the global pollution. III. Trace analysis of polychlorinated biphenyls (PCB) by ECD glass capillary gas chromatography in environmental samples of different trophic levels," Fresnius A. Anal. Chem., $\underline{304}$, 337-349 (1980)

4. Ballschmiter, K., Zell, M., "Baseline studies of the global pollution. I. Occurrence of organohalogens in pristine European and Antarctic aquatic environments," Intern. J. Environ. Anal. Chem., $\underline{8}$, 15-35 (1980)

5. Ballschmiter, K., Buchert, H., Bihler, S., Zell, M., "Baseline studies of the global pollution. IV. The pattern of pollution by organo-chlorine compounds in the North Atlantic as accumulated by fish," Fresnius A. Anal. Chem., $\underline{306}$, 323-339.(1981)

6. Swain, W., "Chlorinated organic residues in fish, water, and precipitation from the vicinity of Isle Royale, Lake Superior," J. Great Lakes Res., $\underline{4}$, 398-407 (1978)

7. Reinert, R. E., "Pesticide concentrations in Great Lakes fish," Pest. Monit. J., $\underline{3}$, 233-240 (1970).

8. Konasewich, D., Traversy, W., Zar, H., "International Joint Commission, Water Quality Board, Appdx. E, Windsor, Ontario, Canada (1978)

9. Veith, G. D., Kuehl, D. W., Puglisi, F. A., Glass, G. E., Eaton, J. G., "Residues of PCBs and DDT in the western Lake Superior ecosystem," Arch. Envir. Cont. Toxic., $\underline{5}$, 587-599.

10. Horn, E. G., Hetling, L. J., Tofflemire, T. J., "The problem of PCBs in the Hudson River system," Annals N.Y. Acad. Sciences, $\underline{320}$, 591-609 (1979)

11. Thomann, R. V., St. John, J. P., "The fate of PCBs in the Hudson River ecosystem," Annals N.Y. Acad. Sciences, $\underline{320}$, 610-629.

12. Hartig, J. H., "Highlights of Water Quality and Pollution Control in Michigan," Dept. of Natural Resources, State of Michigan, Publ. No. 4833-9804 (1981).

13. Eisenreich, S. J., Hollod, G. J., Johnson, T. C., Evans, J. E., "PCB and other microcontaminant-sediment interactions in Lake Superior in Contaminants and Sediments, Vol. 1, R. A. Baker (ed.), Ann Arbor Science Publishers, Inc., Ann Arbor, Mich., p. 67-94 (1980).

14. Eisenreich, S. J., unpublished data, University of Minnesota.

15. Ballschmiter, K., Zell, M., Neu, H. J., PCBs: will they never degrade?" Chemosphere, $\underline{7}$, 173 (1978).

16. Hollod, G. J., Eisenreich, S. J., Johnson, T. C., "Sources and sinks of PCBs in Lake Superior," submitted to Environ. Sci. Tech. (1978).

17. Weininger, D., Armstrong, D. E., "Organic contaminants in the Great Lakes in Restoration of Lakes and Inland Waters," EPA 440/5-81-010, pp. 364-372 (1981).

18. Capel, P. D. M.S. Thesis, University of Minnesota, Minneapolis (1982).

19. Nie, N. H., Hull, C. H., Jenkins, J. G., Steinbrenner, K., Bent, D. H., Statistical Package for the Social Sciences, 2nd Ed., McGraw-Hill Book Company, N.Y.

20. Albro, P. W., Parker, C. E., "Comparison of the compositions of Aroclor 1242 and Aroclor 1016," J. Chromatog. 169, 161-166.

21. Albro, P. W., Corbett, J. T., Schroeder, J. L., "Quantitative characterization of PCB mixtures (Aroclors 1248, 1254, 1260) by gas chromatography using capillary columns," J. Chromatogr., 205, 103-111 (1981).

22. STORET, U.S. EPA, 1981.

23. Thomann, R. V., Koutaxis, M. T., "Mathematical modeling estimate of environmental exposure due to PCB-contaminated harbor sediments of Waukegan Harbor and North Ditch," U.S. EPA, Ind. Env. Res. Lab., Cinn., Ohio, by HydroQual., Inc., Mahwah, N.J., 109 pp. (1981).

24. Hollod, G. J., Ph.D. thesis, University of Minnesota, Minneapolis (1979).

25. Eisenreich, S. J., Looney, B. B., Capel, P., Hollod, G. J., PCBs in the Lake Superior Atmosphere: 1978-1981," in preparation.

26. Osterroht, C., Smetacek, U., "Vertical transport of chlorinated hydrocarbons by sedimentation of particulate matter in Kiel Bight," Marine Ecol. Prog. Series, 2, 27-34 (1980).

27. Frank, R., Thomas, R. L., Braun, H. E., Rasper, J., Dawson, R., "Organochlorine insecticides and PCB in surficial sediments of Lake Superior (1973)," J. Great Lakes Res., 6, 113-120 (1980).

28. Bahnick, D. A., Markee, T. P., Anderson, C. A., Roubal, R. K., "Chemical loadings to southwestern Lake Superior from red clay erosion and resuspension," J. Great Lakes Res., 4, 186-193 (1978).

29. Thomas, R. L., Dell, C. I., "Sediments of Lake Superior," J. Great Lakes Res., 4, 264-275 (1978).

30. Sydor, M., Stortz, K. R., Swain, W. R., "Identification of contaminants in Lake Superior through LANDSAT I data," J. Great Lakes Res., 4, 142-148 (1978).

31. Evans, J. E., Johnson, T. C., Alexander, E. C., Jr., Lively, R. S., Eisenreich, S. J., "Sedimentation rates and depositional processes in Lake Superior from ^{210}Pb geochronology," J. Great Lakes Res., 7, 299-310 (1981)

32. Eisenreich, S. J., Hollod, G. J., Johnson, T. C., "Atmospheric concentrations and deposition of PCBs to Lake Superior," in <u>Atmospheric Pollutants in Natural Waters</u>, S. J. Eisenreich (ed.), Ann Arbor Science Publishers, Inc., Ann Arbor, Mich., pp. 425-444.

33. Weininger, D., "Accumulation of PCBs by lake trout in Lake Michigan," Ph.D. thesis, University of Wisconsin, Madison (1978).

34. Clayton, J. R., Jr., Pavlou, S. P., Breitner, N. F., "PCBs in coastal marine zooplankton: bioaccumulation by equilibrium partitioning," Environ. Sci. Tech., $\underline{11}$, 676-682 (1977).

35. Harding, L. W., Jr., Phillips, J. H., Jr., "PCBs: transfer from microparticulates to marine phytoplankton and the effects on photosynthesis," Science, $\underline{202}$, 1189-1192 (1978).

36. Karickhoff, S. W., Brown, D. S., Scott, T. A., "Sorption of hydrophobic pollutants on natural sediments," Water Res., $\underline{13}$, 241-248 (1979).

37. Steen, W. C., Paris, D. F., Baughman, G. L., "Partitioning of selected PCBs to natural sediments," Water Res., $\underline{12}$, 655-657 (1978).

38. O'Connor, D. J., Connolly, J. P., "The effect of concentration of adsorbing solids on the partition coefficient," Water Res., $\underline{14}$, 1517 (1980).

39. DiToro, D. M., Horzempa, L. M., Casey, M., Richardson, W. "Exchangeable and non-exchangeable components of PCB adsorption-desorption: adsorbent concentration effects," J. Great Lakes Res., in press (1982).

40. Pavlou, S. P., Dexter, R. N., "Distribution of PCBs in estuarine ecosystems. Testing the concept of equilibrium partitioning in the marine environment," Environ. Sci. Tech., $\underline{13}$, 65-71 (1979).

41. Mackay, D., Paterson, S., Shiu, W. Y., Bobra, A., Billington, J., "Physical chemical properties and behavior of PCBs in <u>Physical Behavior of PCBs in the Great Lakes</u>, D. Mackay (ed.), Ann Arbor Science Publishers, Ann Arbor, Mich. (1982).

42. Doskey, P. V., Andren, A. W., "Modeling the flux of atmospheric PCBs across the air/water interface," Environ. Sci. Technol. $\underline{15}$, 705 (1981).

43. Doskey, P. V., Andren, A. W., "Concentrations of airborne PCBs over Lake Michigan," J. Great Lakes Res., 7, 15-20 (1981).

44. Eisenreich, S. J., Looney, B. B., "Evidence for the Atmospheric Flux of PCBs to Lake Superior" in <u>Physical Behavior of PCBs in the Great Lakes</u>, D. Mackay (ed.), Ann Arbor Science Publishers, Ann Arbor, Mich. (1982).

45. Bierman, V. J., Jr., Swain, W. R., "Mass balance modeling of DDT dynamics in Lakes Michigan and Superior," J. Great Lakes Res., in press (1982).

46. Ballschmiter, K., Zell, M., "Analysis of polychlorinated biphenyls (PCB) by glass capillary gas chromatography," Fresenius Z. Anal. Chem., 302, 20-31 (1980).

CHAPTER 12

THE ROLE OF THE BENTHIC
BOUNDARY IN THE CYCLING OF
PCBs IN THE GREAT LAKES

Brian J. Eadie
Great Lakes Environmental Research Laboratory
National Oceanic and Atmospheric Administration

Clifford P. Rice and William A. Frez
Great Lakes Research Division
University of Michigan

INTRODUCTION

In many regions of the world, high quality fresh water for municipal, agricultural or industrial use is in short supply and constitutes a limiting resource. The Great Lakes Region, on the other hand, has had, and can continue to have, ample supplies of high quality water to aid in its future growth and industrial development. This growth will result in increasing and often conflicting demands on the use of the Great Lakes resource. Over the past few decades, careless use of the Great Lakes has resulted in increased loads of nutrients and toxic substances, along with increased pressure from water usage. This has resulted in a decrease in resource value in terms of water quality (ecosystem structure, drinking water, recreation and filtration problems), along with a degradation of commercial and sports fisheries. Balanced future development and use of the Great Lakes depends on our understanding of the responses of aquatic systems to imposed stresses. Investigations of the cycling, behavior and fate of persistent compounds, such as the polychlorinated biphenyls (PCBs) described in this book, enable us to better understand and eventually to accurately predict these responses.

The great Lakes are particularly susceptible to contamination by synthetic materials, primarily because of the heavy industrialization of the region (leading to high

---

GLERL Contribution No. 311

loads) and the slow flushing times of the lakes. Slow flushing causes compound residence times to be scaled to internal removal process rates. The disposition of synthetic organics in aquatic environments depends strongly on their solubility. Chlorobiphenyls range in solubility from ∿1 ppm to ∿1 ppb (1) and, in general, the group of compounds can be classified as hydrophobic, characterized by a strong tendency to sorb onto particles and settle. For compounds characterized by slow decomposition rates, such as highly chlorinated PCB isomers, settling into the benthic boundary layer (BBL) and burial are a prime removal mechanism. This paper discusses available information on the processes of recent sediment accumulation, resuspension and transport within the BBL and the importance of these processes to the residence times of compounds in the Great Lakes.

In this work, calculations based on analyses of materials collected in sediment traps in southern Lake Michigan indicate a significant reentrainment of sediment bound PCB. Approximately 14% of the current water column inventory of particle associated PCB can be accounted for by winter resuspension. Reentrainment of pore water associated PCB is small, but may be significant if associated with soluble natural organics or colloidal material. The net effect of resuspension is to increase the residence time for contaminants in the active ecosystem.

THE BENTHIC BOUNDARY LAYER

The benthic boundary layer is usually described as a relatively thin layer extending above and into the lake sediments. Within the boundary layer, there are measurable gradients of chemical species, particles and heat. Most notable from our observations in Lake Michigan are profiles of total suspended matter (TSM) and particle flux (2).

During the stratified period (June-November), a layer characterized by high TSM (presumably from local resuspension) is generated within the bottom waters of the shelf/slope region of the lake. This localized region of high TSM, often called a nepheloid layer, may be thought of as the water column portion of the BBL. This near-bottom layer is a region of active transport of particulate matter from nearshore out into the deep depositional basins. The nepheloid region appears to be enhanced periodically through localized resuspension in the deep basins. In agreement with our work in Lake Michigan, old transmissometry data from the other four Great Lakes indicate that the nepheloid region is a common feature during the period of stratification (3).

After thermal stratification breaks down, deep mixing resuspends copious quantities of surficial sediments and the particulate material associated with the BBL is reintroduced into the water column. The impact of this is most easily seen in the seasonal cycles of silica (4,5) and plutonium (6). A major unanswered question is what effect this massive reentrainment process has on the cycling and fate of hydrophobic organic compounds, such as the PCBs. In this paper we attempt to scale this process for Lake Michigan.

Although this book focuses on physical processes, it must be mentioned that the BBL also serves as the primary residence and food source for the organisms comprising the base of the benthic food web. Food web transfer of "sediment bound contaminants" apparently reaches the top lake carnivores and eventually man. Thus, food chain transfer of PCBs needs to be considered in any realistic system modeling effort.

CHARACTERISTICS OF THE SEDIMENT COMPONENT OF THE BBL

Over the past decade, many investigators have commented upon the delicacy of the structure of the sediments at the sediment-water interface, and during collection, a great deal of effort is spent on protecting this interface. Surficial material, often called "floc," is enriched in organic matter and hydrated iron oxides, and appears somewhat gelatinous (7). There are at least three sources for this material: the settling products of internal lake productivity; terrigeneous material loaded into the lake by atmospheric, tributary or erosional processes and the products of the very active benthic organisms who live within the upper few centimeters of the sediments. In areas of sediment accumulation, the benthic organisms, along with physical resuspension, tend to rapidly homogenize the upper few centimeters of sediments, leading to average residence times of 20 years or more for particles within the mixed zone in Lake Huron (8). Typically, the upper few centimeters of sediment are unconsolidated and fine grained. In an environment similar to Lake Michigan depositional basins (Goderich Basin, Southern Lake Huron), the porosity (pore water volume/total volume) was generally greater than 0.8 (9).

In non-depositional regions of the lake, accumulation of recently deposited floc is a transient phenomenon although a small amount of new material is incorporated into the sediment matrix (10).

CHARACTERISTICS OF THE WATER COLUMN COMPONENT OF THE BBL (NEPHELOID LAYER)

Sediment trap profiles and water samples collected near the bottom have allowed us to characterize the nepheloid layer in southern Lake Michigan (2). Characteristic profiles of total suspended matter (TSM) and particle size distribution are illustrated in Figure 1. Sampling at the same slope base station five times during the stratified period showed a wide range in the mass of TSM in the bottom few meters, indicating considerable activity in the region. The particle size distribution shows that the mass increase near the bottom is primarily due to an increase in large (16-60 μm) particles. During the unstratified period, the near-bottom TSM concentrations are almost identical for the three collections (which may be a consequence of our ability to sample only during relatively calm periods) and the near-surface concentrations are higher ($\bar{x}$ = 1.8 ± 0.4 vs 1.3 ± 0.4 mg/L) for the stratified period. Particle size distributions during this period generally show a relatively high concentration of fine (2-8 μm) particles. These fine grained particles are high in organic carbon and are the materials that eventually collect the lakes' depositional basins. They are also, presumably, the vehicle upon which hydrophobic contaminants are transported.

Our results indicate that the BBL is an active region in which localized resuspension and transport play an, as yet, incompletely understood role in the cycling of contaminants within the Great Lakes. It must be remembered that the particle inventory of the water column is relatively small. In Lake Michigan, an average value of TSM would be approximately 1 mg/L, which means that the entire particle inventory can be replaced by the reentrainment of the upper 1-2 mm of sediments.

PCBs IN THE BBL

The only information on PCBs in the boundary layer are in the form of mixed surficial sediment concentrations, primarily in the sediment maps reported by Frank et al. (10,11,12), Frank and Thomas (this volume) and Armstrong (this volume). Eisenreich et al. (13) determined the sedimentary profile of PCB in several Lake Superior box cores sectioned at 0.5 cm intervals and compared their results (which they believe contained undisturbed interfacial floc) to much lower values reported by Glooschenko et al. (14) and Vieth et al. (15) and later by Frank et al. (12). These differences have been partially attributed to the very high

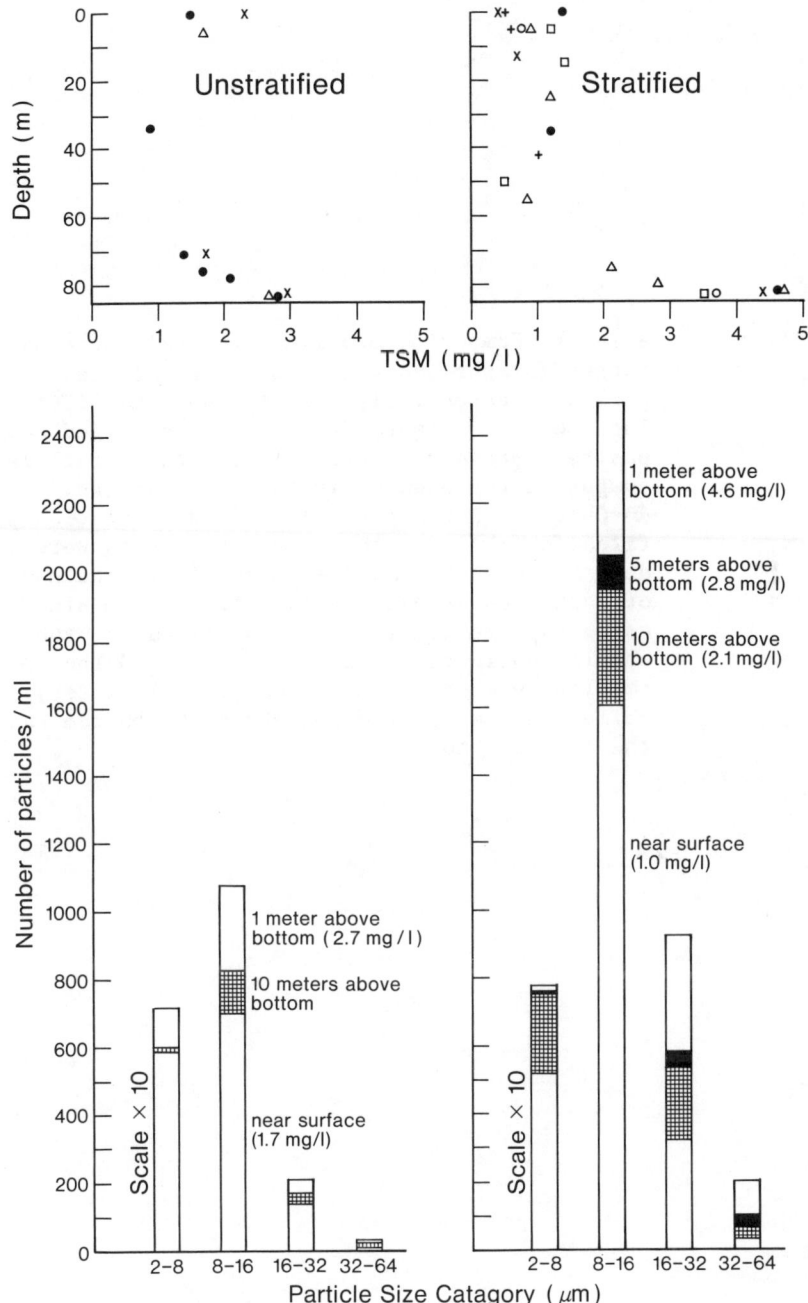

(Caption See Page 218)

Figure 1. a and b—Composite profiles of total suspended matter for the unstratified and stratified periods (respectively) at station 7/8. (For location, see Figure 3.) The unstratified composite represents three sets and the stratified composite represents five sets of samples. c and d—Characteristic particle size distribution profiles for the same location. Size data were determined on a HIAC instrument (2). The number of particles at each depth (within each size category) are cumulative. In all cases, the samples collected near the surface had the smallest number of particles per milliliter, while the sample from 1 m above the bottom had the greatest number.

concentrations that appear to occur at the surface of the sediment.

Water column values for total PCBs have been reported to range from approximately 1 to 30 parts per trillion (ppt) (1) with the most recent numbers ranging from 1-10 ppt (this volume). To our knowledge, there are no water column values for PCBs within a few meters of the bottom; however, a series of PCB values from a profile of sediment traps in Lake Michigan is available (16), representing the concentrations on the settling particles. These values, shown in Figure 2, include a self consistent set of values for surficial sediments (0-1 cm), which are again much higher than those recently reported by Frank et al. (10). The traps were located approximately 20 km offshore Grand Haven,

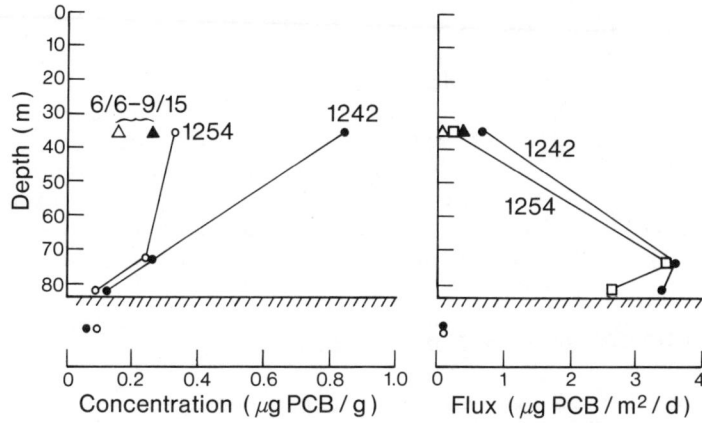

Figure 2. a--Profile of PCB concentrations on particulate matter collected in traps near station 7/8. Open symbols represent PCB 1254, solid symbols represent 1242. Triangles at 35 m represent collection period June 6-September 15; circles represent collection period September 15-October 24. Data in the sediment are for 0-1 cm samples collected at the trap location. b--Profile of PCB fluxes calculated by multiplying PCB concentration by mass flux at each trap depth. Sediment accumulation rate calculated by multiplying local 0-1 cm PCB concentration by the local sediment accumulation rate 0.64 g/m^2/d (18).

Michigan (Figure 3), in 83 m of water in the same general location as the station at which the particle data are reported in Figure 1. Two sets of PCB concentrations are shown in figure 2 for the 35 m traps at station 7; one set for June 6-September 15, the second for September 15-October 24. In both cases, PCB 1242 concentrations are significantly higher than PCB 1254. The reason for the higher concentrations in the fall may be related to the change in predominant particle type from biogenic detritus to calcium carbonate ($CaCO_3$) (17). Deeper in the water column, the PCB concentrations approach those of the local surficial sediments, indicating a dilution of the newly settling material with resuspended sediment.

Fluxes of the PCBs illustrated in the right-hand panel of Figure 2 show a strong resuspension component. The traps placed at 35 m depth accumulate mass at the same rate as the local sediments (2). Local sediment accumulation (0.64 $g/m^2/d$) (18) multiplied by the 0-1 cm sediment concentration of PCBs (see Table 1) is also illustrated. These values agree well with flux measured in our 35 m traps. The bottom two traps show an order of magnitude higher flux, which we find to be characteristic of the BBL. The source of this material is either local surficial sediment resuspension or advective transport of sediments resuspended elsewhere in the basin. This increased flux appears to be confined to the lower 10-20 m of the water column during the period of stratification, but once stratification breaks down, the resuspended surficial sediments move throughout the water column.

REENTRAINMENT OF PCBs FROM SEDIMENTS

The role of lake-scale particle reentrainment on the cycling of PCBs in the Great Lakes is conjectural at this point; in this paper we attempt to make some order of magnitude estimates based on recent field and laboratory data. Figure 4 illustrates the magnitude of increased particle fluxes in the water column. Station 80-7 is very near the station from which the PCB samples were analyzed (in 1978), while stations 4 and 9 are near midlake in the southern basin.

If we ignore the stratified period when the effects of reentrainment of PCB from the BBL appears to be confined to the near bottom region and possibly the nearshore (through upwellings), order of magnitude estimates of the importance of reentrainment can be made. We will first consider the effect of resuspended particles. It appears from Figure 4

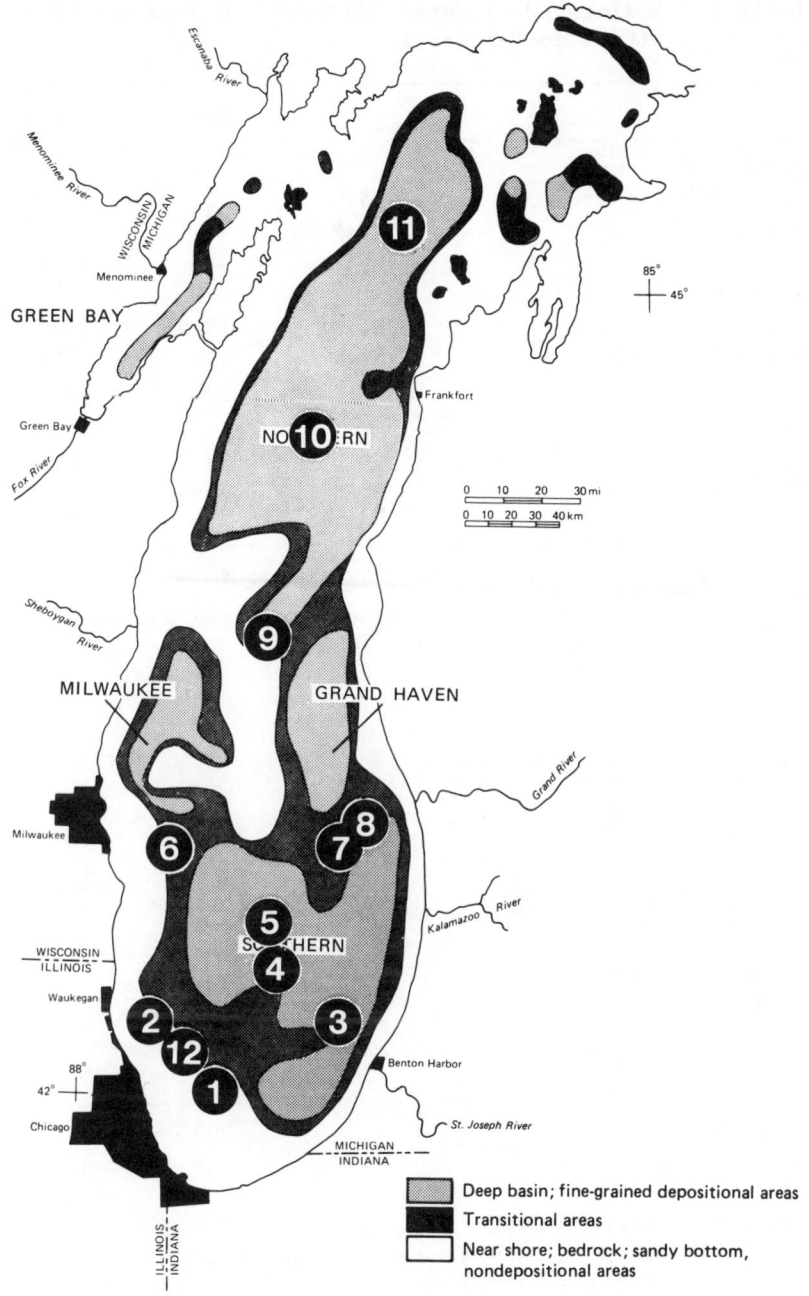

Figure 3. Sediment trap and sampling locations overlain on a map of depositional basins for Lake Michigan (from Cahill, 23).

Table I. Sediment/Pore Water PCB Data and apparent partition coefficients.

	% O.C.	1242	1254	Total
Nearshore* (Sandy)				
Pore water (ng/l)		53.0	6.0	159.0
Sediment (ng/g)	0.7	31.9	31.6	63.5
K		08.	67.	399.
Sta 7 (Clay/silt)				
Pore water (ng/l)		01.0	13.0	214.0
Sediment (ng/g)	1.7	69.0	91.0	160.0
K		43.	00.	747.
Sta 3 (Clay)**				
Pore water (ng/l)		08.0	34.0	342.0
Sediment (ng/g)	3.8	64.4	86.1	150.5
K		09.	32.	440.

* Inshore of Station 7/8 (see Figure 3); 45 m depth

** Off Benton Harbor, Mich.; highest recorded accumulation rate in Lake Michigan; 3 g/m^2/day

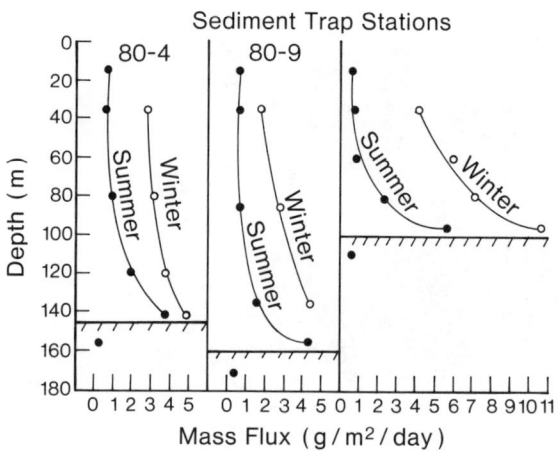

Figure 4. Particle mass flux profiles from three trap profiles in Lake Michigan. (See Figure 3.) Summer profiles represent the period mid-June through mid-November; winter profiles represent the period mid- November through mid-June. Sediment accumulation rates are from local cores measured by Edgington and Robbins (24).

and other data that the winter mass flux (at 35 m) is conservatively estimated at 1 g/m^2/d greater than the flux during the stratified period. From a 200-day period, this represents 200 g/m^2 of resuspended sediment reaching near-surface waters. Assuming the PCB concentration in surficial sediments subject to resuspension to be 100 ng/g, then approximately 20 μg/m^2/year of particle associated PCB would be reintroduced into the water column. In Lake Michigan, the inventory of PCB in one square meter of water column is approximately 425 μg PCB/m^2 (5 μg PCB/m^3 x 85m). Rice et al. (this volume and 19) report that approximately 1/3 of total PCB is associated with particles; therefore the reentrainment represents about 14% of the particulate PCB inventory. Presumably, this material will behave similarly to other particulate PCB and settle back out rapidly.

Along with the particle reentrainment in the resuspension process, there is mixing of the "pore" water associated with the solids. The surficial sediments subject to resuspension are probably 90% water, and a calculation is needed to estimate its importance. If this "pore" water contained only a few ppt PCB as does the lake water, its contribution would be negligible. However, recent work by O'Connor and Connolly (20) supported by some recent laboratory and field work described below indicates that the concentrations are much higher. From published data, O'Connor and Connolly showed that the equilibrium partition coefficients,

$$K = \frac{\text{particle associated concentration}}{\text{dissolved concentration}},$$

of several substances were inversely proportional to the concentration of substrate and that the range of K can vary over several orders of magnitude under substrate conditions encountered in the environment. If true, it would predict pore water concentrations that are much higher than overlying water measurements.

In order to test this in the field, we carefully collected three Shipek samples from sedimentary environments of known different organic carbon content; the surface 1 cm was removed and centrifuged to separate the pore water. These samples were then analyzed for PCBs (Table 1). The data for the dissolved fraction allow us to estimate the importance of entrainment for dissolved material. Assuming the water reentrained is four times the particulate resuspension (vol/vol) or approximately 400 ml/m^2 and has a PCB concentration of 200 ng/l, one can calculate a dissolved PCB flux of 80 ng/m^2/year. This is about 0.4% as large as the particulate reentrainment and less than 0.1% of the water column inventory. The importance of this small load may be increased if the dissolved reentrained PCB is "complexed" with soluble natural organic matter. Such an association would affect the cycling of this fraction and tend to increase its residence time in the active ecosystem. These indicate that reentrainment from sediments can be very important, but more data are needed to get reliable estimates.

Apparent partition coefficients (K) for the three sets of sediment/pore water samples listed in Table 1 averaged 253 (±78) for PCB 1242 and 4933 (±2253) for PCB 1254. For total PCB, the average was 529 (±190) and the values were not related to the organic content of the sediments. These data are several orders of magnitude lower than values observed by Rice et al. (19, this volume) for Lake Michigan water or those calculated from solubility, which are in the range of 10^5. The lower values appear to support the idea that K is a function of substrate concentration (20). A second possibility is that the material we measure in pore water is associated with soluble natural organics and colloidal material that pass through our filters. These submicron particles would be expected to have very high concentrations of hydrophobic contaminants.

Finally, we've attempted to put these fluxes in perspective in Figure 5, which indicates some serious discrepancies. The atmospheric and total load numbers obtained from the literature (Table 2) are seriously out of balance with the estimated sediment accumulations. This indicates either that the lake is far from steady state and PCBs are building up in the water column, in contradiction to the apparent decline of PCBs in fish, or that reported load estimates are not representative of current loads. Reports in this volume indicate that our estimate of the sediment inventory of PCB might be a little high (Armstrong) and that atmospheric loads are probably high by an order of magnitude (Andren). These adjustments would bring the system near to balancing. The conservatively estimated resuspension terms are relatively large and will play a role of increasing importance in the lake budget as other load terms decline in the near future. The reentrainment of sediment associated contaminants will undoubtedly increase the residence time of these materials within the active ecosystem.

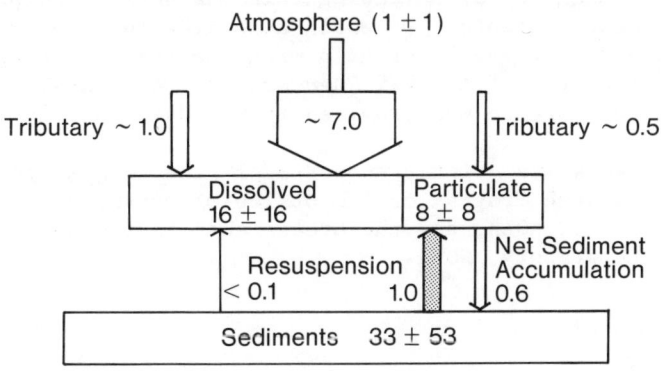

Figure 5. Annual mass fluxes and distribution of PCBs in Lake Michigan. Units for fluxes (arrows) are Mt/yr; units for inventories (boxes) are Mt. The width of the boxes and arrows are proportioned to average inventory and flux values, respectively. Inventory estimates as follows: atmosphere, conc. = 2 ± 2 ng PCB/m3, ~20 Km thick layer; water, conc. = 5 ± 5 ng PCB/liter; sediments from Frank et al. (10). Fluxes are from Table 2 and text. The shaded arrow coming out of the sediment represents the conservatively estimated reentrainment of sediment bound PCB.

Table II. PCB Loads (Mt/Yr) for Lake Michigan

Atmospheric input[1]	6.9
Total load[2]	8.95
Sediment accumulation[3]	0.6
Particulate resuspension	1.0
Dissolved resuspension	0.005

[1] Eisenreich et al. (21)

[2] Murphy et al. (22)

[3] Average sediment accumulation = 10 mg/cm^2/yr; PCB concentration = 100 ng/g

Acknowledgments: The authors would like to express their appreciation to Wayne Gardner, Peter Landrum, John Robbins and Claire Shelske for reviewing this manuscript and providing useful comments. We would also like to thank Richard Chambers and Jerry Bell for help with the sediment trap program and the crew of the R/V *Shenahon* for their assistance. Editorial assistence was provided by Ms. Jeanne Kelley.

This work was jointly supported by the Long Range Effects Research Program of the Office of Marine Pollution and Assessment, NOAA, and the Great Lakes Environmental Research Laboratory, NOAA.

REFERENCES

1. NAS, Polychlorinated biphenyls. National Academy of Sciences Washington, D.C. (1979).

2. Chambers, R.L. and Eadie, B.J., Nepheloid and suspended particulate matter in south-eastern Lake Michigan. Sedimentology 28, 439-447 (1981).

3. Bell, G.L., Chambers, R.L. and Eadie, B.J., Evidence of a concentrated benthic suspended (nepheloid) layer in the Great Lakes. Proc. 23rd Conf. Great Lakes Res., Int. Ass. Great Lakes Res. Program Abs. p. 25 (1980).

4. Conway, H.L., Parker, J.I., Yaguchi, E.M. and Mellinger, D.L., Biological Utilization and Regeneration of Silicon in Lake Michigan. J. Fish Res. Bd. Can. 35, 537-544 (1977).

5. Schelske, C.L., Silica and nitrate depletion as related to rate of eutrophication in Lakes Michigan, Huron, and Superior. pp. 277. In Coupling of land and water systems, A.D. Hasler (ed.), Ecological Studies, Vol. 10, Springer-Verlag New York Inc. (1975).

6. Wahlgern, M.A., Robbins, J.A. and Edgington, D.N., Plutonium in the Great Lakes. In: Transuranics in the Environment, pp. 659-683. (Ed. by W.C. Hanson) publication TIC-22800. U.S. Department of Energy (1980).

7. Freeman, D.H. and Cheung, L.S., A gel partition model for organic desorption from a pond sediment. Science 214, 790-792 (1981).

8. Robbins, J.A., Stratigraphic and dynamic effects of sediment reworking by Great Lakes zoobenthos. Proc. 2nd Internat. Symp. on the Interactions between Sediments and Fresh Water (1982).

9. Robbins, J.A., Sediments of southern Lake Huron: Elemental composition and accumulation rates. EPA-600/3-80-080 (1980).

10. Frank, R., Thomas, R.L., Braun, H.E., Gross, D.L. and Davies, T.T., Organochlorine insecticides and PCB in surficial sediments of Michigan (1975) J. Great Lakes Res. 7(1), 42-50 (1981).

11. Frank, R., Thomas, R.L., Holdrinet, M., Kemp, A.L.W. and Braun, H.E., (1979) Organochlorine insecticides and PCB in surficial sediments (1968) and sediment cores (1976) from Lake Ontario. J. Great Lakes Res. 5, 18-27.

12. Frank, R., Thomas, R.L., Braun, H.E., Rasper, J. and Dawson, R., (1980) Organochlorine insecticides and PCB in surficial sediments of Lake Superior (1973) J. Great Lakes Res. 6, 113-120.

13. Eisenreich, S.J., Hollod, G.J. and Johnson, T.C., Accumulation of polychlorinated biphenyls (PCB's) in surficial Lake Superior sediments. Envir. Sci. Tech. 13, 569-573 (1979).

14. Glooschenko, W.A., Strachan, W.M.J. and Sampson, R.C.J., Distribution of pesticides and polychlorinated biphenyls in water, sediments and seston of the Upper Great Lakes - 1974. Pestic. Monit. J. 10, 61-67 (1976).

15. Veith, G.D., Kuehl, D.W., Puglish, F.A., Glass, G.F. and Eaton, J.G., Residues of PCBs and DDT in the western Lake Superior ecosystem. Arch. Environ. Contam. Toxicol. 5, 487-499 (1977).

16. Rice, C.P., Frez, W.A., Eadie, B.J. and Anderson, M.L., PCB content of settling sediments in Lake Michigan. Submitted to Chemosphere (1982).

17. Strong, A.E. and Eadie, B.J., Satellite observations of calcium carbonate precipitations in the Great Lakes. Limnol. Oceanogr. 23(5), 877-887 (1978).

18. Robbins, J.A. and Edgington, D.N., Determination of recent sedimentation rates in Lake Michigan using PB-210 and CS-137. Geochim. cosmochim. Acta $\underline{39}$,285-304 (1975).

19. Rice, C.P., Eadie, B.J. and Erstfeld, K.M., Enrichement of PCBs in Lake Michigan surface films. J. Great Lakes Res. (In press) (1982).

20. O'Connor, D.J. and Connolly, J.P., The effect of concentration of adsorbing solids on the partition coefficient. Nat. Res. $\underline{14}$, 1517-1523 (1980).

21. Eisenreich, S.J., Looney, B.B. and Thornton, J.D., Airborne organic contaminants in the Great Lakes ecosystem. Envir. Sci. Tech. $\underline{15}$, 30-38 (1981).

22. Murphy, T.J. and Rzeszutko, C.P., Polychlorinated biphenyls in precipitation in the Lake Michigan basin. PA-600/3-78-071 (1978).

23. Cahill, R.A., Geochemistry of recent Lake Michigan sediments. Ill. State Geol. Survey, Circular 517 (1981).

24. Edgington, D.N. and Robbins, J.A., Records of lead deposition in Lake Michigan sediments since 1800. Envir. Sci. Tech. $\underline{10}$, 266-274 (1976).

CHAPTER 13

PCB ACCUMULATION IN SOUTHERN
LAKE MICHIGAN SEDIMENTS: EVAL-
UATION FROM CORE ANALYSIS

David E. Armstrong

Deborah L. Swackhamer
    Water Chemistry Program
    University of Wisconsin-Madison

INTRODUCTION

   The distribution and amounts of PCBs in lake sediments provide a record of PCB accumulation in the lake system. If PCBs entering a lake from all sources are transported rapidly to the bottom sediments and retained without losses by degradation or diffusion, the sediment record may be used to calculate the loading of PCBs to the lake system.
   The residence time in lake water of substances tending to associate with suspended particulate matter is relatively short. For example, the residence time of Pb in natural waters is estimated to be about one year [4]. Similarly, the residence time of the nonpolar chlorinated pesticide DDT in Lake Michigan is apparently about two years [18]. By comparison, the residence time of PCBs in the lake water column with respect to transport to bottom sediments is expected to be short, similar to Pb and DDT. Consequently, the sedimentary PCBs should be a record of recent PCB inputs to the system.
   Losses of PCBs from the lake system may occur through volatilization or degradation. The extent of loss by volatilization is uncertain, but some loss from the lake water is likely, especially for the low chlorine (1 to 3) chlorobiphenyls [2]. However, adsorption by suspended particulate matter and sedimentation reduce volatilization. Chlorobiphenyls (especially the low chlorine isomerides) are biodegradable [9], but degradation is slow and may be relatively unimportant in sediments [8]. Thus, "weathering" by volatilization and degradation probably alters the composition of PCB mixtures (Aroclors) in the environment, but may be relatively unimportant for PCBs deposited in deep basins of lakes. Postdepositional transport by diffusion may also affect PCB distribution and accumulation in

sediments. However, profiles of PCBs in lake sediments
(e.g., [6]) and sediment-water partition coefficients for
PCBs [10] indicate transport by diffusion may be slow
compared to linear sedimentation rates.

The accumulation of PCBs in lake sediments (e.g.
[5,6,12]) demonstrates sediments are a net sink for PCBs.
Although the gross flux to the lake may exceed the net
transport to the sediments, the accumulation of PCBs in
sediments provides a record of the minimum input to the lake
system.

The purpose of this paper is to estimate the accumulation of PCBs in southern Lake Michigan. The depth
distribution of PCBs was measured in four sediment cores
from southern Lake Michigan. In one core (station 18), the
sedimentation rate and mixing depth of the surface sediment
were also calculated from Pb-210 analysis of the core. The
results indicate sediment PCB profiles are determined by a
combination of PCB deposition rate, sedimentation rate, and
mixing depth. Two profiles suggested a recent decrease in
PCB accumulation. At station 18, total PCB accumulation
was 0.052 $\mu g\ cm^{-2}$ and the mass sedimentation rate was
0.016 $g\ cm^{-2}\ yr^{-1}$. Based on an average sedimentation rate
of 0.007 $g\ cm^{-2}\ yr^{-1}$ for southern Lake Michigan, total PCB
accumulation in southern Lake Michigan sediments was
estimated to be 4200 kg ± 20%. This calculation assumed
the PCB concentration in the depositing sediment is uniform
over the southern lake area and would not reflect possible
high concentrations in near-shore sediments affected by
local sources. Comparisons with atmospheric loading estimates
and PCB concentrations in Lake Superior sediments are also
discussed.

METHODS

Sediment Sampling

Sediments were obtained from Lake Michigan with a box
corer which recovers a 9 x 24 x 25.5 cm sample with an
undisturbed surface layer. Cores were sectioned on board
ship at intervals of 0.5 or 1 cm. Samples were stored at
4°C in the laboratory until analysis.

PCB Analysis of Sediments

Sediment samples were extracted using steam distillation [17] with a dichromate digestion [15]. Wet sediment (100 g) was extracted for 4 hours in the presence of hexane. Extracts were passed through a pre-column of anhydrous Na_2SO_4 (5 g), then through a column of 10% deactivated alumina (5 g) over HCl-washed copper grains (2 g) to remove organic interferences and elemental sulfur. Samples from station 18 were also passed through 3% deactivated silica gel (5 g). Spike recovery experiments showed no loss of PCBs in this additional step.

Extracts were concentrated to less than 10 ml in a Kuderna-Danish apparatus using a Snyder column and steam bath. Extracts were brought to final volume in 10 ml volumetrics. The overall procedural recovery is 85-90% for Aroclors 1242, 1248, and 1254.

Extracts were analyzed on a Hewlett Packard 5830 gas chromatograph using a WCOT or SCOT capillary column (60 m; SP 2100 coating) and ^{63}Ni electron capture detection. The conditions were as follows: Carrier gas, H_2, 2 ml/min; makeup gas, $Ar-CH_4$, 20 ml/min; splitless injection; initial column temperature, $50°C$; injection port temperature, $225°C$; detector temperature, $325°C$; multi-ramp temperature program, $180° - 260°$, $1-4°C/min$. The gas chromatographic data was analyzed using a multiple linear regression analysis computer program, COMSTAR [20] to determine the combination and concentrations of Aroclor standards giving the best agreement with the observed areas of PCB peaks in the samples.

Pb-210 Dating of Sediments

Sediment moisture content was obtained by drying a weighed subsample (2-3 g) for 48 hr at $150°C$. The porosity and dry bulk density were calculated using the solids density (2.45 ± 0.054 g/cm^3) for Lake Michigan sediments reported by Robbins and Edgington [13].

The activity of Pb-210 was determined by counting the activity of the daughter (Po-210) self-plated on silver discs from acid extracts of the sediment [13, 16]. The sediment samples had been stored for 1½ years, eliminating the need to measure unsupported Po-210. A spike of Po-208 was added to monitor the recovery of Po-210. Details of the digestion, extraction, plating and counting procedures are given by Talbot [16]. The sedimentation rate and surface sediment mixed layer depth were calculated from the unsupported Pb-210 activity [24, 16].

RESULTS AND DISCUSSION

Comparison Sediment Cores

Four sediment cores from southern Lake Michigan were analyzed for PCB concentration as a function of depth (Figure 1). According to sediment maps reported by Cahill

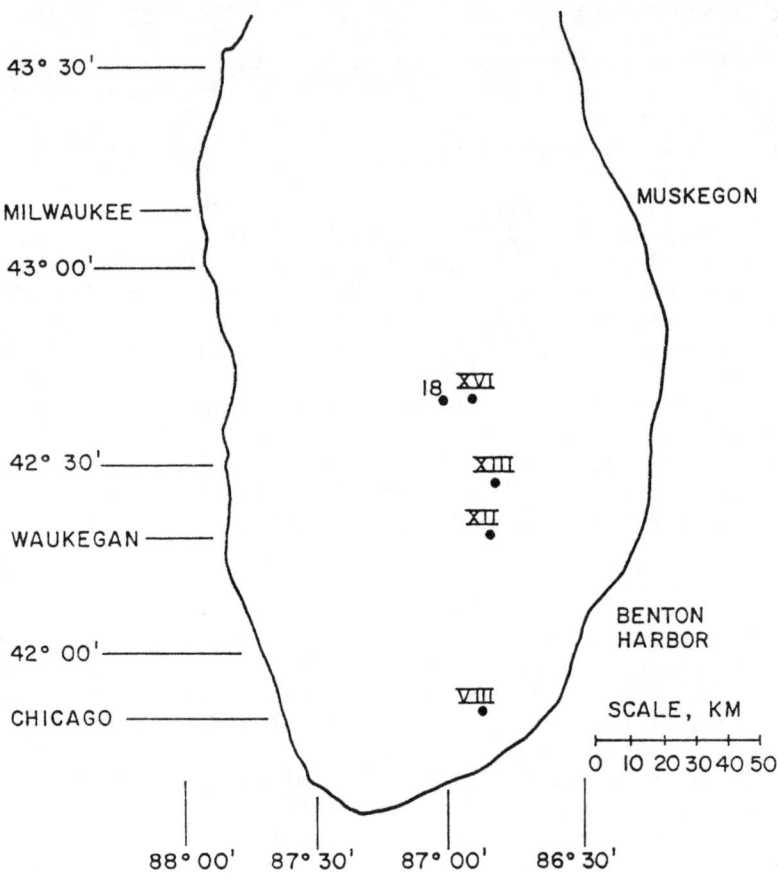

Figure 1. Location of sediment sampling sites in southern Lake Michigan.

Table I. PCB Concentrations and Other Data for Lake Michigan Sediment Cores

Depth Interval (cm)	Water Content (wt %)	Dry Bulk Density (g/cm^3)	PCB Concentration (μg/g)	(μg/cm^3)
Station 18 42°44'N, 87°00'W Depth = 160 meters				
0-1	80.0	0.200	0.091	0.021
1-2	79.4	0.245	0.055	0.013
2-3	77.9	0.245	0.042	0.011
3-4	78.1	0.245	0.019	0.005
4-5	77.8	0.245	0.008	0.002
Station XIII 42°28'N, 86°51'W Depth = 146 meters				
0-1	78.0	0.25	0.103	0.026
1-2	76.1	0.28	0.211	0.059
2-3	73.4	0.32	0.097	0.031
Station XII 42°19'N, 86°51'W Depth = 110 meters				
0-0.5	74.0	0.31	0.104	0.032
0.5-1.5	73.7	0.31	0.164	0.051
1.5-2.5	74.4	0.30	0.106	0.032
Station VIII 41°50'N, 86°50'W Depth = 22 meters				
0-1	65.0	0.44	0.201	0.088
1-2	30.1	1.19	0.014	0.017
2-3	20.0	1.52	0.015	0.023

[1] stations XII, XIII, and 18 are located in the southern basin, a deep basin, fine-grained sediment depositional area, while station VIII is located near the boundary of the transitional and non-depositional zones in the south-east corner of the lake. These differences were apparent in the physical appearance of the sediment cores. Sediments from stations XII, XIII, and 18 were dark-colored muds, while station VIII sediments consisted of a thin layer ($\sim$1 cm) of dark mud overlying sand and clay. Although the composition of sediments within the depositional zone of the southern basin is generally uniform [1] sedimentation rates can vary appreciably among sites within the depositional zone [13].

The concentrations and depth distribution of PCBs varied considerably among the 4 sediment cores (Figure 2). Concentrations were highest in the upper 1 cm at station VIII but decreased abruptly below 1 cm. At station 18, the total PCB concentration decreased gradually with increasing depth while stations XII and XIII exhibited a sub-surface maximum in concentration. The reasons for these differences are explored below.

The PCB concentrations in Figure 2 are plotted on a wt/volume basis (μg PCB/cm^3 of sediment) to facilitate comparisons among sediments of varying porosities. The corresponding mass/mass concentrations (μg PCB/g of sediment) are also given in Table I. Differences in porosity tend to accentuate differences in PCB concentrations expressed on a μg/cm^3 basis. Porosity values were highest at station 18, slightly lower at stations XII and XIII, and lowest at station VIII. The difference in porosities between stations VIII and 18 (0.91 vs. 0.82) corresponds to a factor of almost 2 (0.23 vs. 0.44) in the dry bulk density (g dry sediment/cm^3 of wet sediment). This almost doubles the ratio of concentrations between the two stations (station VIII/station 18) when expressed on a wt/volume basis (4.2) as compared to wt/wt basis (2.2). The effect is most apparent in the 1 to 2 and 2 to 3 cm layers at station VIII. The wt/wt concentration in these sediments (about 0.015 μg/g) is similar to the concentration in the 3 to 5 cm zone at station 18. However, because of the low porosity at station VIII, the mass/volume concentration in the 1 to 3 cm zone (about 0.02 μg/cm^3) is similar to the concentration in the 0 to 1 cm layer at station 18. Thus, expressing concentrations on a mass/volume basis may overemphasize low PCB concentrations in low porosity sediments.

The precision and accuracy of PCB measurements must also be considered in comparing concentration differences among sediment cores. The analytical method involving extraction by steam distillation and measurement by capillary column gas chromatography with electron capture

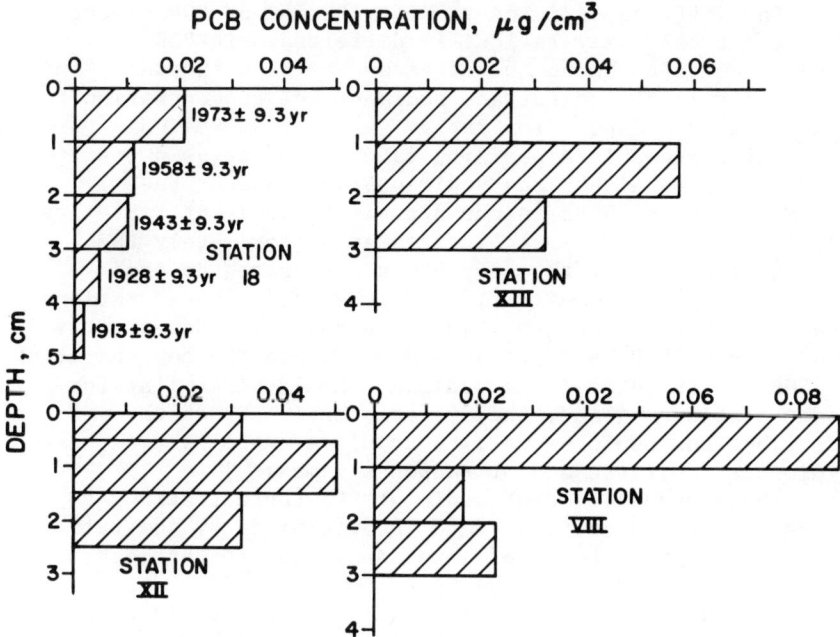

Figure 2. Depth distribution of PCBs in four sediment cores from southern Lake Michigan. The dates for station 18 are the average ages of the respective 1 cm intervals. The uncertainty ($\pm$ 9.3 years) is due to the mixing depth of 1.2 cm at this station.

detection was recently evaluated [15]. Comparison with Soxhlet extraction and analysis of spiked samples indicated the method was accurate. Analysis of duplicate and split samples gave a coefficient of variation of about 20%. The precision is less for low concentrations (0.02 µg/g). Thus, the surface layer concentrations are similar at stations 18, XII and XIII and lower than at station VIII. The decrease in concentration with depth at station 18 and the sub-surface maxima at stations XII and XIII should be significant. However, the uncertainty in the concentrations between 1 and 3 cm at station VIII is relatively high.

The differences in depth distribution of PCBs among the 4 sediment cores may reflect 1) differences in PCB deposition rates among the sites and/or 2) differences in response time of the surface sediment PCB concentration to changes in PCB deposition rate.

Differences in deposition rate can occur as a result of differences in either the sedimentation rate or the concentration of PCBs in the depositing sediment. Stations 18, XII, and XIII are all located at approximately the east-west center of the deep basin designated as the southern basin by Cahill [1]. Furthermore, these stations are not located close to major tributaries. Consequently, the sources of PCBs to these sediments and the concentration of PCBs in the depositing sediment should be similar for these stations. Thus, the differences in PCB depth distribution should be related to differences in sedimentation rates and/or response times.

The response time of sediment PCB concentrations to changes in PCB deposition rate is determined by the ratio of the sediment surface mixed layer depth (S) to the mass sedimentation rate (R) [19]. PCB concentrations at locations with a high S/R ratio will change slowly with changing PCB deposition rate while locations with a low S/R ratio will change more rapidly. Values for S and R were estimated from Pb-210 data for several sites in the southern basin by Robbins and Edgington [13], and this data has been interpolated by Weininger et al. [19]. The interpretation suggests a 50% response time of 10 to 20 years at stations XII and 18, and greater than 20 years at station XIII. However, differences in S and R occur over relatively short distances, and response times obtained for specific sites from the interpolated data may be inaccurate.

The profile at station 18 (decreasing concentration with increasing depth) is consistent with dilution of the deposited sediment by mixing with a surface layer containing lower PCB concentration. Apparently, the PCB concentration in recent sediments is diluted by mixing and the surface layer concentration has not reached the concentration in the depositing sediment. The presence of a surface mixed layer is consistent with the Pb-210 data (Figure 3). The profile is also consistent with a continual increase in PCB deposition rate between about 1928 and the present. Production of PCBs was initiated in 1929 and increased steadily until 1971 [3]. Although the input rate should be declining, evidence of a recent decrease in the PCB deposition rate in the profile from station 18 would be partially masked by mixing of the surface sediment. The average age of the surface 1 cm is 1973 ± 9.3 years (Figure 2).

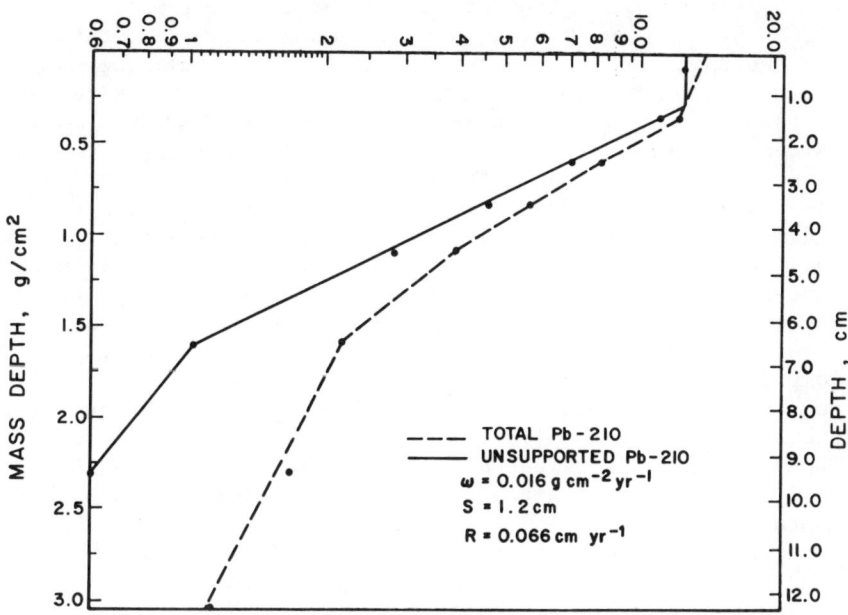

Figure 3. Depth distribution of Pb-210 at station 18. The supported Pb-210 activity was 1.1 pCi/g. The mass sedimentation rate (ω) was calculated from $\ln A = \ln A_m - \lambda(z-S)/\omega$, where A = Pb-210 activity (1 to 8 cm interval); A_m = Pb-210 activity in the mixed layer; λ = decay constant for Pb-210; z = mass depth (g cm^{-2}), calculated from dry bulk density (Db, g cm^{-3}) x depth (cm); S = depth of surface mixed layer, 0.3 g cm^{-2} or 1.2 cm. The linear sedimentation (R) was calculated from ω/Db.

The shapes of the profiles at stations XII and XIII (sub-surface maximum) suggest a recent decrease in PCB deposition rate. This decrease should be most apparent in locations with a short response time (low S/R ratio). However, as discussed above, the response time may be longer at stations XII and XIII than at station 18. One possible explanation for this apparent anomaly remains: the differences in profile shapes are a result of differences in the degree of resolution of the PCB profile due to differences in sedimentation rates among the three stations.

We have used Pb-210 measurements to estimate the mass sedimentation rate (ω), linear sedimentation rate (R) and surface mixed layer depth (S) at station 18 (Figure 3). The values obtained were 0.016 g cm^{-2} yr^{-1} for ω and 1.2 cm for S. We have not been able to date cores XII and XIII using Pb-210. However, PCB measurements can be used to infer sedimentation rates relative to station 18, assuming the sediment deposited at the three stations is identical in PCB concentration. Under this condition, the total areal PCB accumulation over the total PCB depth profile should be directly proportional to the mass sedimentation rate. Comparison of PCB accumulation among the stations (Table II) suggests the respective sedimentation rates at stations XIII and XII are 2.25 and 1.9 times higher than at station 18. These are minimum values because the profiles at stations XIII and XII do not extend to the lower boundary of the PCB contaminated zone.

While this analysis of the PCB profiles is not conclusive, we suggest the following interpretation: 1) the sedimentation rates at stations XIII and XII are considerably higher than at station 18; the upper three layers at stations XIII and XII (2.5 or 3 cm) may represent a time period more recent than the upper 2 layers (2 cm) at station 18; 2) the higher sedimentation rates at stations XIII and XII result in a more detailed resolution of the profiles with respect to time than at station 18. Thus, a recent decrease in PCB deposition rate may have occurred, as indicated at stations XII and XIII, but this decrease is not observed at station 18 because of the slower sedimentation rate, the relatively high mixed layer thickness, and the depth of the core sections.

Table II. PCB Accumulation In Lake Michigan Sediment Cores

Station	Total PCB Accumulation (g/cm^2)	Relative PCB Accumulation
18	0.052	1.0
XIII	0.116	2.3
XII	0.099	1.9
VIII	0.128	2.5

The profile at station VIII (sharp decrease in concentration below 1 cm) reflects a change in sediment type below 1 cm. As mentioned earlier, this station is located in a transitional area where a thin layer of recent, fine-grained sediment overlies sand or clay. Consequently, this core is not directly comparable to the other three cores. The surface sediments in the transitional zone are probably subject to lateral transport to the deep basins. The concentration in the surface sediment at station VIII (0.20 µg/g) is higher than at the other stations (Table I). This may partly reflect dilution by mixing after deposition in the deep basins, but probably also indicates the effect of a local source on sediments temporarily deposited in the transitional area.

Differences in concentration were also observed among the three cores located in the depositional area (Table I). These differences are partly due to porosity differences; the wt/wt concentrations in the surface layer are similar (about 0.10 µg/g). However, the concentrations below 1 cm are lower at station 18. This may be due to greater dilution by mixing of PCB-contaminated sediment with underlying uncontaminated sediment at station 18. According to the model presented by Weininger et al. [19], concentrations in the present surface layer would be diluted about 10% by mixing, if $R = 0.066$ cm yr^{-1}, $S = 1.2$ cm (as indicated by Pb-210 analysis) and PCB deposition has occurred at a constant rate for more than 40 years. The effect of dilution would increase with increasing depth below 1 cm at station 18; dilution would be less important in the upper layers (0 - 3 cm) at stations XII and XIII if the sedimentation rate is higher at these stations (as suggested by the higher PCB accumulation) and S at stations XII and XIII is $\leq$ S at station 18.

The possibility that the differences between station 18 and stations XII and XIII are due in part to analytical error or accuracy of coring and core sectioning can not be discounted completely. However, the concentration in the 1 to 2 cm interval at station 18 (0.055 µg/g) is comparable to the 1 to 2 cm interval concentration at station XVI (0.070 µg/g; the 0 to 1 cm interval was lost) located at about 5 km east of station 18.

Accumulation In Southern Lake Michigan

The total accumulation of PCBs in southern Lake Michigan sediments can be calculated from the accumulation at station 18 by assuming 1) the concentration of PCBs in sediment deposited is constant throughout southern Lake Michigan and 2) accumulation at a given area is proportional to the mass sedimentation for the area.

PCB accumulation at station 18 was measured to be 0.052 $\mu g\ cm^{-2}$ (Tables I and II). The sedimentation rate calculated from Pb-210 data for station 18 was 0.016 $g\ cm^{-2}\ yr^{-1}$ (Figure 3). For a station at about the same location (LM-17), Robbins and Edgington [13] obtained a sedimentation rate of 0.012 $g\ cm^{-2}\ yr^{-1}$. The mean sedimentation rate for southern Lake Michigan, calculated from the thickness of the Waukegan member, is 0.007 $g\ cm^{-2}\ yr^{-1}$ [4]. If PCB accumulation is proportional to mass sedimentation rate, the average accumulation for southern Lake Michigan should be about 0.023 $\mu g\ cm^{-2}$. The area of southern Lake Michigan is about $18.1 \times 10^{13}\ cm^2$ [4]. Thus, the calculated PCB accumulation is about 4200 kg. Allowing ± 20% for analytical error, the estimated total PCB accumulation in southern Lake Michigan is in the range of 3360 to 5040 kg. Using data from surface sediments at about 30 sites distributed over the southern basin Weininger et al. [19] estimated the amount of PCB accumulated in the surface mixed layer (average of 1.16 cm thick) was about 2760 kg. This suggests about 55 to 80% of the sedimentary PCB burden is contained in the surface layer.

The above calculations rely on the accuracy of the data for station 18 as well as the other assumptions. We plan to further evaluate the accuracy of these calculations by measuring sedimentation rates and total PCB accumulation in additional cores.

The estimated amount of PCBs accumulated in southern Lake Michigan is relatively small compared to some estimates of atmospheric input [2, 11]. If PCBs are not lost by volatilization, one estimate [2] of the combined input from particulate deposition, vapor phase transfer, and wet deposition is about 1820 kg/yr for the southern basin. This assumes the southern basin is 28% of the lake area. Thus, only 2 to 3 years of input would be required to accumulate the present PCB burden of the sediments. However, PCBs may be lost by volatilization from the lake water [2], and the net vapor phase and wet deposition input may be lower than the estimated gross input. The estimated input from dry particulate deposition for southern Lake Michigan would be about 335 kg/yr. At this input rate, more than 10 to 15 years could be required to accumulate the present sediment burden of PCBs. Possibly, particulate deposition is the most important route of atmospheric transport.

Comparison of PCB accumulation in Lake Michigan and Lake Superior sediments also is of interest in considering whether atmospheric input is a major source of PCBs to the two lakes. If the areal atmospheric input rate were similar for the two lakes and if PCBs are efficiently scavenged and transported to the bottom sediments by settling particulates, the concentration of PCBs in sediment depositing in the two lakes might also be similar. Some differences would be expected if the flux of particulate matter to the sediments was substantially different for the two lakes. However, the mass sedimentation rates for Lake Superior cores [7] are generally similar to rates in Lake Michigan [13]. As discussed previously, PCB accumulation at a given site should be proportional to the mass sedimentation rate. Consequently, PCB accumulation must be normalized to the mass sedimentation rate to compare different sites. Normalized PCB accumulation values ($\mu g\ g^{-1}$ yr) are obtained by dividing total PCB accumulation ($\mu g\ cm^{-2}$) over the depth of the contaminated zone by the mass sedimentation rate ($g\ cm^{-2}\ yr^{-1}$).

PCB concentrations in Lake Superior sediments were measured by Eisenreich et al. [6]. Differences in methodology of PCB analysis may be a factor in comparing these data to the data for Lake Michigan.

The normalized PCB accumulation value for station 18 in southern Lake Michigan is 3.25 ± 0.65 $\mu g\ g^{-1}$ yr (assuming $\pm$ 20% experimental error). Values calculated from data on Lake Superior [6, 7] are 5.4, 5.2, 2.6, and 3.1 for cores 1 BX, 11 BX, 15 BX, and 26 BX, respectively. The higher values are for stations in the western basin (1 BX and 11 BX). Stations 15 BX and 26 BX are located near the east-west center and in northern and southern sections, respectively. The ranges for the two lakes overlap, and the differences among stations may not be significant because of the degree of uncertainty in both the PCB concentrations and mass sedimentation rates. However, the values suggest local sources may affect the western basin of Lake Superior. Furthermore, the general similarily between the values for central Lake Superior and southern Lake Michigan suggests the areal loading rates of PCBs to these areas of the two lakes are similar.

CONCLUSIONS

PCB accumulation in lake sediments can be used as a conservative measure of the cumulative loading of PCBs to the lake. Losses by volatilization or degradation and postdepositional transport may result in differences between the gross loading to the lake and accumulation in the sediment.

The depth distribution of PCBs in lake sediments provides a record of changes in PCB accumulation with time. However, to interpret depth profiles, information on sedimentation rate and mixing depth is also required.

Analysis of PCB profiles in sediments from the southern basin of Lake Michigan indicates: (1) sediment PCB concentrations have been diluted by mixing of surface sediments; at site investigated in most detail (station 18), concentrations in the surface sediment continue to be diluted by mixing; (2) PCB deposition has occurred in Lake Michigan since about 1928 $\pm$ 9 years; 3) total PCB accumulation in southern Lake Michigan sediments is about 3400 to 5000 kg. Comparison with estimates of atmospheric input indicates some loss of PCBs from the lake probably occurs and suggests particle deposition may be the most important route of net transport of atmospheric PCBs to the lake.

ACKNOWLEDGEMENTS

This investigation was supported in part by NOAA, Office of Sea Grant, through an institutional grant to the University of Wisconsin, and by the Environmental Protection Agency Agreement No. CR-807836-01-1. The authors thank David Liebl and Patricia Polando for assistance with PCB and Pb-210 analyses, respectively. We also thank David Weininger for exchanging ideas about PCB behavior and David Edgington for helpful suggestions concerning sedimentation processes in Lake Michigan.

LITERATURE CITED

1. Cahill, R.A., "Geochemistry of Recent Lake Michigan Sediments" Ill. Geol. Survey, Circular 517 (1981).

2. Doskey, P.V. and Andren, A.W., "Modeling the Flux of Atmospheric Polychlorinated Biphenyls Across the Air-Water Interface" Environ. Sci. Technol., 15, 705 (1981).

3. Durfee, Robert L., "Production and Usage of PCB's in the United States" In National Conference on Polychlorinated Biphenyls. EPA-560/6-75-004 (1976).

4. Edgington, D.N. and Robbins, J.A., "Records of Lead Deposition in Lake Michigan Sediments Since 1800" Environ. Sci. Technol., 10, 266 (1976).

5. Eisenreich, S.J., Holland, G.J. and Johnson, T.C., "Accumulation of Polychlorinated Biphenyls (PCBs) in Surficial Lake Superior Sediments. Atmospheric Deposition" Environ. Sci. Technol., 13, 569 (1979).

6. Eisenreich, S.J., Holland, G.J., Johnson T.C. and Evans, J., "Polychlorinated Biphenyl and Other Microcontaminants-Sediment Interactions in Lake Superior" In Robert A. Baker, ed., Contaminants and Sediments Vol. 1. Ann Arbor Science Publishers, Inc., Ann Arbor, Michigan. (1980).

7. Evans, J.T., Johnson, T.C., Alexander, E.D., Jr., Livey, R.S., and Eisenreich, S.J., "Sedimentation Rates and Depositional Processes In Lake Superior from Pb-210 Geochronology" J. Great Lakes Res., 7, 299 (1981).

8. Flotard, R.D., "The Degradability of PCBs in Lake Mendota Sediments" Ph.D. Thesis, Water Chemistry Program, University of Wisconsin-Madison (1978).

9. Furukawa, K. and Matsumura F., "Microbial Metabolism of Polychlorinated Biphenyls. Studies on the Relative Degradability of Polychlorinated Biphenyl Components by Alkaligenes sp." J. Agric. Food Chem. 24, 251 (1976).

10. Karickhoff, S.W., "Semi-empirical Estimation of Sorption of Hydrophobic Pollutants on Natural Sediments and Soils" Chemosphere 10, 833 (1981).

11. Murphy, T.J. and Rzeszutko, C.P., "Precipitation Inputs of PCBs to Lake Michigan" J. Great Lakes Res. 3, 305 (1977).

12. National Academy of Sciences, "Polychlorinated Biphenyls Committee on Assessment of Polychlorinated Biphneyls in the Environment" National Research Council, Washington, D.C. (1979).

13. Robbins, J.A. and Edgington, D.N., "Determination of Recent Sedimentation Rates in Lake Michigan Using Pb-210 and Cs-137" Geochim. Cosmochim. Acta 39, 285 (1975).

14. Robbins, J.A., Krezokski, J.R. and Mosley, S.C., "Radioactivity in Sediments of the Great Lakes: Post-depositional Redistribution by Deposit-feeding Organisms" Earth Planet Sci. Lett. 36, 325 (1977).

15. Swackhamer, D.P., "The Recovery of PCBs in Sediments by Steam Distillation" M.S. Thesis, Water Chemistry Program, University of Wisconsin-Madison (1981).

16. Talbot, R.W., "Atmospheric Fluxes and Geochemistries of Stable Pb, Pb-210, and Po-210 in Crystal Lake, Wisconsin" Ph.D. Thesis, Water Chemistry Program, University of Wisconsin-Madison (1981).

17. Veith, G.D. and Kiwus, L.M., "An Exhaustive Steam Distillation and Solvent Extraction Unit for Pesticides and Industrial Chemicals" Bull. Environ. Contam. Toxicol. $\underline{17}$, 631 (1977).

18. Weininger, D. and Armstrong, D.E., "Organic Contaminants in the Great Lakes" $\underline{In}$ Restoration of Lakes and Inland Waters. EPA 440/5-81-010 (1980).

19. Weininger, D., Armstrong, D.E. and Swackhamer, D.L., "Application of a Sediment Dynamics Model for estimation of Vertical Burial Rates of PCBs in Southern Lake Michigan" (This book) (1982).

20. Weininger, D. and Burkhard, L., "Analysis of PCB Capillary Column Chromatograms by Comparison to Aroclor Standards Using the Multiple Linear Regression Program COMSTAR" (In Preparation) (1982).

CHAPTER 14

PCBs IN SEDIMENT AND FLUVIAL
SUSPENDED SOLIDS IN THE
GREAT LAKES

R. L. Thomas
    Great Lakes Biolimnology Laboratory
    Canada Centre for Inland Waters

R. Frank
    Provincial Pesticide Testing Laboratory
    Ontario Ministry of Food and Agriculture

INTRODUCTION

    Persistent organochlorine insecticides and industrial chemicals have been found in water, sediment, and fish in the Great Lakes Basin. Increasing documentation has demonstrated their ubiquity throughout the system. PCB's have been found in the ecosystems of all the lakes, yet the extent of their distribution has been best demonstrated in the sediment system, both bottom sediment and sediment introduced by rivers from land drainage. Levels in Great Lakes sediments for example, have been reported (1), (2), (3), as well as results of extensive surveys in individual lakes; Lakes Erie and St. Clair (4), Lake Ontario (5), Lake Huron (6), Bay of Quinte (7), Lake Michigan (8), and Lake Superior (9). In addition, the levels of PCB's associated with suspended solids recovered from the mouths of rivers tributary to the Canadian section of the Great Lakes have been reported (10).

    The present paper is intended to collate and present as a composite the work described by Frank and co-workers, with respect to the levels and distribution of PCB's in the sediment and fluvial suspended solids in the Great Lakes.

METHODS AND MATERIALS

    The original sampling design and general analytical protocols for a major survey of the sediments of the Great Lakes was developed in 1968 and implemented in that year on

Lake Ontario. The Lake Ontario grid was a polyconic projection with a baseline grid dimension of 8 km. All other lakes were on Universal Transverse Mercator grids of either 10 km spacing or alternate samples taken E - W on a 10 km grid (Figure 1). Lake Ontario (216 stations) was sampled in 1968, Lake Huron (174 stations) in 1969, Lake St. Clair (55 stations) in 1970 and 1974, Lake Erie (259 stations) in 1971, the Bay of Quinte (214 stations) in 1972 and 1973, Lake Superior (405 stations) with Georgian Bay (115 stations) and North Channel (55 stations) in 1973, and Lake Michigan, (279 stations) in 1975. All samples were taken with a Shipek sampler and the surficial 3 cms of sediment were recovered and freeze-dried for analysis and storage. Prior to analysis, samples were sieved to pass 20 mesh to remove pebbles and pulverized to pass 100 mesh to fully homogenize the sample.

In addition, selected Benthos core samples from Lakes Huron and Erie were used for PCB analysis in order to evaluate the deposition history of the compounds.

Suspended solids were collected at river mouths by pumping large volume water samples through a continuous flow Westphalia KDD 605 centrifuge. Suspended solids were removed from the centrifuge bowls and freeze-dried for storage and analysis (11).

A 10 gram aliquot of the sample was adjusted to about 50 percent field capacity with respect to moisture control and allowed to stand for 12 hours. A 250 ml hexane:acetone mixture (1:1 v/v) was used to extract samples by shaking for 2 hours on a wrist-action shaker. After filtering the extract, a 100 ml aliquot was mixed with 10 ml saturated NaCl solution and 300 ml water and shaken vigorously for one minute. The hexane phase was passed through a 3 cm layer of sodium sulphate and evaporated just to dryness with a rotary vacuum evaporator (12).

Twenty five grams of activated FLORISIL were introduced into a 22-mm internal diameter (i.d.) chromatography column. After pre-washing with 50 ml of hexane the sample extract was added to the column. The column was eluted successively at the rate of about 5 ml/min with 20 ml of dichloromethane: hexane (20:80 v/v). The eluate was evaporated just to dryness with a rotary vacuum and the residue reconstituted in 5 to 10 ml acetone (13).

A 10 mm i.d. chromatography column was prepared containing glass wool, 1 cm sand, 7.5 cm coconut charcoal, and 1 cm

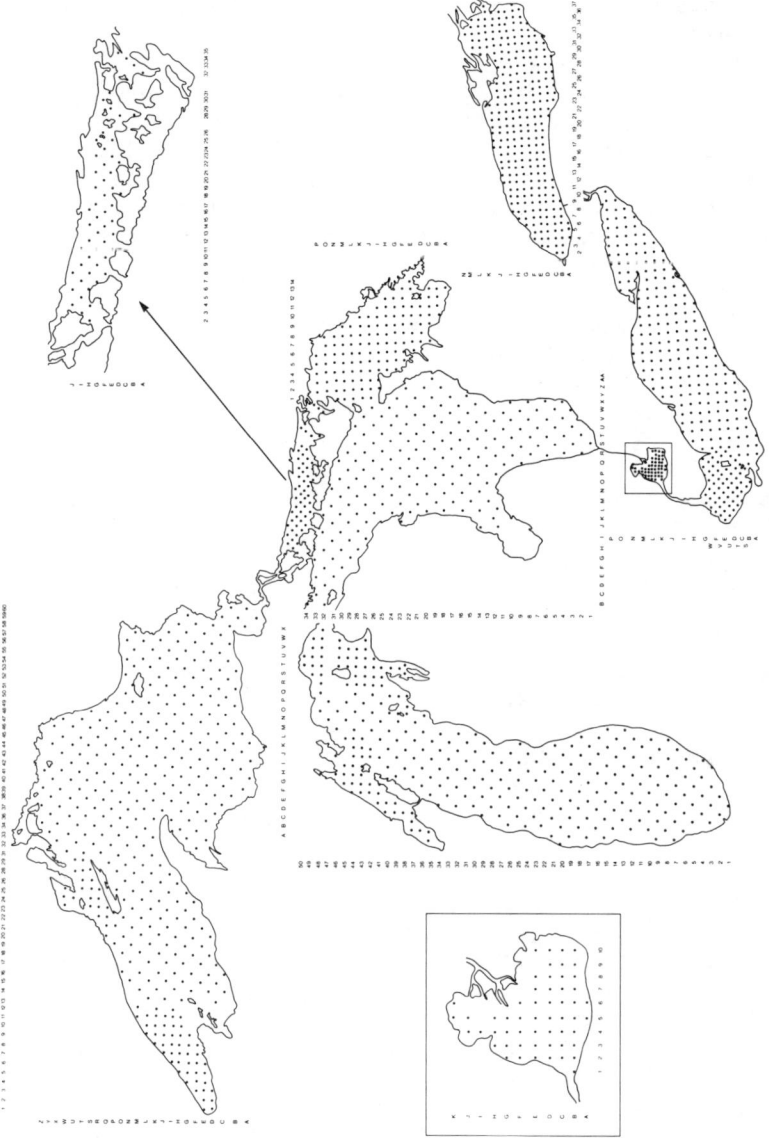

Figure 1. Great Lakes surface sediment sampling grid.

sand. The column was prewashed with acetone:ether (1:3 v/v). The eluate was quantitatively added and the column was eluted with 180 ml of acetone:ether (1:3 v/v) to remove DDT-like compounds and then with 80 ml benzene to elute the PCB fraction. This fraction was evaporated just to dryness with a rotary vacuum and reconstituted in hexane for determination (14),(15).

Micro-Tek Model MT 220 and Tracor 550 gas chromatographs equipped with Ni^{63} electron capture detectors were used. These instruments had 1.8 m columns with i.d. of 3.2 mm. Columns were packed with 1.5 percent OV - 17/2.0 percent OV-210 on 100-200 mesh Gas Chrom Q. Columns were pre-conditioned for 72 hours at 240°C and 30 ml nitrogen per minute. Operating parameters were as follows:

    Nitrogen carrier gas at 60 ml/min
    Injector temperature 220°C
    Column temperature   180°C
    Detector temperature 300°C

PCB levels were estimated against an Aroclor 1254 standard by summing peaks 7, 8, and 10 according to the numbering system of (16).

Recovery measurements were undertaken using samples fortified at 100 ppb and extracted 2 days, 1 week, and 2 weeks after fortification. Recovery percentage for PCB (Aroclor 1254), was 89 percent based on the mean recovery of nine determinations.

RESULTS AND DISCUSSION

Distribution and Levels of PCB - Open Lakes

The distribution of sediment types observed in the Great Lakes was determined during the field phase by the use of low frequency echosounding verified by sample descriptions at the sampling stations. For the purpose of this discussion, two major sedimentary sub-divisions can be made based upon in-lake processes. The first is a zone of non-deposition in the nearshore zones of the lakes and in offshore regions of higher elevations or relatively shallow water. This zone is characterized by the occurrence of bedrock, some rare sand deposits, lag sands and gravels, which veneer the exposed surfaces of glacial tills and glacial tills and glacio-lacustrine clays (17).

The second zone occurs predominantly in the deeper water offshore zones of the lake in which active fine-grained sediment deposition is occurring. These areas are defined as depositional basins, which in some lakes have been split (for statistical and comparative purposes) into sub-basins. Deposits occurring in the depositional basins consist of siltyclays and clays invariably rich in organic matter.

The distribution of the depositional basins and the non-depositional zones is given in Figure 2.

The mean levels for PCB concentrations in Great Lakes sediments are summarized in Table 1, broken down into lake, and lake sectors. Lake Superior shows significantly lower mean values than the other lakes, with Lakes Michigan, Huron, St. Clair, and Georgian Bay showing similar levels which are marginally higher than those observed in Lake Superior. Lakes Erie and Ontario, including the Bay of Quinte, display mean values considerably higher than the Upper Lakes. In all lakes, other than Georgian Bay, higher values are observed in the depositional basin than in the non-depositional zones, reflecting the differences in sediment type and associated sedimentary processes. It is interesting to note that PCB occurs in all zones including the non-depositional zones. This relates to sedimentation in the lakes which may be described very simply. Firstly, studies in Lakes Huron and Ontario, have shown that PCB correlates with mean sediment grain size and organic carbon (5), (6). Concentrations of PCB thus increase with decreasing particle size and hence concentrations increase outward into the depositional basins (See Figure 3). Secondly, sedimentation occurs on a lakewide basis so that under normal conditions some fine-grained sediment with associated PCB's will accumulate on all sedimentary surfaces. In the non-depositional zones, these recently sedimented materials are recovered by sampling and can be observed as an olive-green coloured ooze. Under storm conditions, these fine grained materials are resuspended and subjected to transportation and subsequent redeposition. This cycle of intermittent shallow water resuspension is a continuing process, transporting sediment progressively into the deeper water deposits. Depth of water and length of fetch ultimately determine the final sink such that the materials so deposited will be sedimentologically unique for their location within the physical milieu of the lake. This sorting process has been called sediment focussing and represents an important facet in the understanding of contaminant dispersal in aquatic systems.

Table 1 also shows considerable variations between

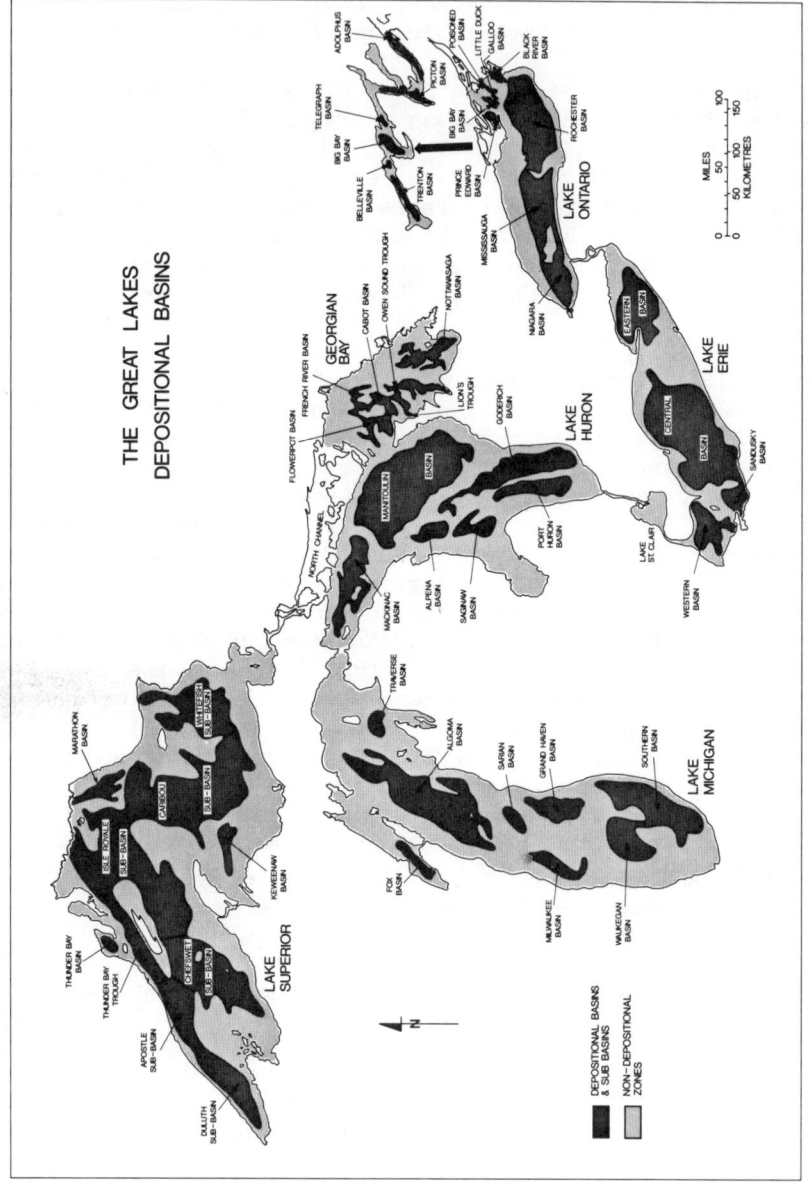

Figure 2. Distribution of the depositional basins and the non-depositional zone in the Great Lakes.

basins within lakes. These variations are in part related to lake processes, but in larger measure may be accounted for by major source regions of PCB's to the lakes. This is best seen with reference to the distribution of PCB's in the lakes, as given in Figure 3, together with an evaluation of basin concentrations summarized in Table 1.

Lake Superior

Lake Superior shows low concentrations throughout. Of the 405 samples analyzed, PCB's were non-detectable (to a level of 2.5 ng/g) in 56 percent and traces (2.5 to 5.0 ng/g) in 22 percent. In the measurable sediments, the mean concentration was 8.5 ng/g. In Figure 3, areas of measurable concentrations (greater than 5 ng/g) occur in a patchy fashion, but predominate in the Duluth and Thunder Bay region of the lake, which may indicate some industrial sources. However, the PCB's in this lake could be attributed to atmospheric precipitation (3), (9). The concentrations quoted here are those given by (9). Reported values ranged from 5 to 290 ng/g in the surface 0 to 0.5 cm of sediment which declined to non-detectable at a depth of 4 cm (3). These residues are an order of magnitude higher than the values reported in this study for offshore stations. Since we are reporting an homogenized surface 3 cms of sediment, then lower values would be expected. It should be noted however that the samples reported here were taken in 1973, whereas the samples referred to above were taken in 1977 (3). Only Aroclor 1254 is reported for the 1973 samples since this was the predominant fraction, whereas in 1977, Aroclor 1242 was dominant and PCB was reported as the sum of Aroclor 1242 and 1254. When residues for Aroclor 1254 were compared for both studies, good agreement was observed at six of the nine comparable deep water stations. The differences may thus not be as great as suggested by an initial comparison of the data sets.

The obvious suppression of concentration levels due to the sampling depth will be greatest in the sediment of Lake Superior, since this lake has the lowest sedimentation rate of all the Great Lakes. Additionally, organic carbon is low and the process of sediment mixing by physical and biological processes will be less pronounced here than in the other lakes. Concentrations reported for the other lakes will thus more closely approximate the concentrations at the immediate surface of the sediment. However, it should be recognized that the initial design of the Great Lakes sediment sampling program was aimed at describing texture, mineralogy, and gross geochemistry of the sediments rather

Table I. Mean values and standard deviations for PCB's (ng/g) by lake, non-depositional zones (NDZ) and depositional basins for the surficial 3 cms of sediment in the Great Lakes.

Lake Superior	Mean	S.D.	Lake Michigan	Mean	S.D.
Whole Lake	3.3	5.7	Whole Lake	9.7	15.7
Total Basin	4.8	5.5	Total Basin	17.3	23.9
NDZ	3.9	2.1	NDZ	6.3	8.1
Sectors:-			Sectors:-		
Duluth	8.6	13.7	Fox	73.5	78.9
Apostle	5.0	2.2	Southern	17.1	11.1
Chefswet	3.3	1.3	Milwaukee	29.2	23.1
Thunder Bay	5.7	3.6	Waukegan	19.5	13.2
Trough	5.5	2.9	Grand Haven	17.1	23.1
Isle Royale	4.5	2.2	Sarian	7.9	-
Caribou	3.7	1.6	Algoma	10.1	10.6
Marathon	6.4	7.3	Traverse	2.5	-
Whitefish	4.4	3.0			
Keweenaw	3.1	1.3	Georgian Bay		
			Whole Lake	11.2	10.7
Lake Huron			Total Basin	11.1	8.1
Whole Lake	12.8	10.3	NDZ	11.2	13.2
Total Basin	15.4	12.8	Sectors:-		
NDZ	10.7	7.3	Nottawasaga	19.6	12.3
Sectors:-			Owen Sound		
Mackinac	17.9	14.9	Trough	9.1	6.1
Manitoulin	12.2	5.7	Lion's Trough	18.0	8.5
Alpena	8.5	0.7	Cabot	15.5	3.7
Saginaw	33.0	38.3	French River	24.0	24.0
Port Huron	17.0	5.2	Flowerpot	9.1	2.6
Goderich	18.6	14.0	North Channel	8.2	4.0
Lake St. Clair			Lake Erie		
Whole Lake 1970	19.1	8.9	Whole Lake	94.6	113.6
			Total Basin	115.2	114.8
Whole Lake 1974	9.9	6.3	NDZ	64.0	105.1
			Sectors:-		
			Western	251.7	156.0
Lake Ontario			Sandusky	106.9	46.0
			Central	74.1	55.7
Whole Lake	57.5	56.2	Eastern	85.6	85.2
Total Basin	85.3	57.0			
NDZ	28.1	34.7	Bay of Quinte		
Sectors:-					
Niagara	89.1	68.3	Whole Bay	50.6	53.6
Mississauga	77.1	50.6			
Rochester	89.4	56.6			

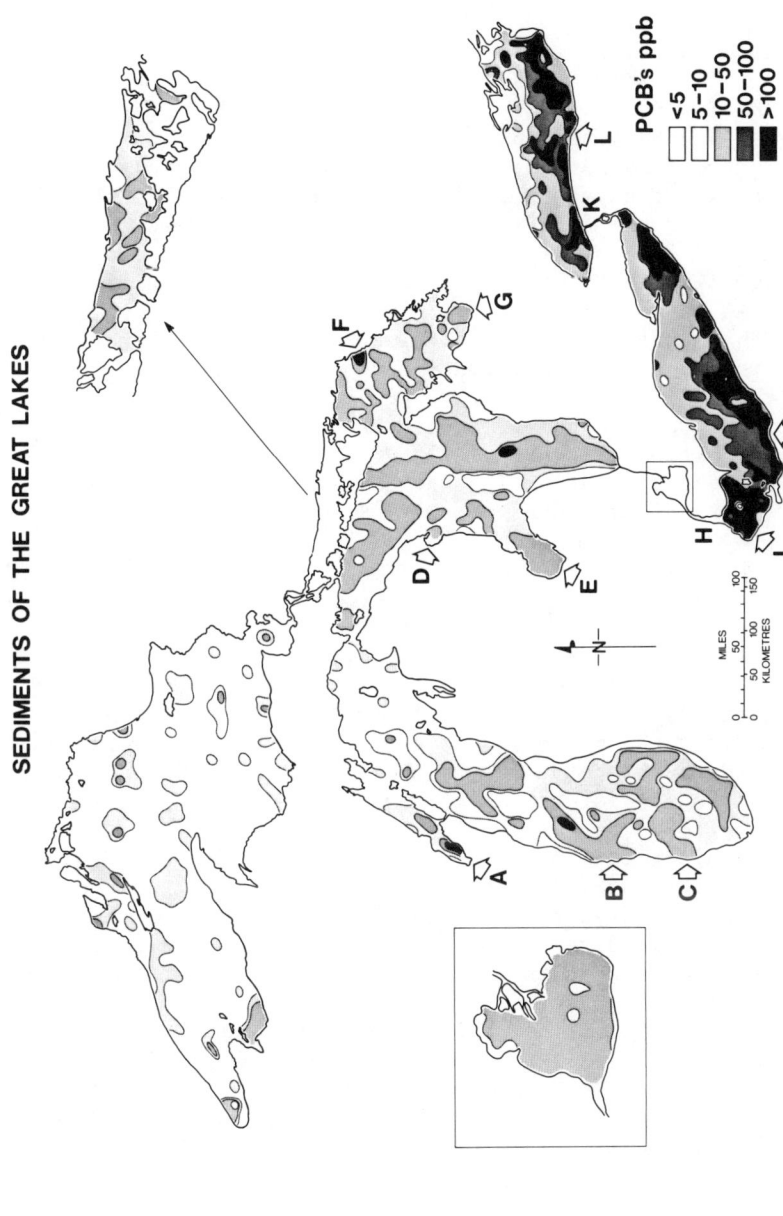

Figure 3. Distribution of PCBs in ng/g in the surface 3 cms of sediment in the Great Lakes.

than a design specific to recent contaminants. Three centimeters was selected as a good integration of recent processes and was maintained for all lakes to ensure consistency of procedure during the entire survey.

Lake Michigan

PCB's were detected in 79.6 percent of 279 sediments analyzed and 67.4 percent contained between 2 to 20 ng/g. The mean residue for the whole lake was 9.7 ng/g with a non-depositional zone value of 6.3 ng/g, as compared to the 17.3 ng/g mean for the depositional basins. The highest mean residues of 73.5 ng/g occurred in the Fox Basin followed by 29.2 ng/g in the Milwaukee Basin (Table 1). The distribution of PCB is shown in Figure 3, in which it can be seen that, other than possibly the Fox, Milwaukee, and Southern Basins, there is not a good conformation to the basin structure of the lake (Figure 2). The compound is distributed at low levels throughout the lakes with three obvious inferred sources on the western shoreline. These are, from north to south, the Fox River, Milwaukee and Waukeegan, shown as A,B, and C respectively in Figure 3.

Lake Huron

All 174 samples contained PCB residues in the range of 3 to 90 ng/g. Unlike Lake Michigan, mean residues for the non-depositional zone (11 ng/g) was similar to the 15 ng/g observed for the basins. Among the basins, the Saginaw Basin contained the highest mean value (33 ng/g), compared to the lowest (9 ng/g) for the Alpena Basin.

The distribution of PCB in Lake Huron shows only a general relationship to the basin structure of the lake which is masked by nearshore areas of higher values apparently associated with direct sources. These occur at Alpena and Saginaw Bay (D and E in Figure 3). Normalization of the PCB concentrations by quartz correction confirmed these two sources (6). It further showed that the loadings from Saginaw Bay were dispersed around the "thumb of Michigan" into southern Lake Huron and into the St. Clair River. PCB from this source is thus probably the most significant source to Lake St. Clair immediately downstream.

Georgian Bay, North Channel

All sediment samples from Georgian Bay and North Channel

contained PCB's with mean concentrations of 13 and 8 ng/g respectively. Mean residue levels in the non-depositional zones and basin of Georgian Bay were identical at 11 ng/g. Among the Georgian Bay basins, highest values were observed in the French River Basin (24 ng/g) and lowest values in the Owen Sound Trough and Flowerpot Basin with a mean concentration of 9 ng/g. The distribution (Figure 3) does not replicate the basin distribution, but two sources can be inferred in the region of the French River outlet and Collingwood, F and G respectively in Figure 3. There are no obvious sources to North Channel and the general distribution may be related to atmospheric deposition and run-off from the St. Mary's River.

Lake St. Clair

PCB's at detectable levels were found in all 55 samples from Lake St. Clair, both in 1970 and 1974. Mean levels for 1970 were 19 ng/g, but had declined to 10 ng/g in 1974. The major source is believed to be Saginaw Bay and the decline is thought to be due to the voluntary restraint in commercial usage of PCB starting in 1971. An evaluation of trace metal and other organochlorine levels in 1970 and 1974 in Lake St. Clair has been reported (18) and showed that the decline is due to sediment re-suspension and onward transmission to the Detroit River, with replacement by new sediment from Lake Huron containing lower concentrations of PCB.

Lake Erie

The mean level of PCB in Lake Erie for 259 samples was 95 ng/g, some five times higher than observed in Lake St. Clair (Table 1). Mean PCB residues in the decompositional basins were almost double that of the non-depositional zone. The value of 64 ng/g in the non-depositional zone, however, was three times the value observed in Lake St. Clair. The highest mean basin value (252 ng/g) was seen in the Western Basin, closest to the Detroit River, followed by the Sandusky Basin (107 ng/g), the Eastern Basin (86 ng/g) and the Central Basin, with 74 ng/g. It has been shown that the Western Basin is statistically significantly higher in PCB's when compared to the other basins (4).

The distribution of PCB's is shown in Figure 3. This diagram shows high values in the Western Basin, the southern portion of the Central Basin and the deep water region of the Eastern Basin. The distribution indicates major sources from the Detroit River, the Maumee River and from Cleveland (H, 1 and J in Figure 3, respectively).

Lake Ontario

216 samples were analyzed for PCB's and give a mean whole-lake value of 58 ng/g. PCB concentrations in the depositional basins (85 ng/g) were almost three times the 28 ng/g mean value for the non-depositional zone. All three individual basins showed very similar mean values between 77 and 89 ng/g (Table 1). The distribution of PCB's in Lake Ontario (Figure 3) shows a major influence from the Niagara River (K on Figure 3), extending out into the Niagara Basin, along the south shore of the Mississauga Basin and extending into the Rochester Basin. A possible additional source to the Rochester Basin may be the Genesee River, point L on Figure 3. This distribution reflects the major circulation pattern of the lake relative to an input source from the Niagara River and is similar to the mercury distribution (19).

In the Bay of Quinte, 214 samples were analyzed and gave a mean value of 48 ng/g. Little variation was noted in the Bay, with basins giving a mean value of 50 ng/g, as compared to 47 ng/g in the non-depositional zone. Individual basins gave mean values ranging from 33 to 74 ng/g.

Loadings

The annual rates of sediment loading to the depositional basins of Lakes Superior, Huron, Erie, Ontario and Georgian Bay have been determined (20) (21), (22), (23). No such information is, as yet, available for Lake Michigan, the North Channel and the Bay of Quinte. Using the annual loading values referred to above, together with basin mean conconcentrations as given in this study, crude estimates of annual PCB loadings have been compiled and are given in Table 2. It should be noted that this figure is for an integrated 3 cm subsample and assumes no sedimentation in the non-depositional zone. Since this non-depositional zone may be regarded as an intermittent supply zone to the depositional basins, then it can be assumed that on an average annual basis, this PCB is incorporated into the basin loadings. The integration of the surface 3 cms of surface sediment will likely produce a value lower than the true PCB accumulation rate.

Sediment Cores

PCB's were analyzed in cores taken from Lakes Huron,

Table 11. Annual loading of PCB's to the Depositional Basins of the Great Lakes. (Excluding Georgian Bay and Lake Michigan)

Lake	Mean Annual Sediment Loads metric tons x 10^3	Mean Annual PCB Loads metric tons
Superior	6,029	0.028
Huron	2,880	0.0149
Erie	14,400	1.6589
Ontario	4,040	0.3520

Erie, and Georgian Bay. The chronology for Lake Erie is probably good due to the high rate of sediment accumulation relative to the units sub-sampled in the core. For Lake Erie, PCB's were first detected in the core increment dated 1956-1958 and increased to the surface with highest values between 1969-1971. This chronology is compatible with the known production figures for PCB, which showed a dramatic increase in 1954. The situation in Lake Huron and Georgian Bay is far less clear due to much lower sedimentation rates in which the upper 2 cms of sediment represents time spans ranging from 1924 to 1974 and 1943 to 1971 in five cores. In the optimum case (core V-9, Lake Huron), PCB first appears in the time span ranging from 1936 to 1947. Much finer sub-sampling of these cores is required to provide the needed resolution.

TRIBUTARY LOADINGS OF PCB'S

Rural Sources

Although PCB's have not been used in agricultural pursuits, they are nevertheless found in the rural environment. Release to the environment from electrical transformers has occurred as a result of spreading spent oils on rural roads to reduce dust and as a carrier in the spraying of herbicides to control brush and trees (24).

PCB's are found in rain water and were present in all 11 agricultural watersheds studied (25). The aerial fallout of PCB on particulate matter and in precipitation in southern Ontario has been variably estimated to range from 0.2 to 0.95 g/ha annum (26). In the Grand River basin this could vary

Table III. Inputs into the Grand and Saugeen Rivers from Rural Activities and Losses at the Mouth

Year	Location	PCB's Grand River		PCB's Saugeen River	
		Amount (kg)	Concentration	Amount (kg)	Concentration
1975-76	Entry - Rural (Water + SS)	128	4.6 ng/L	63	2.9 ng/L
	Mouth - SS	18.7	56 µg/kg	7.0	36 µg/kg
	- Water	15.8	5.8 ng/L	0.5	0.2 ng/L
	- Total	34.5		7.5	
1976-77	Entry - Rural	48	3.4 ng/L	25	1.3 ng/L
	Mouth - SS	35.1	70 µg/kg	10.3	63 µg/kg
	- Water	7.6	3.7 ng/L	0.6	0.3 ng/L
	- Total	42.7		10.9	
	- Sediment		25 µg/kg		< 0.1 µg/kg

from 136 kg to 3.4 metric tons and in the Saugeen River basin from 80 kg to 2.0 metric tons. Much of these deposits could be absorbed and held by soil particles with only small portions finding their way into streams and rivers.

Agricultural Watersheds

Losses of PCB to streams were identified in 11 agricultural watersheds (27). Unit area loading varied from 97 to 269 mg/ha in 1975 to 1976 and 25 to 119 mg/ha in 1976 to 1977. A similar magnitude in loadings was evident in the Grand and Saugeen Rivers. The decline between 1975 and 1977 was unexplained, but may have been due to actions taken to reduce losses of PCB to the environment from (a) open ended uses and (b) disposal of contaminated spent oils.

Grand and Saugeen River Basins

Estimated losses of PCB to the Grand and Saugeen rivers appears in Table 3. The losses to the Grand River were roughly twice as high as those to the Saugeen River, but the latter basin is only a little over half the area (59 percent) of the former.

Concentrations of PCB in water and on supended solids from the two river basins also appear in Table 3. The similar orders of magnitude in both years for both water and suspended solid phase of the two systems, would suggest a general overall contamination for the basins.

Within the rural watersheds PCB's accumulated in sediment, invertebrates, and in small fish (Table 4). Biomagnification between 300 and 600 times the concentration in water was observed.

Discharges at the mouth of the Grand River appear in Table 3. The amounts associated with suspended solids represented 54 and 82 percent of the total discharge. Based on the amounts entering the river system from rural areas, the discharges at the mouth represented 27 and 89 percent. Residues in sediments measured by the Ontario Ministry of Environment (unpublished) during the study period would suggest that considerable losses to the river system occurred in the environs of major cities, giving rise to a much larger input than herein estimated from only the rural part of the basin.

In the Saugeen River discharges into Lake Huron, 93 and 94 percent of the PCB was carried on the suspended solids

Table IV. PCB concentrations in biota of 11 agricultural watersheds

| | PCB's | | |
Item	Mean	Range	Magnification
Water + suspended solids (ng/L)	38	< 2-200	
Bottom sediment (µg/kg)	15	< 2-8	395
Invertebrates (µg/kg)	12	4-35	316
Small fish (µg/kg)		18-52	605

respectively in the two one-year periods. Of the inputs from rural areas to the system, 13 and 48 percent appeared in discharge waters at the mouth.

The water in both river systems was alkaline; in the Grand the pH ranged from 7.7 to 8.8 and in the Saugeen 7.1 to 8.5. Water temperatures in both ranged from 0 to 26°C. Biological activity was high, especially in the lower reaches of the Grand River. In addition, this river has a complex series of weirs and dams which interrupt the water flow leading to settling and storage of suspended solids. In the lower reaches of the Grand River, the velocity of flow slows, leading to further deposition of solids. The resident time for water would appear to be 2 to 3 days during spring runoff periods, a time when the river is at its highest, and 5 to 7 days during the low summer flow period. The movement of suspended solids can be quite variable and depend on storm events and stream flow velocities.

Canadian Tributaries

PCB residues were found to be ubiquitous in suspended solids collected in all streams sampled on the Canadian side of the Great Lakes.

Spring Flow 1974

PCB residues on solids collected from 43 streams ranged from 20 to 1000 g/kg with the highest being found in Sixteen

Mile Creek (Oakville Creek). Half of the streams had residues at or above the 100 µg/kg level and the rest were below this concentration. The highest mean levels (186 µg/kg) were among streams along the north shore of Lake Ontario; levels were only fractionally lower (172 µg/kg) for streams along the southwest shoreline (Table 5).

Spring Flow 1975

Residue levels of PCB were considerably lower in the 1975 sampling period than in 1974. The highest residue occurred in Muddy Creek, Lake Erie (330 µg/kg). Only 3 of 36 suspended solids had residue levels above 100 µg/kg, and 32 of the remaining 33 were between 10 and 100 µg/kg. Solids from the Ausable River contained only 8 µg/kg PCB.

Spring Flow 1977

Residue levels in the 42 streams sampled in 1977 showed a further decline over those sampled in 1975. While five streams had levels above 100 µg/kg PCB in suspended solids, none were below 10 µg/kg. More variation was observed in concentrations. The highest mean residue levels were observed in the Niagara and Detroit Rivers sections (128 and 395 µg/kg).

Monitored Streams 1974-76

The mean residues of PCB on suspended solids of the five monitored streams entering Lake Ontario ranged from 58 to 275 µg/kg. The highest concentrations occurred in the Niagara and Humber Rivers (275 and 224 µg/kg respectively). There was considerable fluctuation from month to month and from stream to stream with individual readings ranging from 20 to 600 µg/kg. PCB residues in the other five monitored streams entering Lake Erie, St. Clair and Lake Huron were uniformly similar throughout the year and all but two individual sample readings were below 100 µg/kg.

Lake System

In 1974 to 1975, the highest mean residues of PCB were found in streams entering Lake Ontario and especially the north and southwest shorelines. By 1977 residues had declined markedly in these sections of the lake.

Table V. Mean levels of PCB's in the suspended solids of Canadian streams entering the Great Lakes, 1974 to 1977. (The number in parentheses after the locations is the number of samples used).

	PCB's	
Locations	Arith. Mean (µg/kg)	Coeff. of Variation (%)
Monitored Streams 1974 - 75		
Bronte Creek (5)	68	59
Humber R. (10)	224	79
Credit R. (4)	62	45
Welland Canal (7)	91	42
Niagara R. (4)	273	78
Grand R. (12)	54	35
Cedar Cr. (10)	62	31
Thames R. (9)	68	53
Saugeen R. (5)	36	28
Nottawasaga R. (6)	41	44
Total (72) & Means	93	144
Lakes & Lake Sections 1974-75		
Bay of Quinte (6)	63	56
L. Ontario N. (20)	166	132
L. Ontario S.W. (12)	159	137
L. Erie (13)	65	143
L. St. Clair (3)	33	18
L. Huron (10)	30	43
Georgian Bay (5)	116	122
N. Channel & St. Mary's R. (6)	32	88
L. Superior (3)	23	25
Lakes & Lake Sections 1977		
Bay of Quinte (1)	8	-
L. Ontario N. (8)	31	168
L. Ontario S. (4)	27	33
L. Erie (13)	81	193
L. St. Clair (1)	91	-
L. Huron (2)	33	110
Georgian Bay (13)	24	46

Lake Loadings

Long term mean annual suspended solids loadings derived from the Ontario Ministry of the Environment's river monitoring network were calculated (28). Individual stream concentrations for PCB were used to calculate individual stream loadings, which were then summarized into a regional breakdown. These loadings are given in Table 6. Considerable differences are observed when the regional loadings are compared to regional concentrations given in Table 5. This in large measure is a function of the total loading of suspended solids for regions of varying dimensions and varying PCB uses and losses (e.g. north shore Lake Erie, as compared to southwestern shore, Lake Ontario). Intensity of usage is still reflected by solids with higher concentrations, yet in order to truly reflect mass balances, measured loadings from streams would be an essential component of any future study.

Great Lakes

PCB's have been used in a wide variety of industrial and commercial applications and their disposal over the years has resulted in an untold number of possible sources, including many hundreds of landfill sites.

A. Point Sources

Both industrial and municipal wastewaters have been found to contain PCB's. Wastewater PCB levels have been examined in some jurisdictions, and measured loads range from several to hundreds of kilograms per year. Twenty-six large sewage treatment plants in Ontario discharged a total of about 250 kg/yr of PCB's. One industry in Ontario was found to be discharging 7 kg/yr into Lake Ontario (25).

B. Diffuse Sources

Studies indicate that between 5 and 50 metric tons/yr of PCB's are deposited directly on to the water surface of the Great Lakes from the atmosphere (25). The monitored total tributary PCB load to the Great Lakes is between 490 and 770 kg/yr. These values include tributaries with forested or agricultural watersheds, again implying atmospheric sources (Table 7). The loading of PCB's from urban areas is about 310 kg/yr for the Great Lakes Basin. PCB distribution in sediments indicates that urban areas represent a major PCB contribution to the lakes.

Table VI. Crude loadings calculated from concentrations of contaminants in suspended solids (1974-1977) and mean annual discharges between 1964 and 1972-1974.

Lake	No. of Streams Used in Calculations	Suspended Solids Loading (tonnes/yr)	Crude Loadings PCB (g)
Ontario			
Bay of Quinte	7	50,676	1,873
North Shore	23	89,529	1,293
Southwest Shore	12	8,352	1,520
Niagara R. Trib.	5	1,406	141
Erie			
North Shore	18	154,135	5,914
Detroit R. Trib.	3	12,047	1,024
St. Clair			
Whole Lake	3	249,919	8,423
St. Clair R.	1	704	64.1
Huron			
East Shore	8	113,875	3,754
Georgian Bay	13	34,684	1,638
North Channel	1	7,777	241
St. Marys R. Trib.	3	4,608	100
Superior			
North Shore	3	390,460	7,785

Table VII. Comparative figures on loadings to Great Lakes.

Lake	Tributary Suspended Solids x 10 metric tons/yr.	PCB's	
		PLUARG Study (kg/yr)	Canadian Trib. (kg/yr)
Superior	1,378	33	21
Michigan	706	61	0
Huron	1,053	13	37
St. Clair	1,400	–	47
Erie	6,532	530	273
Ontario	1,597	140	51
Totals	12,666	777	429

CONCLUSIONS

The occurrence of PCB's on sediment particles is ubiquitous throughout the Great Lakes. It occurs on particles in all the Canadian tributaries which fact reflects a widely dispersed origin from the atmosphere and amplified by point and diffuse sources from the industrial regions within the basin. Levels observed within the lakes reflect these sources, with widely distributed low level concentrations (as seen in Lake Superior), originating from the atmosphere, whilst higher concentrations in the other lakes are due to amplification from industrial source regions. The intensity of urban/industrial development in the Lower Lakes is reflected by the higher concentrations observed in Lakes Erie and Ontario and southern Lake Michigan. The predominant regions from which these diffuse industrial sources predominate have been individually identified. In-lake processes which result in the sorting and size grading of sediment result in concentration gradients for PCB's closely approximating the textural fabric of the lake sediments. The concentration gradients further provide an insight into the derivation, transport, and final deposition of the PCB's. Together with some knowledge of the rate of sediment accumulation in the sink areas, crude annual loading of PCB can be provided.

REFERENCES

1. Glooschenko, W.A., Strachan, W.M.J., & Sampson, R.J.C. "Distribution of pesticides and polychlorinated biphenyls in water, sediments and seston of the Upper Great Lakes - 1974". Pestic. Monit. J., 10. 61. (1976).
2. Veith, G.D., Kuehl, D.W., Puglish, F.A., Glass, G.R., & Eaton, J.G. "Residues of PCB's and DDT in the western Lake Superior ecosystem". Arch. Environm. Contam. Toxicol., 5. 487. (1977).
3. Eisenreich, S.J., Hollod, G.J., & Johnson, T.C. "Accumulation of polychlorinated biphenyls (PCB's) in surficial Lake Superior sediments. Atmospheric deposition". Environm. Sci. Technol., 13. 569. (1979).
4. Frank, R., Holdrinet, M., Braun, M.E., Thomas, R.L., Kemp, A.L.W., & Jaquet, J.M. "Organochlorine insecticides and PCB's in sediment of Lake St. Clair (1970 and 1974) and Lake Erie (1971)". Sci. Total Environm., 8. 205. (1977).
5. Frank, R., Thomas, R.L., Holdrinet, M., Kemp, A.L.W., & Braun, H.E. "Organochlorine insecticides and PCB in surficial sediments (1968) and sediment cores (1976) from Lake Ontario". J. Great Lakes Res., 5. 18. (1979).

6. Frank, R., Thomas, R.L., Holdrinet, M., Kemp, A.L.W., Braun, H.E., & Dawson, R. "Organochlorine insecticides and PCB in the sediments of Lake Huron (1969) and Georgian Bay and North Channel (1973)". Sci. Total Environm., <u>13</u>. 101. (1979).
7. Frank, R., Thomas, R.L., Holdrinet, M., & Damiani, V. "PCB residues in bottom sediments from the Bay of Quinte, Lake Ontario, 1972-73". J. Great Lakes Res., <u>6</u>. 371. (1980).
8. Frank, R., Thomas, R.L., Braun, H.E., Gross, D.L., & Davies, T.T. "Organochlorine insecticides and PCB in surficial sediments of Lake Michigan (1975)". J. Great Lakes Res., <u>7</u>. 42. (1981).
9. Frank, R., Thomas, R.L., Braun, H.E., Raspher, J., & Dawson, R. "Organochlorine insecticides and PCB in the surficial sediments of Lake Superior (1973)". J. Great Lakes Res., (in press). (1982).
10. Frank, R., Thomas, R.L., Holdrinet, M., McMillan, R.K., Braun, H.E., & Dawson, R. "Organochlorine residues in suspended solids collected from mouths of Canadian streams flowing into the Great Lakes, 1974-1977". J. Great Lakes Res., (in press). (1982).
11. Thomas, R.L., & McMillan, R.K. "Recovery and analysis of suspended solids in some Canadian rivers tributary to the Great Lakes". (In preparation) (1982).
12. Chilba, M., & Morley, H.V. "Factors influencing extraction of aldrin and dieldrin residues from different soil types". J. Agric. Food Chem., <u>16</u>. 916. (1968).
13. Mills, A., Bang, A., Kamps, R., & Burke, A. "Elution solvent system for Florisil column clean-up in organochlorine pesticide residue analysis". J. Assoc. Off. Anal. Chem., <u>55</u>. 39. (1972).
14. Berg, O.W., Diosady, P.L., & Rees, G.A.V. "Column chromatographic separation of polychlorinated biphenyls from chlorinated hydrocarbon pesticides and their subsequent gas chromatographic quantitation in terms of derivatives". Bull. Environm. Contam. Toxicol., <u>7</u>. 338. (1972).
15. Holdrinet, M. "Determination and confirmation of hexachlorobenzene in fatty samples in the presence of other halogenated hydrocarbon pesticides and PCB's". J. Assoc. Off. Anal. Chem., <u>57</u>. 580. (1974).
16. Reynolds, L.N. "Pesticide residue analysis in the presence of polychlorinated biphenyls (PCB's)". Residue Rev., <u>34</u>. 27. (1971).
17. Thomas, R.L., Kemp, A.L.W., & Lewis, C.F.M. "Distribution, composition and characteristics of the surficial sediments of Lake Ontario". J. Sediment. Petrol., <u>42</u>. 66. (1972).
18. Thomas, R.L., Jaquet, J.-M., & Mudroch, A. "Sedimentation processes and associated changes in surface

sediment trace metal concentration in Lake St. Clair, 1970-1974". Proc. Internat. Conf. Heavy Metals Environm., Toronto, 1975, 691. (1977).
19. Thomas, R.L. "The distribution of mercury in the sediments of Lake Ontario". Can. J. Earth Sci., $\underline{9}$. 636. (1972).
20. Kemp, A.L.W., & Harper, N.S. "Sedimentation rates and a sediment budget for Lake Ontario". J. Great Lakes Res., $\underline{2}$. 324. (1977).
21. Kemp, A.L.W., & Harper, N.S. "Sedimentation rates in Lake Huron and Georgian Bay". J. Great Lakes Res., $\underline{3}$. 215. (1977).
22. Kemp, A.L.W., MacInnis, G.A., & Harper, N.S. "Sedimentation rates and a revised sediment budget for Lake Erie". J. Great Lakes Res., $\underline{3}$. 221. (1977).
23. Kemp, A.L.W., Dell, C.I., & Harper, N.S. "Sedimentation rates and a sediment budget for Lake Superior". J. Great Lakes Res., $\underline{4}$. 276. (1978).
24. Scott, W.S. "Use of oil sorbant materials for small herbicide spills". Bull. Environm. Contam. Toxicol., $\underline{23}$. 123. (1979).
25. PLUARG. "Environmental management strategy for the Great Lakes system". Rept. Internat. Ref. Group Great Lakes Pollut. Land Use Activities, I.J.C., Windsor, Ontario, Canada.
26. Sanderson, M., & LaValle, P.D. "Surface loadings from pollutants in precipitation in southern Ontario: some climatic and statistical aspects". J. Great Lakes Res., $\underline{5}$. 52. (1979).
27. Frank, R., Braun, H.E., & Holdrinet, M. "Residues from past uses of organochlorine insecticides and PCB's in waters draining eleven agricultural watersheds in southern Ontario, Canada, 1975-1977". Sci. Total Environm., $\underline{20}$. 255. (1981).
28. Ongley, E.D. "Land use, water quality and river mouth loadings, a selective overview for southern Ontario". Submitted to PLUARG, Task Group D (CANADA), Activity 2.1, Windsor, Ontario, Canada. 110 pp. (1978).

CHAPTER 15

ESTIMATING BIOCONCENTRATION
POTENTIAL FROM OCTANOL/WATER
PARTITION COEFFICIENTS

G. D. Veith
Environmental Research Laboratory-Duluth
6201 Congdon Boulevard
Duluth, Minnesota 55804

P. Kosian
Center for Lake Superior Environmental Studies
University of Wisconsin-Superior
Superior, Wisconsin 54880

Veith et al. [1] presented a structure-activity relationship for estimating the bioconcentration potential of chemicals in fish from the n-octanol/water partition coefficient (P or K_{ow}) which can be measured or computed from structure. The relationship was based on more than 50 tests from five general classes of chemicals and summarized as follows:

$$\log BCF = 0.85 \log P - 0.70 \quad R^2 = 0.897$$

The dependance of bioconcentration on the partition coefficient is comparable to many other SARs involving transport through biological membranes. Although the coefficient of 0.85 is commonly observed, it is only an empirically derived value resulting from the linear regression model imposed.

Mackay [2] derived a mechanistic relationship between bioconcentration and the partition coefficient in which the slope is 1.0 rather than 0.85. This linear model applied to the data of Veith et al. [1] resulted in the SAR

$$K_B = 0.048 \ K_{ow} \quad R2 = 0.95$$

where K_B is the bioconcentration factor. This relationship yields a standard error of 0.25 in log K_B which is approximately a factor of two in K_B. This is adequate to screen chemicals to identify potentially bioaccumulative chemicals from those which do not produce residues and to provide an order of magnitude estimation of the bioaccumulation factor of chemicals under test conditions.

Although present funding does not permit continued testing of chemicals specifically from determination of bioconcentration factors, an on-going literature review has been conducted since 1979 to provide additional data for the model. Moreover, residues have been determined at the termination of 30-day embryo-larval toxicity tests being conducted at the Environmental Research Laboratory-Duluth. In these tests, the bioconcentration factor is simply the slope of the least-squares fit of residues in fish versus test concentration in the water. Only no-effect concentrations in these toxicity tests were used in determining the 30-day bioconcentration factors.

Bioconcentration factors and log P values for 122 bioaccumulation tests are presented in Table 1. The SAR was evaluated by regression analysis using different classes of chemicals and species of fish. The slope, intercept, number of data points, and R^2 for seven sets of data are summarized in Table 2. When only fathead minnow data on non-ionic chlorinated hydrocarbons were considered, the regression equation changed only slightly from Veith et al. [1] and R^2 increased from 0.90 o 0.92. Including only fathead minnow data on all classes of chemicals (54 tests), the SAR was indistinguishable from the Veith SAR. By including bluegill with fathead minnow data, the 88 tests lowered the slope from 0.85 to 0.81 and increased the intercept from -0.70 to -0.49. Adding the 9 tests with guppies did not alter the equation. Finally, examining the data of all 13 species of freshwater and marine fish and all classes of chemicals, the relationship between log BCF and log P is summarized as:

$$\log BCF = 0.79 \log P - 0.40 \quad N = 122 \quad R^2 = 0.86$$

The increase of the intercept from -0.70 to -0.40 is the only significant change in expanding the data base to 122 tests. This increase is associated with decrease in slope to 0.85 to 0.79 which means the predicted results are quite similar. In fact, the predicted BCF for a chemical with log

Table I. Summary of Bioconcentration Factors for Chemicals in Fish

Chemical	Ref	Log P	Log BCF	Species	Time	Tissue
Lindane	1	3.85	2.67	Fathead	304D	W. body
Atrazine	1	2.63	0.90	Fathead	276D	W. body
Heptachlor	1	5.44	4.30	Fathead	276D	W. body
2-Ethylhexiphthalate	1	4.20	2.93	Fathead	56D	W. body
DASC-3	1	1.00	0.32	Bluegill	30D	W. body
DASC-4	1	1.00	0.32	Bluegill	30D	W. body
NTS-1	1	1.00	0.66	Bluegill	35D	W. body
BSB	1	1.00	0.32	Bluegill	50D	W. body
FWA-2-A	1	1.80	0.32	Bluegill	105D	W. body
FWA-3-A	1	1.48	0.32	Bluegill	105D	W. body
FWA-4-A	1	1.20	0.32	Bluegill	105D	W. body
Nitrobenzene	1	2.93	1.18	Fathead	28D	W. body
P-Nitrophenol	1	1.91	1.88	Fathead	28D	W. body
Naphthalene	1	3.59	2.63	Fathead	28D	W. body
Chlorobenzene	1	3.79	2.65	Fathead	28D	W. body
2,4,5-Trichlorophenol	1	3.72	3.28	Fathead	28D	W. body
Endrin	1	4.56	3.66	Fathead	300D	W. body
1,1,2,2-Tetrachloroethylene	1	2.88	2.06	Rainbow Trout	4D	Muscle
Hexachlorobenzene	1	6.18	4.37	Rainbow Trout	4D	Muscle
P-Biphenylphenyl Ether	1	5.55	3.22	Rainbow Trout	4D	Muscle
Diphenyl Ether	1	4.21	2.77	Rainbow Trout	4D	Muscle
Carbon Tetrachloride	1	2.64	1.72	Rainbow Trout	4D	Muscle
P-Dichlorobenzene	1	3.38	2.81	Rainbow Trout	4D	Muscle
Biphenyl	1	4.09	3.12	Rainbow Trout	4D	Muscle

Table I (Continued)

Chemical	Ref	Log P	Log BCF	Species	Time	Tissue
Chloropyrifos	1	4.82	2.67	Mosquito Fish	35D	W. body
Endrin	1	4.56	3.17	Mosquito Fish	35D	W. body
2,5,6-Trichloropyridinol	1	1.35	0.49	Mosquito Fish	35D	W. body
Fluorene	1	4.38	3.11	Fathead	28D	W. body
Dibenzofuran	1	4.12	3.13	Fathead	28D	W. body
2-Chlorophenanthrene	1	5.16	3.63	Fathead	28D	W. body
Phenanthrene	1	4.46	3.42	Fathead	4D	W. body
2-Methylphenanthrene	1	4.86	3.48	Fathead	4D	W. body
Heptachlor	1	5.44	3.98	Fathead	32D	W. body
Heptachloroepoxide	1	5.40	4.16	Fathead	32D	W. body
P,P'-DDE	1	5.69	4.71	Fathead	32D	W. body
Pentachlorophenol	1	2.97	2.89	Fathead	32D	W. body
Hexabromobiphenyl	1	6.39	4.26	Fathead	32D	W. body
Methoxychlor	1	4.30	3.92	Fathead	32D	W. body
Mirex	1	6.89	4.26	Fathead	32D	W. body
Hexabromocyclododecane	1	5.81	4.26	Fathead	32D	W. body
Heptachloronorbornene	1	5.28	4.05	Fathead	32D	W. body
Hexachloronorbornadiene	1	5.28	3.81	Fathead	32D	W. body
1,2-Dichlorobenzene	8	3.40	1.95	Bluegill	14D	W. body
1,3-Dichlorobenzene	8	3.44	1.82	Bluegill	14D	W. body
1,4-Dichlorobenzene	8	3.37	1.78	Bluegill	14D	W. body
1,2,3,5-Tetrachlorobenzene	8	4.46	3.26	Bluegill	28D	W. body
Pentachlorobenzene	8	4.94	3.53	Bluegill	28D	W. body
Carbon Tetrachloride	8	2.73	1.48	Bluegill	21D	W. body

Table I (Continued)

Chemical	Ref	Log P	Log BCF	Species	Time	Tissue
Chloroform	8	1.90	0.78	Bluegill	14D	W. body
1,2-Dichloroethane	8	1.45	0.30	Bluegill	14D	W. body
1,1,1-Trichloroethane	8	2.47	0.95	Bluegill	20D	W. body
1,1,2,2-Tetrachloroethane	8	2.39	0.90	Bluegill	14D	W. body
Pentachloroethane	8	3.21	1.83	Bluegill	14D	W. body
Hexachloroethane	8	3.93	2.14	Bluegill	28D	W. body
Bis(2-Chloroethyl)Ether	8	1.12	1.04	Bluegill	14D	W. body
1,1,2-Trichloroethylene	8	2.42	1.23	Bluegill	14D	W. body
Tetrachloroethylene	8	2.53	1.69	Bluegill	21D	W. body
Isophorone	8	1.67	0.84	Bluegill	14D	W. body
N-Nitrosophenylamine	8	3.13	2.34	Bluegill	14D	W. body
2-Chlorophenol	8	2.16	2.33	Bluegill	28D	W. body
2,4-Dimethylphenol	8	2.16	2.33	Bluegill	28D	W. body
Butylbenylphthalate	8	4.05	2.89	Bluegill	21D	W. body
Dimethylphthalate	8	1.61	1.76	Bluegill	21D	W. body
Diethylphthalate	8	2.70	2.07	Bluegill	21D	W. body
Alkyl Benzene Sulfonate	9	1.59	2.02	Bluegill	21D	W. body
Alkyl Benzene Sulfonate	9	1.59	1.56	Bluegill	21D	Muscle
Naphthalene	17	3.59	1.90	Coho Salmon	35D	Muscle
2-Methylnaphthalene	17	3.84	2.28	Coho Salmon	35D	Muscle
1-Methylnaphthalene	17	3.84	2.11	Coho Salmon	35D	Muscle
Hexachlorocyclohexane	11	3.85	2.15	Guppy	4D	W. body
Hexachlorocyclohexane	12	3.85	2.70	Guppy	4D	W. body
Endrin	13	4.56	4.02	Flagfish	30D	W. body

Table I (Continued)

Chemical	Ref	Log P	Log BCF	Species	Time	Tissue
Endrin	13	4.56	4.18	Flagfish	65D	W. body
Endrin	13	4.56	3.85	Flagfish	110D	W. body
Al254	14	6.47	4.60	Brook Trout	118D	W. body
Al254	15	6.47	4.43	Spot	56D	W. body
1,4-Dichlorobenzene	16	3.37	1.96	Guppy	19D	W. body
1,2,3-Trichlorobenzene	16	4.20	2.81	Guppy	19D	W. body
1,3,5-Trichlorobenzene	16	4.20	2.85	Guppy	19D	W. body
1,2,3,5-Tetrachlorobenzene	16	4.46	3.56	Guppy	19D	W. body
Pentachlorobenzene	16	4.94	4.11	Guppy	19D	W. body
Hexachlorobenzene	16	6.18	4.16	Guppy	19D	W. body
Aroclor 1016	1	5.88	4.63	Fathead	32D	W. body
Aroclor 1248	1	6.11	4.85	Fathead	32D	W. body
Aroclor 1254	1	6.47	5.00	Fathead	32D	W. body
Aroclor 1260	1	6.91	5.29	Fathead	32D	W. body
Chlorodane	1	6.00	4.58	Fathead	32D	W. body
Octachlorostyrene	1	6.29	4.52	Fathead	32D	W. body
P,P-DDT	1	5.75	4.47	Fathead	32D	W. body
O,P'-DDT	1	5.75	4.57	Fathead	32D	W. body
Hexachlorobenzene	1	6.18	4.27	Fathead	32D	W. body
1,2,4-Trichlorobenzene	1	4.23	3.32	Fathead	32D	W. body
Lindane	1	3.85	2.26	Fathead	32D	W. body
5-Bromoindole	1	2.97	1.15	Fathead	32D	W. body
2,4,6-Tribromoanisol	1	4.48	2.94	Fathead	32D	W. body
N-Phenyl-2-Naphylamine	1	4.38	2.17	Fathead	32D	W. body

Table I (Continued)

Chemical	Ref	Log P	Log BCF	Species	Time	Tissue
Tricresyl-Phosphate	1	3.42	2.22	Fathead	32D	W. body
Diphenyl Amine	1	3.42	1.48	Fathead	32D	W. body
Toluene	1	3.16	1.96	Fathead	32D	W. body
Tetrachloroethylene	2	2.53	1.79	Fathead	30D	W. body
1,1,2,2-Tetrachloroethane	2	2.39	0.91	Fathead	30D	W. body
Pentachloroethane	2	3.21	1.78	Fathead	30D	W. body
Hexachloroethane	2	3.93	2.85	Fathead	30D	W. body
1,3-Dichlorobenzene	2	3.44	1.99	Fathead	30D	W. body
1,4-Dichlorobenzene	2	3.37	2.05	Fathead	30D	W. body
1,2,4-Trichlorobenzene	2	4.52	2.60	Fathead	30D	W. body
1,2,3,4-Tetrachlorobenzene	2	4.46	3.41	Fathead	30D	W. body
Hexachlorobenzene	2	6.18	4.37	Fathead	30D	W. body
Hexachloro 1,3-Butadiene	2	5.10	3.84	Fathead	30D	W. body
Acridine	3	3.30	2.10	Fathead	17D	W. body
Toxaphene	4	5.28	3.64	Sheepshead Minnow	4D	W. body
Toxaphene	4	5.28	3.59	Pinfish	4D	W. body
Pentachlorophenol	10	2.97	1.11	Bluegill	8D	Muscle
Imidan	5	2.83	0.90	Fathead	2D	W. body
Imidan	5	2.83	1.04	Channel Catfish	2D	W. body
Imidan	5	2.83	0.90	Bluegill	2D	W. body
Diazinon	6	1.92	2.18	Topmouth Gudgeon	7D	W. body
Diazinon	6	1.92	1.56	Silver Crucian Car	7D	W. body
Diazinon	6	1.92	1.81	Carp	7D	W. body
Diazinon	6	1.92	1.24	Guppy	7D	W. body

Table I (Continued)

Chemical	Ref	Log P	Log BCF	Species	Time	Tissue
Endrin	7	4.56	3.21	Channel Catfish	55D	W. body
Acenaphthene	8	3.92	2.59	Bluegill	28D	W. body

References: (1) Veith, G. D., D. L. DeFoe, and B. V. Bergstedt. 1979. Measuring and estimating the bioconcentration factor of chemicals in fish. J. Fish. Res. Board Can. 36: 1040-1048. (2) ERL-D, unpublished data. (3) Southworth, G. R., B. R. Parkhurst, and J. J. Beauchamp. 1979. Accumulation of acridine from water, food, and sediment by the fathead minnow (Pimephales promelas). Water, Air and Soil Pollut. 12: 331-341. (4) Schimmel, S. C., J. M. Patrick, Jr., and J. Forester. 1977. Uptake and toxicity of toxaphene in several estuarine organisms. Arch. Environ. Contam. Toxicol. 5: 353-367. (5) Julin, A. M., and H. O. Sanders. 1977. Toxicity and accumulation of the insecticide imidan in freshwater invertebrates and fishes. Trans. Am. Fish. Soc. 4: 386-392. (6) Kanazawa, J. 1978. Bioconcentration ratio of diazinon by freshwater fish and snail. Bull. Environ. Contam. Toxicol. 20: 613-617. (7) Argyle, R. L., G. C. Williams, and H. K. Dupree. 1973. Endrin uptake and release by fingerling channel catfish (Ictalurus punctatus). J. Fish. Res. Board Can. 30: 1743-1744. (8) Veith, G. D., K. J. Macek, S. R. Petrocelli, and J. Carroll. 1980. An evaluation of using partition coefficients and water solubility to estimate bioconcentration factors for organic chemicals in fish. Aquatic Toxicol. : 116-129. (9) Kimerle, R. A., K. J. Macek, B. H. Sleight, III, and M. E. Burrows. 1981. Bioconcentration of linear alkylbenzene sulfonate (LAS) in bluegill (Lepomis macrochirus). Water Res. 15: 251-256. (10) Pruitt, G. W., and B. J. Grantham. 1977. Accumulation and elimination of pentachlorophenol by the bluegill (Lepomis macrochirus). Trans. Am. Fish. Soc. 5: 462-465. (11) Canton, J. H., P. A. Greve, W. Slooff, and G. J. van Esch. 1975. Toxicity, accumulation and elimination studies of α-hexachlorocyclohexane (α-HCH) with freshwater

Table I (Continued)

organisms of different trophic levels. Water Res. 9: 1163-1169. (12) Canton, J. H., R. C. C. Wegman, T. J. A. Vulto, C. H. Verhoef, and G. J. van Esch. 1978. Toxicity-, accumulation- and elimination studies of α-hexachlorocyclohexane (α-HCH) with saltwater organisms of different trophic levels. Water Res. 12: 687-690. (13) Hermanutz, R. O. 1978. Endrin and malathion toxicity to flagfish (Jordanella floridae). Arch. Environ. Contam. Toxicol. 7: 159-168. (14) Mauck, W. L., P. M. Mehrle, and F. L. Mayer. 1978. Effects of the polychlorinated biphenyl Aroclor® 1254 on growth, survival, and bone development in brook trout (Salvelinus fontinalis). J. Fish. Res. Board Can. 35: 1084-1088. (15) Hansen, D. J., P. R. Parrish, J. I. Lowe, A. J. Wilson, Jr., and P. D. Wilson. 1971. Chronic toxicity, uptake, and retention of Aroclor® 1254 in two estuarine fishes. Bull. Environ. Contam. Toxicol. 6: 113-119. (16) Koneman, H. 1979. Quantitative structure-activity relationships for kinetics and toxicity of aquatic pollutants and their mixtures in fish. Chapter 2, Toxicokinetics in fish: Accumulation and elimination of six chlorobenzenes by guppies. (17) Roubal, W. T., S. I. Stranahan, and D. C. Malins. 1978. The accumulation of low molecular weight aromatic hydrocarbons of crude oil by coho salmon (Oncorhynchus kisutch) and starry flounder (Platichthys stellatus). Arch. Environ. Contam. Toxicol. 7: 237-244.

Table II. Regression Analysis of Bioconcentration Data Sets: Log BCF = a Log P + b

Data Set	Slope (a)	Intercept (b)	N	R^2
Veith et al. (1979) mixed speces	0.85	−0.70	59	0.90
Fathead minnow, non-ionic chlorinated hydrocarbons	0.86	−0.80	16	0.92
All fathead minnow data > 20 day exposure	0.85	−0.63	50	0.86
All fathead minnow data	0.86	−0.72	54	0.86
All fathead minnow and bluegill data	0.81	−0.49	88	0.87
All fathead minnow, bluegill, guppy data	0.81	−0.49	97	0.87
All species, all chemicals	0.79	−0.40	122	0.86

P = 5.00 is 3548.1 by both the original equation with 59 tests and the expanded set of 122 tests.

If the linear model proposed by Mackay [2] were fitted to this data, the coefficient would be increased from 0.048 to 0.183 $\pm$ 0.389.

On final result of this continuing evolution of the structure-bioaccumulation relationship is that the prediction limits of the regression did not significantly decrease by more than doubling the number of data points. Figures 1-3 present plots of the log BCF vs log P data smallest and most similar data set, the data set for only one species of fish, and, the entire set of all species and all chemicals. The dashed lines are the 95 percent prediction limits (the 95 percent confidence limit is extremely narrow) and it is quite apparent that the order of magnitude limit remains in all sets. This analysis suggests that the predictability is limited by the variance in the measured values themselves and/or the model and will not likely be improved by generating new data with similar inherent errors. Although it must be emphasized that the SAR (with order of magnitude accuracy) can only be used to screen chemicals, the data do show that the single SAR is useful for a wide variety of species and a wide range of chemical properties.

The structure-bioaccumulation relationship has one limitation which has not been thoroughly investigated. The equation implies that the bioconcentration factor will increase without bounds as log P increases. However, increases in log P is often accompanied by an increase in molecular volume which eventually is sufficiently large to inhibit membrane permeability. Polychlorinated paraffins (Chlorafin* etc.) have molecular weights in excess of 700 and do not appear to bioconcentrate in fish. We have also tested hexabromobenzene in a conventional 28-day test with no evidence of residues in fish other than what would be expected from simple adsorption on the surface of the fish. Consequently, there is evidence of a steric cutoff however undefined at this time. A rule-of-thumb discussed by several investigators is that the log BCF/log P relationship should be used with caution for chemicals with a molecular weight greater than 600.

REFERENCES

1. Veith, G.D., DeFoe, D.L. and Bergstedt, B.V., "Measuring and Estimating the Bioconcentration Factor of Chemicals in Fish". J. Fish. Res. Board Can. 36 1040 (1979).
2. Mackay, D. "Correlation of Bioconcentration Factors". Environ. Sci. Technol. 16, 274 (1982).

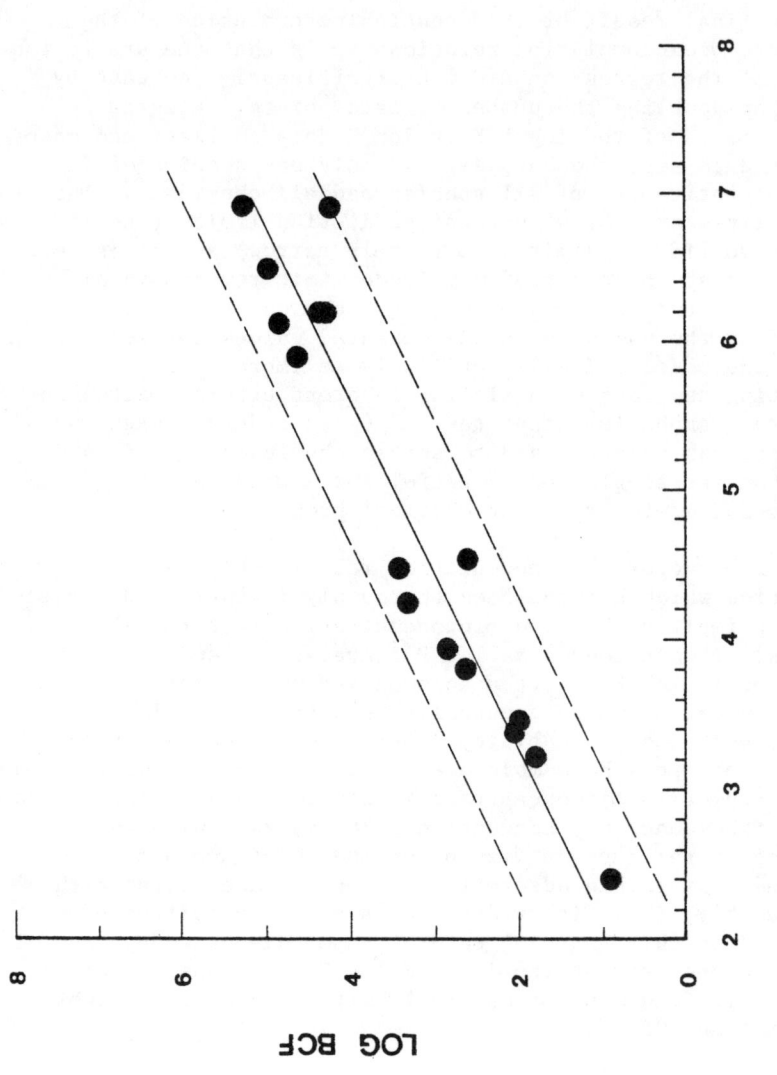

Figure 1. Relationship between log P and log BCF for nonionic chlorinated hydrocarbons and fathead minnows

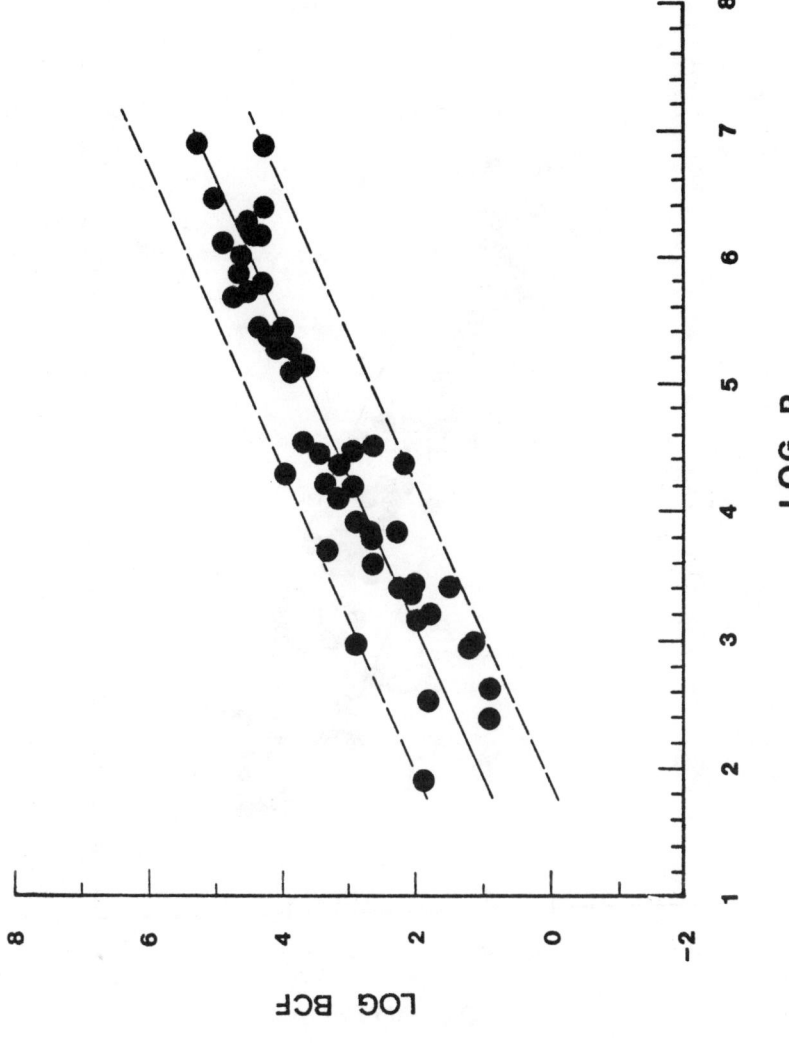

Figure 2. Relationship between log P and log BCF for all chemicals and fathead minnows

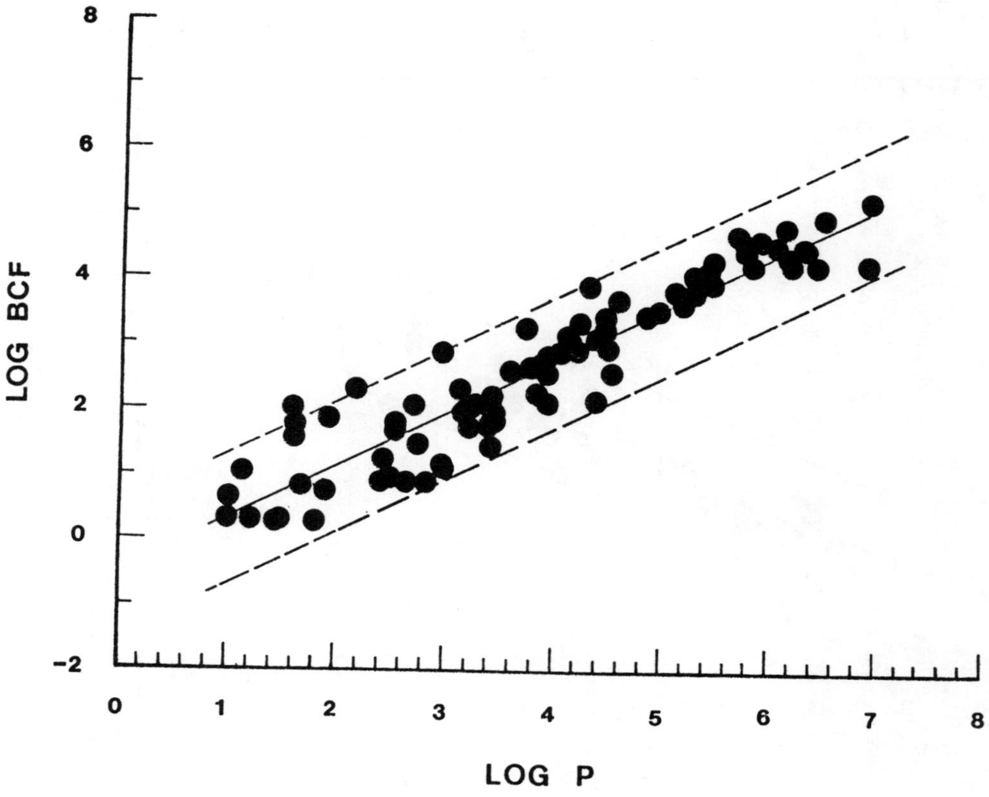

Figure 3. Relationship between log P and log BCF for all chemicals and all fish species

CHAPTER 16

STEADY STATE MODELING OF
TOXIC CHEMICALS-THEORY AND
APPLICATION TO PCBs IN THE
GREAT LAKES AND SAGINAW BAY

Robert V. Thomann
Manhattan College

John A. Mueller
Manhattan College

INTRODUCTION-WHY STEADY STATE?

Understanding the fate of a chemical such as PCBs in a natural water system must fundamentally recognize the dynamic relationship between all phases (atmospheric, water column and sediments). For PCBs in particular, the relevant questions of environmental control often involve time variable issues such as "How long will it take before the concentration of PCBs in a body of water will decline to a given concentration after cessation or reduction of an external input?" Also, if the load history of the chemical is time variable (as it is for PCBs in the Great Lakes due to long term changes in the atmospheric input or direct point and non-point loadings), then the ability to fully describe the resulting concentrations in the water column and sediment must, of necessity, require a time variable calculation framework.

As a result, time variable computational frameworks of the interactions between sediments, water column solids and a toxic chemical have been developed [1,2,3,4]. These models incorporate the mechanisms of solids settling, sediment resuspension, net sedimentation and transfer to the deep sediments, interstitial diffusion of dissolved toxicant and other decay and transfer mechanisms. Other work on modeling the fate of chemicals has assumed a steady state condition and calculated equilibrium concentrations [5,6,7,8,9]. With the availability of time variable modeling frameworks and the significance of the time variable issues associated with control of PCBs especially in the Great Lakes, one could legitimately ask, "What is the usefulness of steady state equilibrium calculations? Why steady state? Why not do all calculations with a time variable model and if steady state is obtained, let such a condition be determined from the time variable framework?" Steady state calculations, however, offer several advantages.

(1) The specification of the various parameters (e.g. resuspension velocity and surface sediment solids concentration for the interactive time variable toxicant models) is not unique and various combinations of the parameters, within reported ranges, yield similar calibrations to observed data. Steady state models, in general, do not require as detailed a specification of parameters as do time variable models and, in some instances, less parameters need be determined.

(2) Time variable calculations, because of the complexity of the computation tend to obscure (although not always) some of the fundamental mechanisms that are operative in a problem context

(3) a need exists for rapid, less expensive, and easy to compute models to describe the fate of toxic chemicals; such a need is partially met by a steady state framework

(4) the analytical solutions to steady state models provide useful insights into relationships between water column and sediment toxicant levels that are not obvious from time variable calculations.

Consequently, it is desirable that a less complex model be developed that provides increased understanding of the processes that are operative and, to the degree possible, permits unambiguous estimation of model parameters. A simplified steady state model provides such an approach. Since the analysis framework is steady state, the time variable questions discussed above cannot be answered. Furthermore, the time for the entire system of water column and sediment to reach steady state may be long. A steady state framework, therefore, is a simplified first approximation to chemical fate that does not address issues related to temporal questions but does provide a useful model for estimating chemical concentrations in a simpler way.

In the research reported on herein, the theoretical framework for a steady state model of systems such as the Great Lakes is derived, principal insights into the mechanisms that result from this derivation are presented with specific reference to PCBs as a special case and finally the application of the model to Saginaw Bay and the Great Lakes as a whole is presented.

THEORY

Suspended Solids Model

Since it is known that many chemicals, such as PCBs, sorb to suspended particulate matter, the first step in the development of the overall model is the mass balance of suspended solids. In this work, a single class of solids is considered and is intended to incorporate inorganic solids and organic particulates, including detrital material and phytoplankton solids.

Consider then the steady state model for suspended solids in a multi-dimensional system. It is assumed that the bed sediment is stationary; a reasonable assumption for the Great Lakes and associated bays and harbors but, of course, a questionable assumption for highly transient solids movements in river systems where bed load transport is significant. Figure 1 shows a definition sketch and Table 1 is a list of all variables and coefficients and associated units. The mass balance equation for the solids in segments #1 and 1s, using backward finite differences, is:

$$0 = W_1 - Q_{12}M_1 + E'_{12}(M_2 - M_1) - w_{a1}AM_1 + w_{rs1}AM_{1s} \quad (1)$$

$$0 = w_{a1}AM_1 - w_{rs1}AM_{1s} - w_{s1}AM_{1s} \quad (2)$$

where the numerical subscripts refer to segment locations.

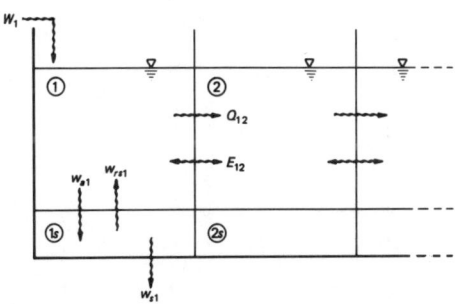

Fig.1. Definition sketch for steady state suspended solids model

A similar set of equations can be written for segments 2 and 2s. There are several aspects of even this simple formulation that result in some practical difficulty. The designation of an interactive sediment layer is not precise

TABLE 1 DEFINITION OF TERMS

Symbol	Definition	Units[1]
A	Interfacial area	$[L^2]$
c_d	Dissolved toxicant concentration porosity corrected	$[M_T/L^3_{s+w}]$
c'_d	Dissolved toxicant concentration	$[M_T/L^3_w]$
c_g	Gas phase toxicant concentration	$[M_T/L^3_g]$
$c_{p,T}$	Particulate or total toxicant conc.	$[M_T/L^3_{s+w}]$
E'	Bulk mixing/dispersion coefficient	$[L^3_{s+w}/T]$
$f_{d,p}$	Fraction of total toxicant concentration in dissolved or particulate form	
H	Water column or sediment depth	$[L]$
H_e	Henry's constant for toxicant	$[M_T/L^3_g \div M_T/L^3_w]$
K	Overall loss rate of total toxicant	$[1/T]$
K_d	Loss rate of dissolved toxicant	$[1/T]$
K_L	Sediment-water diffusive transfer coefficient	$[L/T]$
K_p	Decay rate of particulate toxicant	$[1/T]$
k_ℓ	Overall volatilization transfer coefficient	$[L/T]$
M	Solids concentration	$[M_s/L^3_{s+w}]$
r	Toxicant concentration on solids	$[M_T/M_s]$
Q	Advective flow transport	$[L^3_{s+w}/T]$

Symbol	Definition	Units[1]
V	Volume of water column or sediment segment	$[L^3_{s+w}]$
W	Mass input of solids	$[M_s/T]$
$W_{d,p,T}$	Mass input of dissolved, particulate and total toxicant respectively	$[M_T/T]$
w_a	Particulate settling velocity	$[L/T]$
w_d	Net dissolved sedimentation velocity	$[L/T]$
w_n	Net removal velocity of solids from water column	$[L/T]$
w_{rs}	Particulate resuspension velocity	$[L/T]$
w_s	Net particulate sedimentation velocity	$[L/T]$
w_T	Net removal velocity of toxicant from water column	$[L/T]$
π	Toxicant-solids partition coefficient	$[M_T/M_s \div M_T/L^3_w]$
π'	Partition coefficient, porosity corrected	$[M_T/M_s \div M_T/L^3_{s+w}]$
ϕ	Porosity	$[L^3_w/L^3_{s+w}]$

[1] $[M_{s,T}]$ = Mass of solids or toxicant
 $[L]$ = Length
 $[L^3_{s+w}]$ = Volume of solids and water
 $[L^3_w]$ = Volume of water
 $[T]$ = Time

so that the sediment solids concentration that is effectively interacting with the water column is not well specified. The velocity terms involving settling, resuspension, and sedimentation are also not uniquely specified and can have a wide range depending on solids type, and hydrological and meteorological conditions. However, observed data on input loads of solids and resulting in-lake suspended solids concentrations in the water column can be used to narrow the range of specification of the model parameters and provide a "tighter" analysis.

From Eq. (1) and Eq. (2), a single equation for the water column can be obtained as:

$$0 = W_1 - Q_{12}M_1 + E'_{12}(M_2 - M_1) - w_{n1}AM_1 \qquad (3)$$

where w_{n1} is the net loss of solids from water column segment #1 and is given by:

$$w_{n1} = \frac{w_{a1} w_{s1}}{w_{rs1} + w_{s1}} \qquad (4)$$

Note that Eq. (3) states that for a multi-dimensional water body,
the sediment interaction can be substituted out of the equation set and incorporated in the net solids loss parameter, w_{n1}. A tracer substance, such as chlorides, can be used to estimate the net advective flow and the dispersion coefficients in Eq. (3). With data on input solids, W, and observed solids concentration, M, in the water column, Eq. (3) then permits direct estimation of the net loss of solids, w_n. A mass balance around the sediment segment then yields

$$w_{s1}AM_{1s} = w_{n1}AM_1 \qquad (5)$$

which simply states that the solids flux into the sediment layer from the water column is balanced by the solids flux leaving the sediment segment due to net sedimentation. In principle, the net sedimentation velocity can be measured or estimated so that the solids concentration in the sediment could be calculated from Eq. (5). In practice, however, the spatial variability in the sedimentation velocity and the uncertainty in the sediment solids concentration that interacts with the water column make it difficult to separate the two quantities. This is partially a reflection of the occurrence of nepholoid layers or "fluff" layers that are at the boundary between the water column and the bed sediment and the high degree of spatial heterogeneity in zones of deposition, scour or erosion. As a consequence, in the absence of detailed data on w_{sl} or M_{ls}, only the net flux to the sediment from the water column can be estimated and is given by $w_{nl}AM_1$. This assumes, of course, that all terms in the water column equation (3) are known with more certainty than the sedimentation velocity or the sediment solids.

As noted, preliminary estimates of w_{ni} may be obtained from solving Eq. (3) directly with given water column solids concentrations. These estimates, however, may be subject to wide variations, including inconsistency in net sedimentation as a result of small changes in water column concentrations. The procedure finally adopted in this research was to obtain estimates of w_{ni} from trial and error calculations of the water column model by calibration to the observed suspended solids concentrations. A single set of net loss rates is then obtained that balances mass rates in the water column and sediment but does not require specification of the settling or resuspension velocities.

Toxicant Model

The basic objectives of the steady state toxicant model are a) incorporate first principles of chemical fate and then search for simplifications and b) utilize in a consistent way the field data for calibration.

The development of a mathematical model for the physical-chemical fate of a toxic substance in water must include the following features:

1. sorption-desorption mechanism of the chemical with the suspended particulates in the water column

2. similar mechanisms with solids in the sediments

3. loss of the chemical due to mechanisms such as biodegradation, volatilization, chemical and biochemical reactions, photolysis

4. transport of the toxicant due to advective flow transport, and dispersion and mixing

5. settling and resuspension mechanisms between sediment and water column

6. direct inclusion of external inputs that may be subject to environmental control.

Figure 2 shows these principal features in a schematic form. The model proceeds from these general features by development of the basic mass balance equations for the toxicant. As noted previously, for the sake of simplicity, only a single solids class is considered, i.e., a distinction is not made between sands, silts, or organic particles. It is assumed also that laboratory data are available for the following:

1) rates of degradation, photolysis, hydrolysis
2) sediment microbial degradation rates
3) partition coefficients for ambient solids concentrations in the water column and sediment.

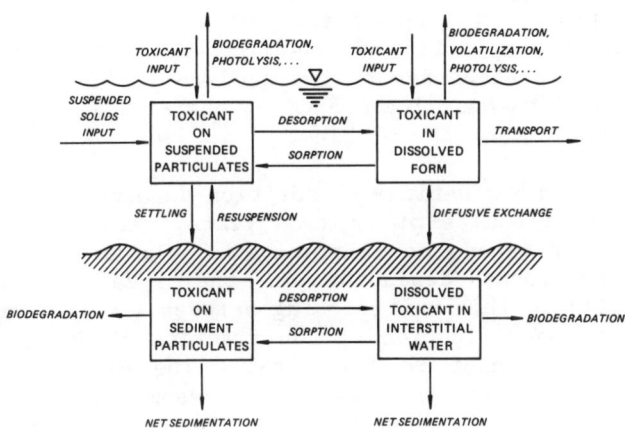

Fig.2. Principal components of physical-chemical model of toxic chemicals

The concentration of a toxicant in the water column and in the sediment is given by a mass balance around each of the finite segments of the water body. The toxicant is composed of two forms: 1) the dissolved form, c'_d, where "dissolved" is considered in an operational manner, i.e. all toxicant passing, for example, a 0.45 µm filter and 2) the particulate form, c_p, i.e. the toxicant sorbed onto particulate matter in the water column or sediment. The total toxicant concentration is then

$$c_T = c_p + c_d \tag{6}$$

where

$$c_d = \phi c'_d$$

Note that since the dissolved toxicant concentration is the mass of toxicant per volume of water and the total toxicant concentration the mass of toxicant per volume of water plus solids, the porosity of the volume must be introduced to maintain a consistent mass balance. Recognizing the concentration as mass per volume of solids plus water permits a consistent equation set between the water column and the sediment. The quantity, c_d, therefore represents the porosity corrected dissolved form of the toxicant.

A full set of equations incorporating all kinetic interactions of sorption and desorption and other physio-chemical interactions can be derived for the dissolved and particulate forms in the water column and sediment [10]. The application of these equations is considerably simplified by assuming that the reaction kinetics if sorption-desorption tend to be "fast" (on the order of minutes-hours) compared to the kinetics inherent in other mechanisms of the problem.

The "fast" kinetics of sorption-desorption indicate that for time scales of days to years, there will be a virtually continuous equilibration of the dissolved and particulate forms depending on the local solids concentration. This partitioning between the two components permits the specification of the fraction of dissolved and particulate toxicant to the total. The dissolved and particulate toxicant are, therefore, assumed to be always in a "local equilibrium" with each other. A partition coefficient can then be defined as follows:

$$\pi' = \pi/\phi = r/c_d \tag{7}$$

From Eqs. (6) and (7), the fraction of the total toxicant that is dissolved is given by

$$f_d = (1 + \pi'M)^{-1} \tag{8}$$

Also, the particulate toxicant as a fraction of total toxicant is given by

$$f_p = \frac{\pi'M}{1 + \pi'M} \tag{9}$$

The local equilibrium assumption, therefore, permits specification at all times and places of the fractions of the total toxicant in the dissolved and particulate forms. Attention can then be focused solely on the mass balance equation for the total toxicant.

It can be shown [10] that the total toxicant in the water column is given by

$$0 = W_{T1} - Q_{12} c_{T1} + E'_{12}(c_{T2} - c_{T1}) - w_{T1} A c_{T1} + k_{\ell 1} A c_g / H_e \tag{10}$$

where the net removal rate from the water column, w_{T1}, is

$$w_{T1} = w_{a1} f_{p1} + K_1 H_1 + K_{L1} f_{d1}/\phi_1$$
$$- \alpha_{T1} [w_{rs1} f_{p1s} + K_{L1} f_{d1s}/\phi_{1s}] \tag{10a}$$

and

$$\alpha_{T1} = \frac{w_{a1} f_{p1} + K_{L1} f'_{d1}}{(w_{rs1} + w_{s1}) f_{p1s} + (w_{d1s} + K_{L1}/\phi_{1s}) f_{d1s} + K_{1s} H_{1s}} \tag{10b}$$

and

$$K_1 = K_{d1} f_{d1} + (k_{\ell 1}/H_1) f_{d1}/\phi_1 + K_{p1} f_{p1} \tag{10c}$$

$$K_{1s} = K_{d1s} f_{d1s} + K_{p1s} f_{p1s} \tag{10d}$$

Eq. (10) is informative since it is in a form similar to standard multi-dimensional water quality models where all the complex sediment interactions are included in the net removal rate of the toxicant, w_{T1}. This provides for an immediate simplification since now only the water column need be considered. The concentration of the toxicant in the water column can be computed from the set of equations represented by Eqs. (10) and the sediment concentration can be calculated from

$$c_{T1s} = \alpha_{T1} c_{T1} \tag{11}$$

PCBs – A Special Case

A useful first approximation can be obtained for chemicals, such as PCBs, that are highly sorbed to solids and for which sediment decay and diffusive exchange processes are small. If then it is assumed that the following are zero:

a) sediment diffusive exchange of PCBs, i.e. $K_{L1} = 0$

b) decay of particulate and dissolved PCBs, i.e. $K_{1s} = K_{1d} = K_{1p} = 0$

c) downward flux of dissolved PCBs in the sediment, i.e. $w_{d1s} = 0$,

then it can be shown that

$$w_{T1} = w_{n1} f_{p1} + k_{\ell 1} f_{d1}/\phi_1 \qquad (12)$$

This result indicates that the net loss rate of PCBs from the water column - as a first approximation - is given solely in terms of the partition information and solids concentration (to give f_{p1}), the net loss rate of solids from the water column from a solids balance (to give w_{n1}) and the exchange rate due to volatilization ($k_{\ell 1}$). This makes Eq. (12) extremely useful for first estimates since it requires a minimum number of parameter specifications. Also it can be shown that for this case

$$r_{1s} = r_1 \qquad (13)$$

These results indicate that for PCBs, if the net solids loss to the sediment, w_{n1} can be obtained and information is available on the partition coefficient, then the entire steady state PCB problem can be approached as a first approximation through the simple equations (12) and (13). For the case of no net sedimentation, then $w_{n1} = 0$ and again $r_{1s} = r_1$.

APPLICATION: SAGINAW BAY PCB MODEL

Saginaw Bay affords an opportunity to test the simple steady state model for PCBs. The basic data base and the overall properties of the Bay have been summarized [3]. This application uses the steady-state model previously discussed that incoporates horizontal transport, net solids settling and resuspension and sedimentation, and the interaction of the solids with PCBs.

Segmentation & Transport Coefficients

The segmentation is from [3] and includes segment volumes, areas, and depths. The dispersive exchange coefficients and the flows between each of the segments and Lake Huron were obtained from a calibration analysis using measured chloride concentrations. The temporally averaged flows and exchange coefficients that were used in the steady-state computation for the five segments in the water column are as shown in Figure 3.

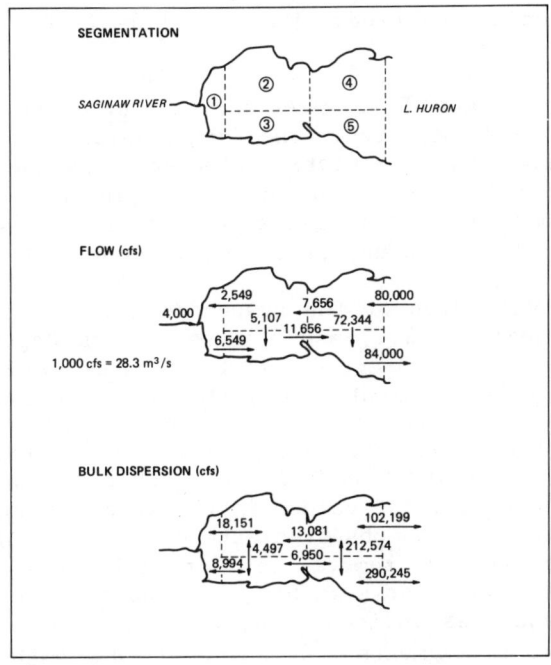

Fig.3. Segmentation for Saginaw Bay and estimated long-term average transport coefficients

Suspended Solids Calibration

The Saginaw River and smaller tributaries to the Bay, shoreline erosion, atmospheric fallout and phytoplankton biomass are the components of solids loads used to calibrate the solids in Saginaw Bay. Long-term estimates of the Saginaw River load are given by [11], whereas the contributions of other tributary drainage areas were estimated from average flows and a long-term average suspended solids concentration estimated for the Saginaw River. Bank erosion values were derived from county by county erosion volumes in [12] and proportioned to Saginaw Bay on the basis of shoreline length. Volumes of eroded material were converted to mass loadings by assuming a porosity of approximately 60% and a specific gravity of 2.65. To account for immediate settling of heavier fractions, a 50% reduction was used to obtain the final estimates. Phytoplankton biomass was obtained from algal model simulations. The principal sources of solids for the Bay as a whole are the Saginaw River (49% of total input) and the phytoplankton production of solids (35% of total input).

Boundary values of 4.3 mg/ℓ for segment 4 and 5.5 mg/ℓ for segment 5 were selected. The former value is thought to be more associated with advective flow entering the Bay from Lake Huron and the latter value more representative of observed segment 5 concentrations leaving the Bay with the net advective flow (see net circulation, Figure 3).

Net removal rates of the suspended solids were then assigned to segments 2,4, and 5 where solids deposition zones are either documented to occur (segment 2) or estimated to occur (segments 4 and 5). No net removal rates were assigned to segments 1 and 3 since sedimentation appeared to be minimal there on the basis of sediment solids and PCB data in sediment cores. The result are shown in Figure 4. THe calculated values in the water column are in good agreement with the observed data of 1976 through 1979, when values of the net removal rate of 12.7, 13.8, and 9.7 meters/year are used for segments 2,4, and 5 respectively. Sensitivity analyses indicate that the differences in the net loss rates in segments 2,4, and 5 are not significant. Equally acceptable calibrations are obtained for values between 10 and 20 m/yr. for segments 2,4, and 5.

With the estimated net removal rate of solids w_n from the water column, the flux of solids into the bed is calculated as $w_{ni} \cdot AM_i [M_s/T]$. This is equal to the sedimentation flux which is calculated as $w_{si} \cdot AM_{is}$. Since both w_{si} and M_{is} vary with sediment depth, no single value can be specified. However, from Eq. (6) the relationship between w_{si} and M_{is} is unique. If w_{si} is selected, then M_{is} is determined. Log-log plots of the relationships between the sedimentation rate and the computed and observed (from [13]) bed solids concentration are included in the bottom of Figure 4.

These results indicate the utility of the simple steady state solids balance. The water column solids data are known with some accuracy and do not exhibit marked spatial gradients. Note that the maximum spatial differences in the average water column suspended solids is about a factor of four. In contrast, the spatial heterogeneity of the

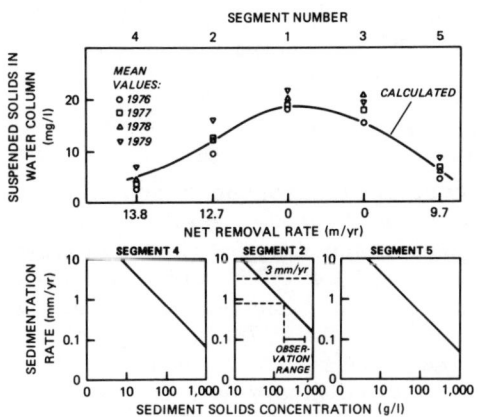

Fig.4. Saginaw Bay solids calibration

sediment is quite marked with regions of deposition, scour and no apparent net deposition. Sediment solids may then vary markedly in the given segment horizontally, but, more importantly, vertically. Boundary layers of sediment solids or "fluff" layers may be available for interaction with the surface water column at concentrations less than sediment data from cores. Conversely, estimated net sedimentation velocities are often cited only for those regions of deposition and not over an area equivalent to a model segment of Saginaw Bay. The calculation discussed above provides a good estimate of the net flux to the sediment over the entire segment area. The trade off between net sedimentation over the entire segment area and sediment solids concentrations is shown in the lower figures of Figure 4. If, as noted above, the solids data from the sediment cores are used, then the net sedimentation velocity varies from 0.25 to 0.8 mm/yr. or almost one order of magnitude less than the 3 mm/yr. previously cited. If on the other hand, an average net sedimentation over the entire area of segment 2 is fixed at say 3 mm/yr., then the sediment solids concentration that is consistent with that sedimentation velocity is about 45,000 mg/ℓ or one order of magnitude less than the average sediment solids in the top 5 cm of the cores. The results indicate, therefore, that with only the net flux of solids to the sediment known with some confidence, then it is not possible to uniquely specify the net sedimentation or boundary layer sediment solids. Additional tracers (of which the radionuclides or PCBs are examples) would provide additional information that could aid in specifying the net sedimentation and sediment solids concentations.

PCB Calibration

With the horizontal transport and net loss rate of suspended solids calibrated, analysis of the PCB concentrations can proceed. Total PCB loadings of 350 kg/yr. were estimated for 1979 [11], the first year for which total PCB field data were available. The Saginaw River load is approximately 75% of the total load and atmospheric sources contribute an additional 25%. Although open lake concentrations are reported to be in the 1 ng/ℓ range, the boundary condition was selected as 10 ng/ℓ - the value needed to calibrate observed data in segments 4 and 5.

Partition coefficients were selected on the basis of observed dissolved and particulate fractions and values of 10,000, 50,000, and 100,000 µg/kg per µg/ℓ were selected for segments 1, 2 and 3, and 4 and 5, respectively. With the partition coefficients selected, the removal rates of total PCBs are then calculated as the particulate fraction of the suspended solids net settling rate (see Eq. (12) for $k_{\ell i} = 0$). Thus, in this calculation, exchange of PCBs with the atmosphere, diffusive exchange with the sediment and decay processes are zero. Therefore, the net total PCBs removal rates for segments 2, 4, and 5 are 4.8, 4.7 and 3.5 m/yr., respectively.

With the loads, boundary conditions and net removal rates described above, together with the horizontal transport, the steady state model is used to calculate total PCB concentrations in the water column. The top panel of Figure 5 shows the agreement between calculated values and data observed in 1979. Dissolved and particulate fractions also agree well with observed data, as noted in the next two panels of the figure. The bottom panel displays the particulate PCBs per unit weight of solids and, again, agreement between observed means and calculated values is good for the water column.

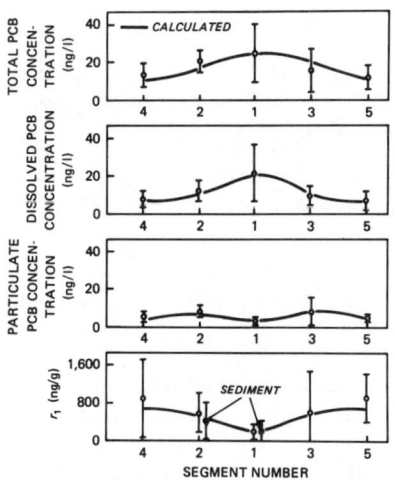

Fig.5. Components of PCBs in water column and sediment, Saginaw Bay, 1979. Data, mean ± 1 standard deviation

In the previous theoretical section of this paper, it was shown that, for depositional areas, and areas where settling and resuspension were equal, the PCBs per unit weight of solids in the water column (r_1) is equal to the sediment toxicant concentration (r_{1s}). The sediment data also shown in the bottom panel of Figure 5 are segment averaged sediment concentrations for 1979, provided by the Grosse Ile Laboratory. Agreement between calculated and observed means is good for segments 1 and 2. It is hypothesized that segment 3 may be a net erosion zone (see suspended solids calibration, Figure 4) in which case the assumption that $r_{1s}=r_1$ is not appropriate.

A mass balance of total PCBs is shown in Figure 6 for the Saginaw River and atmospheric loads, net settling fluxes and boundary fluxes. As noted in the lower right panel, approximately 30% of the total PCBs entering Saginaw Bay from external loads is incorporated into the sediments of the Bay and approximately 70% is exchanged with Lake Huron.

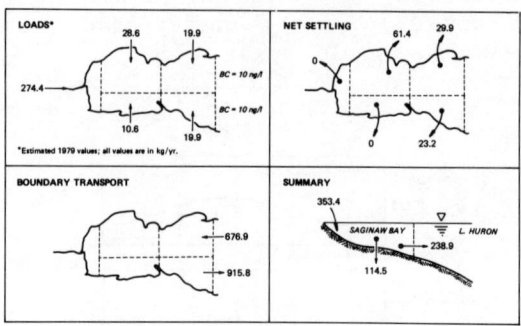

Fig.6. Mass balance of total PCBs for Saginaw Bay, long-term average.

The separate effects of the external PCB loads (Saginaw River and atmospheric) and the boundary conditions are illustrated in Figure 7. The total PCB due to both external loads and boundary conditions is compared with observed data in the top panel of the figure, where the peak concentration in segment 1 is seen to be approximately 24 ng/ℓ. Of the 24 ng/ℓ, approximately 6 ng/ℓ is due to the boundary condition (center panel) and the remaining 18 ng/ℓ is the effect of the loads. Thus, complete removal of the Saginaw River

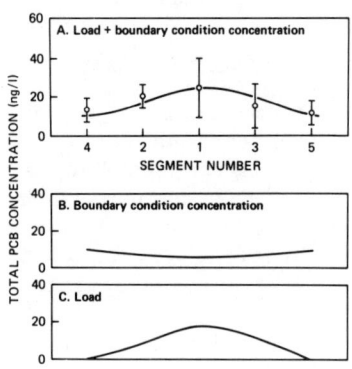

Fig.7. Effects of external loads and boundary conditions on total PCB average concentrations in water column of Saginaw Bay

loads and maintenance of the boundary at 10 ng/ℓ would result in at least a 75% reduction in the segment 1 PCB concentration under this new steady-state condition. Additional reduction would occur since some significant fraction of the boundary concentration is probably caused by the loads. Therefore, reducing the boundary concentration to lower open Lake Huron levels would reduce the concentration in segment 1. If, for example, the boundary decreased to a value of 5 ng/ℓ, the concentration in segment 1, under the no-load situation, would be approximately 3 ng/ℓ.

APPLICATION: GREAT LAKES MODEL OF PCBs

The steady state modeling framework can also be applied on a Great Lakes scale to the distribution of PCBs in the water column and sediment. The details of the Great Lakes model and a review of the time variable version of the model are given in [10,4]. In that work, emphasis was placed on the dynamic behavior of the Lakes and regions of the Lakes to time variable inputs such as the external loading of plutonium-239. Three solids classes were used and because of the time variable nature of the problem, settling and resuspension velocities needed to be specified. In the calculation here, the steady state model is used to estimate the distribution of PCBs with a minimum specification of parameters.

In calculating the distribution of PCBs on a Great Lakes scale, uncertainty exists in two important areas:

a) external input of PCBs due to atmospheric, point and tributary sources, and
b) the extent of volatilization of PCBs from the dissolved phase in the water column to the gaseous phase in the atmosphere.

This steady state PCB model, as noted below, attempts to address this uncertainty by assigning a range to both these aspects of uncertainty, thereby attempting to bracket the appropriate water column and sediment PCB concentrations. The purpose of this steady state Great Lakes PCB model calculation is, therefore, to demonstrate the practical usefulness of the steady state framework for PCBs in spite of the uncertainties in inputs and mechanisms. The practical usefulness results principally from an analysis that uses a well-defined reproducible procedure and that does not require specification of solids resuspension or settling velocities.

Figure 8 shows the segmentation used for the Great Lakes model. With this segmentation, and estimates of net advective flows and dispersion between segments, the transport of the system was estimated [10]. The steady state application then proceeded as follows.

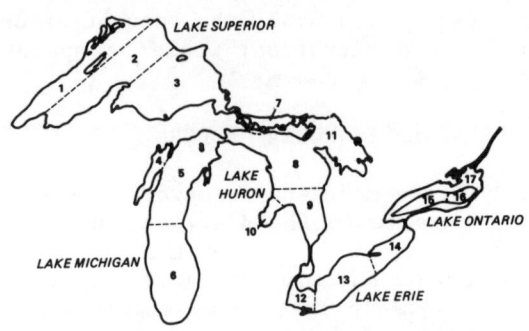

Fig.8. Model segmentation of Great Lakes.

1) With estimates of external solids loading to each segment and associated steady state concentrations of water column suspended solids, the net loss of solids was estimated for each segment. This is the term w_n (m/yr) in Eq. 3.

2) A single partition coefficient for PCBs was specified at 100,000 µg/kg÷µg/ℓ for all segments.

3) With the net loss of solids, and the fraction of PCBs in the particulate phase f_p, (Eq. 9), the net loss rate of PCBs from the water column due to the net loss of solids was calculated as $w_n f_p$.

4) Since the volatilization of PCBs may represent a sink to the atmosphere from the Great Lakes waters, exchange rates of zero (no volatilization) and 0.1 m/d were used. For the latter rate, the total loss of PCBs (net loss due to settling and loss due to volatilization) was then calculated using Eq. 12.

5) For an upper and lower bound in the external load of PCBs as given in [4], the resulting water column concentration of PCBs was then calculated using the steady state modeling framework. Four conditions were thus

calculated:

    i and ii) Upper bound PCBs load with and without volatilization
    iii and iv) Lower bound PCBs load with and without volatilization.

6) Sediment concentrations of PCBs were calculated using Eq. 13, i.e. the PCBs on the solids in the sediment were taken as equal to the PCBs on the solids in the water column.

Table 2 shows the solids concentration (M) utilized for each segment and the net loss of solids, w_n, that results from solution of a simple solids mass balance equation for a given set of external solids loading. As noted, w_n varies from low values of 4.5-7.3 m/yr. for Lake Huron to a maximum value of 250 m/yr. for southern Lake Michigan. These net loss rates of solids depend, of course, on the external solids loading and the associated water column suspended solids concentration. For Lake Michigan, a relatively high external loading of heavier solids from shore line erosion settles out at an elevated rate to result in relatively low levels of open lake suspended solids concentrations. The net loss rates, shown in Table 6, are not necessarily a unique specification since other choices of external solids loading and solids concentrations would result in a different set of values for w_n. Nevertheless, the values shown in Table 2 represent an internally consistent set of loss rates for a given loading and concentration. The net loss rates of PCBs for the case of no volatilization ($k_\ell=0$) are also shown in Table 2. The range of the net loss of PCBs for this case of $k_\ell=0$ is from 1-100 m/yr depending on the loss rate of solids and the associated fraction of PCBs in the particulate form. Note that the lowest loss rate is for Lake Huron while the highest rate is for the eastern basin of Lake Erie due principally to the high solids concentration and high net loss of solids in that basin.

A volatilization transfer rate was obtained using the general approach of [14]. In that work, using 1254 PCB as an example, a Henry's constant (H_e) of 0.114 mmHg/mg/ℓ and a 5m/sec wind, one can calculate that $k_\ell=0.11$ m/d and $k_g=71.4$ m/d so that the exchange process is essentially liquid film controlled and that the overall transfer coefficient, k_ℓ is about 0.1 m/d. If one assumes that the equivalent saturation level of PCBs in the water in equilibrium with the atmosphere is small, then the exchange of PCBs is given as a net sink to the atmosphere only and the source term in

TABLE 2
STEADY STATE PARAMETERS FOR NET LOSS OF SOLIDS (w_n) AND OF PCBs (w_T)

Lake	Segment	H (m)	M (mg/ℓ)	w_n (m/yr)	$w_T^{(1)}$ (m/yr) k_ℓ=0.0	k_ℓ=0.1 m/d
Superior	1	131.0	3.0	76.8	17.7	45.8
	2	163.6	2.0	80.9	13.5	43.9
	3	121.0	2.0	58.1	9.7	40.1
Michigan	4	14.0	5.0	79.7	26.6	50.9
	5	102.6	2.0	160.0	26.7	57.1
	6	74.0	2.0	250.6	41.9	72.3
Huron	7	22.0	2.0	62.0	10.4	40.8
	8	60.9	3.2	7.3	1.8	4.5
	9	56.9	3.3	4.5	1.1	28.6
	10	6.0	7.9	5.9	2.6	23.0
	11	58.1	2.7	118.7	25.3	54.0
Erie	12	7.6	19.0	176.8	115.9	128.5
	13	18.3	24.0	32.7	23.1	33.8
	14	28.0	16.0	111.3	68.6	82.6
Ontario	15	104.6	3.0	213.2	49.4	77.4
	16	136.1	3.5	88.8	23.1	50.1
	17	33.8	5.5	38.9	13.8	37.4

(1) $w_T = f_p w_n + f_d k_\ell$ $f_d = (1+\pi M)^{-1}$

$f_p = \pi M/(1+\pi M)$ $\pi = 100{,}000$ μg/kg per μg/ℓ

Eq. 10, $k_{\ell 1} A(c_g/H_e)$ can be neglected. Table 2 shows the resulting net loss rates of PCBs for each segment for k_ℓ=0.1 m/d. For Lakes Superior and Michigan, the loss rate including volatilization is increased by a factor of about 2-4. For Lake Huron and Saginaw Bay (#8,9,and 10), the increase in the loss rate is greater than a factor of 10. This generally reflects the higher level of PCBs in the dissolved form in these Lakes. For Lake Erie, however, the increase in the loss rate, due to volatilization, is proportionately less because of the higher fraction of PCBs in the particulate form. Note that the overall range of net loss of PCBs when volatilization has been included is now reduced to a factor of about six (23-128 m/yr).

The estimated range of external loading of PCBs is shown in Table 3 [4]. These loadings are believed to

TABLE 3
ESTIMATED RANGE OF TOTAL PCB LOADINGS [4]
(kg/yr)

Lake	Atmospheric (1)	Tributary (2)	Mun. & Ind. (3)	Total	Atmosph. Load as % of Total
Superior	755–7550	630–1890	5– 60	1390–9500	28–92
Michigan	530–5310	460–1380	70– 700	1060–7390	20–91
Huron	500–5030	680–1760	10– 130	1190–6920	21–88
Erie	230–2290	230– 690	220–2180	680–5160	7–84
Ontario	180–1830	330– 990	130–1260	640–4080	7–80

(1) Atmospheric loading range: precipitation, 10–100 ng/ℓ; dry deposition, $1.2 \cdot 10^{-6}$ to $1.2 \cdot 10^{-5}$ g/m^2-yr
(2) Tributary loading @ 10–30 ng/ℓ, except Saginaw Bay where tributary input data were directly available
(3) Municipal & Industrial direct point source loading @ 0.1–1.0 μg/ℓ @ municipal direct point source flows

bracket the current external input and as noted were obtained by applying a range of conditions to each of the principal sources of PCBs. With these upper and lower loading levels applied to each model segment and the associated net loss rate of PCBs given in Table 2, the PCBs concentrations in each water column segment are computed easily from Eq.(10). As noted above, four cases are computed for the upper and lower load levels and with and without volatilization.

Figure 9 shows the calculated water column concentrations and Figure 10 shows the calculated sediment concentrations with a comparison to observed ranges of sediment PCB data. Figure 9 indicates that the water column concentrations are generally below 10 ng/ℓ a concentration that appears reasonable based on some reported data (see [10]). Figure 10, however, is more definitive and indicates that the simple steady state computation provides a goood estimate of the sediment PCB concentration. In general, the upper load level with no volatilization tends to overestimate the observed sediment PCB data while a combination of a lower load level with no volatilization or an upper load level with volatilization appears to represent the data equally well. Individual lakes, however, may have specific combinations that are unique to that lake. For example, the

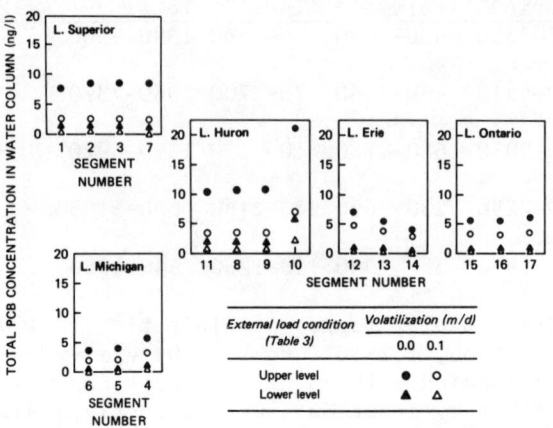

Fig.9. Calculated Great Lakes total PCB concentrations, steady state.

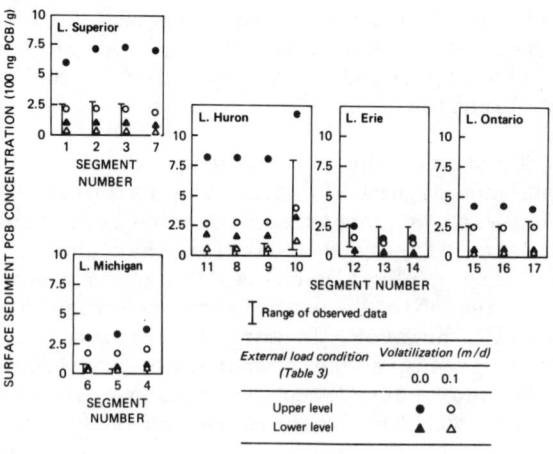

Fig.10. Calculated and observed Great Lakes PCB sediment concentrations, steady state.

best calibration to the sediment data of Lake Michigan is with a lower level of PCB loading while for Lake Erie, the higher loading levels appear more appropriate. In general, however, the simple steady state calculation provides a good first calibration to the observed sediment PCB data. Such data, together with load uncertainty, are not sufficiently detailed to determine what loading level or volatilization rate is more representative. The calculation indicates that, at best, one can only describe the range of external loading and the range of volatilization rates within which it is believed the "true" values fall.

Figure 11 shows the mass balance for the upper load level case and indicates the significance of the volatilization mechanism on the net downward flux of PCBs to the Great Lakes sediments. For Lake Superior, the flux to the sediment for this loading case varies significantly (8-25 kg/d), depending on the volatilization. For Lake Erie, however, the effect of volatilization on the sediment flux is less and varies from 11-17 kg/d. The difference between lakes again reflects varying solids levels and surface area for loss of flux due to volatilization. The change in the flux of PCBs to downstream lakes is not significant and indicates that the mechanisms of net settling and volatilization markedly exceed the mechanism of transport and flushing out of the lake.

Ranges of probable PCB fluxes can be obtained for each of the Lakes from the mass balances of the steady state model and the range of load and degree of volatilization. These ranges are shown in Table 4. The numbers shown in Table 4 were arrived at by evaluation of the comparisons of the model calculation to the sediment data as shown in Figure 10. If the loading was clearly outside the range of the data (as, for example, for Lake Superior upper level load without volatilization) that sediment flux was not included.

The steady state model, therefore, provides an approximate bound on the various PCBs fluxes with estimates of external PCB loads, net solids deposition, partition coefficient and volatilization rate. The caveats regarding the steady state assumption must, however, be recognized. The fluxes shown in Figure 11 and Table 4 and the concentrations calculated for the water column and sediment, assume that the external PCB load has been entering the Lakes for a time sufficiently long to reach equilibrium.

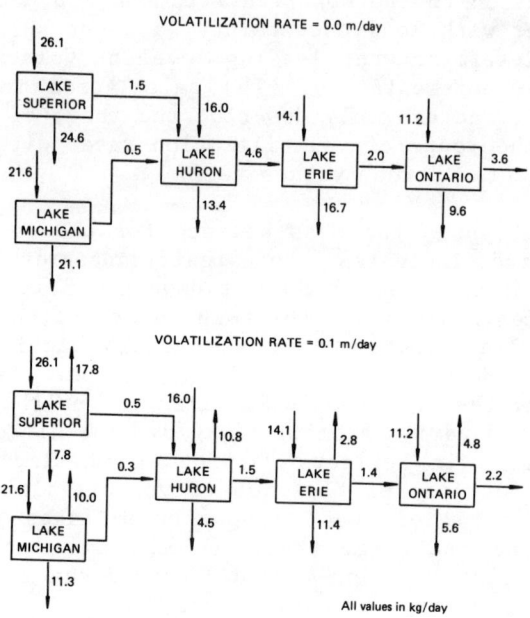

Fig.11. Lakewide mass balance of PCBs for upper level of external load (Table 3) and selected volatilization rates

TABLE 4

PROBABLE PCB FLUXES FOR GREAT LAKES
kg/day

Lake	External Load	To Sediment	To Atmosphere	To Downstream
Superior	3.8–26	1.1– 7.8	0–17.8	0.1–0.5
Michigan	2.9–21.6	1.5–11.3	0–10.0	<0.1–0.3
Huron	3.3–16.0	0.9– 4.5	0–10.8	0.3–1.5
Erie	1.9–14.1	2.5–16.7	0– 2.8	0.3–2.0
Ontario	1.8–11.2	0.9– 5.6	0– 4.8	0.4–2.2

For the water column, this is in the order of several years while for the sediment it may be as long as ten years or greater [10,4]. Nevertheless, it is believed that the steady state model calculations described herein approximately bound the concentrations and fluxes under contemporary (e.g. within 5-10 years) conditions.

SUMMARY AND CONCLUSIONS

Steady State models of the physio-chemical fate of chemicals, such as PCBs, in the aquatic environment can provide easily obtained insight into the principal pathways and resulting concentration of the chemical. For PCBs, and to first approximation, the steady state model indicates that: a) the sediment layers can be "disconnected" from the water column and the resulting water column equations easily solved in a classical fashion, b) resuspension and settling velocities need not be individually specified; only the net loss rate of solids and the partition coefficient need be known, c) the net loss of PCBs is approximately the loss due to particulate settling and loss due to volatilization, d) the concentration of PCBs on the solids in the water column is equal to the concentration on the solids in the sediment. The steady state modeling framework, therefore, results in an easily understood computation and provides a rapid and inexpensive approach to obtaining the spatial distribution of PCBs. However, questions related to response times of a water body following external load changes must be addressed with a time variable modeling framework.

The application of a five segment steady state PCB model to Saginaw Bay indicated that, to first approximation, the theoretical framework could be calibrated to observed data in the water column and sediment. Assuming volatilization of PCBs in Saginaw Bay is small, net removal velocities of PCBs from the water column ranged from zero (for areas with no net deposition of solids) to an average of 4.3 m/yr. for the depositional zones. For the inner Bay, about 80% of the total observed PCBs is due to external loads and 20% due to exchange of PCBs with Lake Huron. Of the approximate 2.1 lb. of PCB/d entering the Bay from external sources, it is estimated from the steady state model that about 1.9 lb/d enters Lake Huron and about 0.2 lb/d is deposited into the sediments of the depositional zones of the Bay.

The PCB steady state model of the Great Lakes, utilizing seventeen water column segments, approximately reproduces observed sediment PCB data with a reasonable range of external loading with or without volatilization of PCBs. However, given the uncertainty in the external load of PCBs to the Lakes, the sediment data are not sufficient to determine the magnitude of the volatilization of PCBs from the Lakes to the atmosphere (or vice versa). The results of the comparison to the sediment data indicate that the percent of external load of PCBs deposited to the sediments varies from better than 90% without volatilization to about 25% with volatilization. Lake Erie because of its smaller surface area and high solids deposition is not as sensitive to the volatilization of PCBs and the net deposition to the sediment varies from 70-90% of the external load (2.5-16.7 kg/d). Generally, downstream transfer of PCBs represents a smaller percentage of the external load indicating that the mechanisms of net settling with and without volatilization exceed the transport out of the Lake.

ACKNOWLEDGEMENTS

This research was supported under a Cooperative Agreement between the U.S. EPA Large Lakes Resarch Station at Grosse Ile, Michigan and Manhattan College (CR 07853-01). Special thanks are due to William Richardson, Project Officer and Nelson Thomas, Chief, Water Quality Branch of the EPA Duluth ERL for their continuing input, and encouragement, and specifically for the compilation of data and free exchange of information. Appreciation is also gratefully extended to Mr. K. Subburamu, Research Assistant at Manhattan College for his efforts in the model calculations and to Ms. Cynthia O'Donnell for her patient typing of the manuscript.

REFERENCES

1. O'Connor, D.J., J.A. Mueller, and K.J. Farley, 1982. Distribution of Kepone in the James River estuary, 20 pp. Submitted for publication.

2. Onishi, Y. (1977). Finite element models for sediment and contaminant transport in surface waters-transport of sediments and radionuclides in the Clinch River, Battelle, Northwest Laboratory, Richland, Washington, Pub. BNWL-2227.

3. Richardson, W.L., V.E. Smith, and R.Wethington, 1982. Dynamic mass balance of PCB and suspended solids in Saginaw Bay-a case study. Proceedings of Workshop on Physical Behavior of PCBs in the Great Lakes. D. Mackay (Ed.)

4. Thomann, R.V., and D.M. DiToro, 1982. Preliminary physical- chemical model of fate of toxicants in the Great Lakes. Submitted for publication.

5. Smith, et al., 1977. Environmental pathways of selected chemicals in freshwater systems, Part I: Background and experimental procedures. EPA-600/7-77-113, U.S. EPA, Washington, D.C., 81 pp.

6. Lassiter, R.R., G.L. Baughman, and L.A.Burns, 1978. Fate of toxic organic substances in the aquatic environment, pp. 219-245. In: Jorgensen, S.E. (ed.) State-of-the-Art in Ecological Modelling, Vol. 7. Int. Soc. Ecol. mod., Copenhagen.

7. Thomann, R.V., J.P. St.John, 1979. The fate of PCBs in the Hudson River ecosystem. Annals of N.Y. Academy of Science, N.Y., Vol. 320:610-629.

8. Eadie, B.J., M.J. McCormick, C.Rice, P. LeVan, M. Simmins, 1981. An equilibrium model for the partitioning of synthetic organic compounds incorporating first-order decomposition. Great Lakes Env. Res. Lab, NOAA, Tech. Memo ERL, GLERL-37, 33 pp.

9. Mackay, D. and S. Paterson, 1981. Calculating fugacity. Environmental Science and Technology 15(9):1006-1014.

10. Thomann, R.V., 1982. Development of a physio-chemical model of toxic substances in the Great Lakes. Final draft report prepared for U.S. EPA Large Lakes Resarch Station, Grosse Ile, Michigan.

11. U.S. Environmental Protection Agency, 1982. Private communication. Large Lakes Research Station, Grosse Ile, Michigan.

12. Monteith, T.J. and W.C. Sonzogni, 1976. U.S. Great Lakes Shoreline Erosion Loadings, International Joint Commission, Windsor, Ontario, 211 pp.

13. Robbins, J.A., 1980. Sediments of Saginaw Bay, Lake Huron: Composition,geochronology, and trace element accumulation rates. U.S. EPA, ERL, ORD, Duluth,Minn.

14. O'Connor, D.J. 1980. Modeling of toxic substances in natural water systems, Manhattan College Summer Institute Notes.

CHAPTER 17

MODEL SIMULATION OF PCB
DYNAMICS IN LAKE MICHIGAN

Paul W. Rodgers
  DePaul University
  Large Lakes Research Station
  9311 Groh Road
  Grosse Ile, Michigan 48138

INTRODUCTION

PCBs (polychlorinated biphenyls) were first commercially marketed in 1929 in the United States. However, it was not until 1959 that these highly stable chlorinated hydrocarbons were brought in bulk to the Lake Michigan basin in the form of hydraulic fluids manufactured by Monsanto Chemical Company. During the 1960's, evidence mounted that PCBs could cause functional failures in animals, induce general symptoms of toxicity, and could be potentially carcinogenic. These findings are particularly alarming because of the persistence of these compounds in the environment, and their known bioaccumulation in the fatty tissue of organisms that inadvertently consume them. In response to this evidence, Monsanto restricted its sales of PCBs to closed systems in 1971, and by 1977 had ceased production altogether.

In 1971, the Environmental Protection Agency found mean concentrations of PCBs (Aroclor 1254) in Lake Michigan fish ranging from 2.7 ppm in the herbivorous rainbow smelt to 15 ppm in the lake's top predatory fish, the lake trout. PCB concentrations in Lake Michigan coho salmon were two to three times higher than coho from Lake Huron, and ten times the concentrations observed in the coho from Lakes Erie and Superior. In 1973, the U.S. Food and Drug Administration established the temporary threshold limit for human consumption of PCBs in fish at 5 ppm. Fish sale restrictions and fish consumption warnings were subsequently issued in Lake Michigan.

By 1974 fish concentrations of PCBs had not dropped, and the necessity of quantifying the extent of the problem became clear. In 1975, the Johnson Motors Division of Outboard Marine Corporation in Waukegan, Illinois was found to be discharging PCBs to Waukegan Harbor and to a tributary of Lake Michigan known as the "North Ditch" [1]. The relative importance of this source, and the necessity of understanding the mobility of PCBs required the quantification of other potential sources. Investigation of influencing chemical and biological processes related to the fate of PCBs was also essential. The aim of this research was to make use of the "best available estimates" for the loads and processes which influence the distribution of PCBs in Lake Michigan, and to develop a time-variable physicochemical model of total PCBs which would simulate changes in concentration in the water column and in the underlying sediments. Potentially these simulations of lake response would serve to assess management alternatives and could ultimately be coupled to a fish-PCB model which requires these concentrations as an input. The present uncertainty in model inputs and in relevant processes makes it necessary for model output to be flexible. In fact, an intent of model output was to facilitate reevaluation of the expectations of system response as estimates of inputs and representation of predominant processes subsequently become more certain. An analysis of historical loads is also examined as a model application.

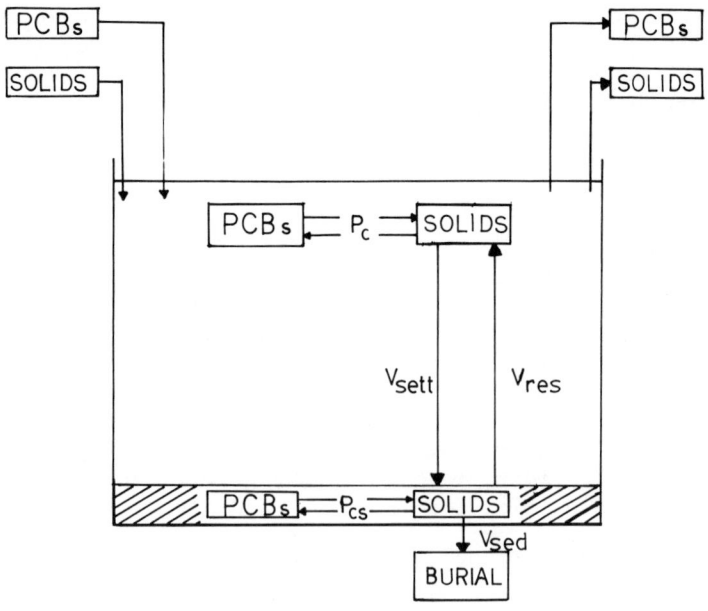

Figure 1. Schematic diagram of PCB dynamics in the water column and sediments of Lake Michigan.

MODEL STRUCTURE

A schematic diagram of the Lake Michigan PCB model is presented in Figure 1. The two state variables, solids and total PCBs, are simulated in two physical compartments, the water column and the underlying sediment layer. The solids are subject to settling from the water column, subsequent resuspension from the sediment layer to the water column, and sedimentation from the active sediment layer to buried, inactive sediments. The dynamics of PCBs are closely associated with the dynamics of the solids via partitioning of PCBs between the solids and the water.

Four mass balance equations are required to describe mathematically the model structure;

1) Solids in the water column, SSW:

$$V_w \frac{d[SSW]}{dt} = W_s - Q[SSW] - v_{sett}[SSW] A + v_{res}[SSSD] A$$

2) Solids in the sediment layer, SSSD:

$$V_{SD} \frac{d[SSSD]}{dt} = V_{sett}[SSW] A - V_{res}[SSSD] A - V_{sed}[SSSD] A$$

3) PCB in water column, PW:

$$V_w \frac{d[PW]}{dt} = W_p - Q[PW] - V_{sett} \frac{P_c[SSW] A}{(1 + P_c[SSW])} [PW]$$

$$- k \frac{[PW]}{(1 + P_c[SSW])} V_w + V_{res} \frac{P_{cs}[SSSD] A}{(1 + P_{cs}[SSSD])} [PSD]$$

4) PCB in sediment layer, PSD:

$$V_{SD} \frac{d[PSD]}{dt} = V_{sett} \frac{P_c[SSW] A}{(1 + P_c[SSW])} [PW] - k \frac{[PSD]}{(1 + P_{cs}[SSSD])} V_{SD}$$

$$- V_{res} \frac{P_{cs}[SSSD] A}{(1 + P_{cs}[SSSD])} [PSD] - V_{sed} \frac{P_{cs}[SSSD] A}{(1 + P_{cs}[SSSD])} [PSD]$$

Terms are defined as:
- V_w - Volume of water column   (m³)
- V_{SD} - Volume of sediment layer   (m³)
- W_s - Solids load to Lake Michigan   (g/yr)

W_p - PCB load to Lake Michigan (g/yr)
V_{sett} - First order settling rate (m/yr)
V_{res} - First order resuspension rate (m/yr)
V_{sed} - First order sedimentation rate (m/yr)
k - General first order loss rate (1/yr)
A - Area of water-sediment interface (m^2)
Q - Mean annual flow (m^3/yr)
P_c - Partitioning coefficient in water (m^3/g)
P_{cs} - Partitioning coefficient in sediment layer (m^3/g)

These equations are presented and examined by Schnoor et al. [2] assuming steady-state conditions. Since the lake response over time is of interest the solutions to the above equations are obtained using a 4th order Runge-Kutta approximation.

Physically, Lake Michigan is simulated as a homogeneous one segment water column and a one segment sediment layer. Observations by all investigators point, however, to the heterogeneous distribution of PCBs and solids. The lack of sufficiently well defined spatial data, and the present inability to simulate the depositional pattern of Lake Michigan have made development of a more highly segmented model of little additional advantage, although models by Thomann et al. [3] have begun to include greater segmentation. As data becomes more abundant, additional segmentation would be advantageous in describing the observed distribution of PCBs in Lake Michigan. The design of this model is intended only to describe lake-wide responses.

The general first order loss rate, k, incorporates all potential loss mechanisms other than settling into one term. This is done only to investigate the potential magnitude of the losses, given the observed conditions. Our current knowledge of loss processes outside of the laboratory is limited. The present controversy over volatilization of PCBs has been examined for several years, and remains to be resolved.

The model structure provides that the partitioning coefficient in the sediment may be different from that in the water column. This aspect of the model was introduced in response to observations that partitioning is a function of solids concentration [4,5].

The model structure represents both settling and resuspension instead of a net loss rate. This aspect is necessary to simulate the time variable behavior of a partitioning chemical like PCB, but it is not required in forecasting the steady-state response of a particular load. However, the overall or net loss rate of solids from the water column, ω, can be derived from steady-state solutions of Equations 1 and 2 as:

$$\omega = \frac{V_{sett} \cdot V_{sed}}{V_{res} + V_{sed}} \qquad (5)$$

where,

ω = overall loss rate of solids from the   (m/yr)
water column

DETERMINATION OF COEFFICIENTS

In determining the values of the model coefficients, the goal is not only to simulate previous field observations of the state-variables, but also to have all coefficients reflect the contemporary knowledge of the processes they represent. This latter character of the coefficients can be referred to as external consistency. Table I has listed the model coefficients and their calibrated value. Before examining model output, some discussion of the evidence supporting the value of each coefficient can be made. Additional support for certain calibrated coefficients will be found in comparing model output with field observations.

Table I. Model Coefficients

Symbol	Value	Units
V_{sett}	430 (141)*	m/yr
V_{res}	0.002 (.0004)	m/yr
V_{sed}	0.000375	m/yr
k	0.0	1/yr
P_c	500,000	ℓ/kg
P_{cs}	1000	ℓ/kg

*Values presented in parenthesis represent possible refinements in calibration obtained from plutonium modeling.

The partitioning coefficient for PCBs has been observed to be sensitive to the solution's solids concentrations and an empirical function has been suggested by O'Connor and Connolly [5] as:

$$\pi = \beta m^\alpha \tag{6}$$

where,

π = specific partitioning coefficient
β = partitioning coefficient at a solids concentration of 1 mg/ℓ
m = specific solids concentration
α = the slope of a log-log plot of π vs. m

This formulation strongly suggests that π should be greater in the Lake Michigan water column where solids concentrations are 1 mg/ℓ than in the sediment layer where the solids concentration is greater than $2 \cdot 10^5$ mg/ℓ. Experimental data from Horzempa and DiToro [4] suggests that based on solids concentrations indicative of the open waters of Lake Michigan the partitioning coefficient should be approximately in the range of $8.3 \cdot 10^4 - 2.6 \cdot 10^5$ ℓ/kg. A similar analysis for the sediment layer yields an expected range for the partitioning coefficient of 55-1200 ℓ/kg. These estimates are based on solids concentrations only. Additional factors, such as the particular water chemistry, may also affect the magnitude of the partitioning. Field data from Lake Michigan observed by Rice [6] indicates a partitioning of PCBs of $8 \cdot 10^5$ ℓ/kg for sub-surface samples. Eisenreich [7] has observed partitioning of PCBs in Lake Superior of about $2.5 \cdot 10^5$ ℓ/kg. Both of these reported field values were for samples having a solids concentration in the vicinity of 1 mg/ℓ.

The values of the coefficients which affect the dynamics of the solids (V_{sett}, V_{res}, and V_{sed}) are difficult to substantiate. However, there are a number of observations which can fix the ranges these coefficients can be expected to lie within and the overall loss rate (ω) required of these coefficients can be determined with some certainty.

Examination of the above coefficients is enhanced by examination of the steady-state solutions of Equations 1 and 2. From such an analysis, the following function can be derived:

$$V_{sed} = \frac{W - [SSW] \cdot Q}{[SSSD] \cdot A} \tag{7}$$

By definition, W, [SSW], and [SSSD] do not change over time on a long term, annual basis. Equation 7 suggests that given estimates of steady-state loading and solids concentration, the sedimentation rate, V_{sed}, can be determined.

Setting the solids concentration in the water column between 1.0 - 1.7 mg/ℓ as an estimate of an annual open-lake average and 2.0 - 2.75 · 10^5 mg/ℓ as the sediment average, permits calculation of the probable range within which V_{sed} lies. The values for the other constituents of Equation 7 as well as the loadings, morphometry and hydrology required by the model may be found in Table II. The solids simulations in the model is concerned only with the fine grained particles which impact the deep portions of the lake, and not with larger particles whose impact is most evident in the near-shore zones. The resulting range for V_{sed} is 2.9 - 4.0 · 10^{-4} m/yr. This range can be compared to an examination of data by Edgington and Robbins [11] that suggests the area averaged sedimentation rate, V_{sed}, is approximately 3 · 10^{-4} m/yr in the southern basin of Lake Michigan.

Table II. Model Loads and Inputs

LOADS

Source	Solids (metric tons/yr)	PCBs (metric tons/yr)
Tributary	706,540	1.65**
Atmospheric	900,000	3.9-5.9***
Direct Pt	140,000	-
Erosion (Fine Grain)	2,900,000	-
TOTAL	4,646,540*	5.55-7.55

MORPHOLOGY

Surface Area	=	5.78 · 10^{10} m^2
Lake Volume	=	4.976 · 10^{12} m^3
Sediment Volume	=	2.89 · 10^9 m^3
Hydraulic Retention Time	=	110.3 yrs.

*Sonzogni, et al. [8].
**Murphy and Rzeszutko [9].
***Murphy 1982, Atmospheric load represents a sum of a wet estimate (.85 - 1.275 metric tons/yr) and a dryfall estimate (3.083-4.625 metric tons/yr) derived from experimental work.

Once the approximate value of V_{sed} is estimated, values for V_{sett} and V_{res} remain to be estimated. However, these coefficients are not independent of one another. Further steady-state analysis yields the following functional relationship:

$$V_{res} = \frac{[SSW] \cdot V_{sett} (1 - \frac{V_{sed}[SSSD]}{V_{sett}[SSW]})}{[SSSD]} \tag{8}$$

Note that given the value of V_{sed} from Equation 7, fixing either of the remaining two coefficients (V_{sett} or V_{res}) results in the direct determination of the third. These observations remove one degree of freedom from the variability within the model calibration.

Presently, data that might yield a quantitative determination of either of the coefficients (V_{sett} or V_{res}) is difficult to obtain. Preliminary analysis of depositional rates based on sediment trap data from Lake Michigan [12] indicates a range for V_{sett} of approximately 150 to 600 m/yr, with variability apparently due to temporal and spatial observations, which cannot be reflected in a model utilizing an annual time scale. The calibrated values for V_{sed}, V_{sett}, and V_{res} reported in Table I yield an overall loss rate (ω), based on Equation 5, of 68 m/yr. Given the estimate of solids loading used in the model and the observed solids concentrations, the value of ω is fixed. However, there are many combinations of the coefficients in Equation 5 which would yield the same overall loss rate. As will be discussed in the model sensitivity analysis, the specific combination of coefficients has no effect on the predicted steady-state concentrations in the water column, but is influential in determining the response time of the system. The value of ω may be compared to the net loss rate of 73 m/yr calibrated in the Lake Michigan phytoplankton model [31]. Field data in support of this net loss value is found in the mean deposition rate observed by Edgington and Robbins [11] of 7 $mg/cm^2/yr$, which suggests a net loss rate of 70 m/yr, if the annual water column concentration of solids is approximately 1 mg/ℓ.

MODEL OUTPUT

The PCB loadings reported in Table II are based on limited experimental data and a number of necessary assumptions. Although the entire range of the load estimate was used in observing model performance, the lower estimate of

5.55 metric tons/yr was used in the calibration and sensitivity analysis, because this value represents the most recent analytical assessments.

Due to the paucity of PCB data reflecting the loads and ambient conditions in Lake Michigan, it is difficult to judge model output or to use comparison of model output to field data as a primary calibration tool. In recognition of these inherent aspects, model output was utilized primarily for comparing the effects of various input estimates. For example, Figure 2 relates the steady-state response in PCB concentration to a range of loads for three different partitioning coefficients. The results demonstrate that PCB concentration increases with increasing loading and decreasing partitioning coefficient. Also evident, is the observation that various estimates in the magnitude of the partitioning coefficient may or may not have profound effects on lake response. The difference in concentration response for partition coefficients between $5 \cdot 10^5$ and $5 \cdot 10^6$ ℓ/kg is not striking. However, the model predicts a significant difference, if the partition coefficient is as small as $5 \cdot 10^4$ ℓ/kg. The point at which the estimate for the partitioning coefficient becomes sensitive is ultimately a function of the solids concentration.

Model output further indicates that the partitioning of PCBs onto solids in the water column is only about 37 per-

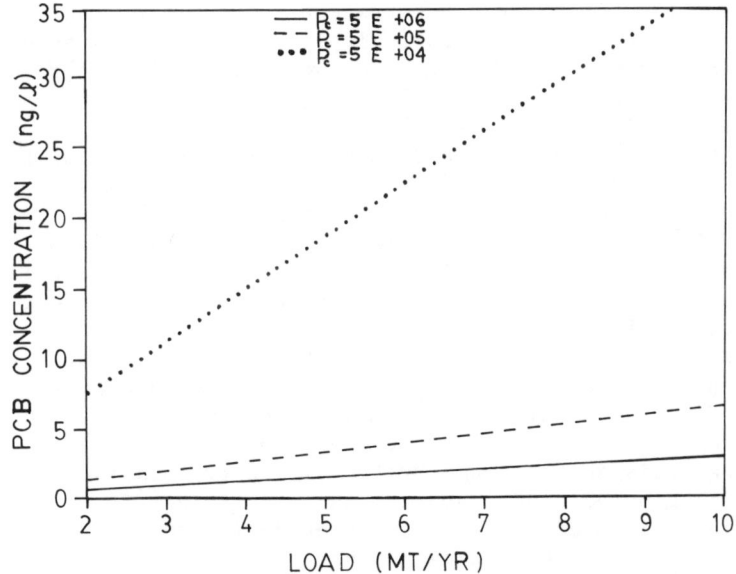

Figure 2. Steady-state response in total PCB concentration as a function of loading and the value of the partition coefficient in the water column (ℓ/kg).

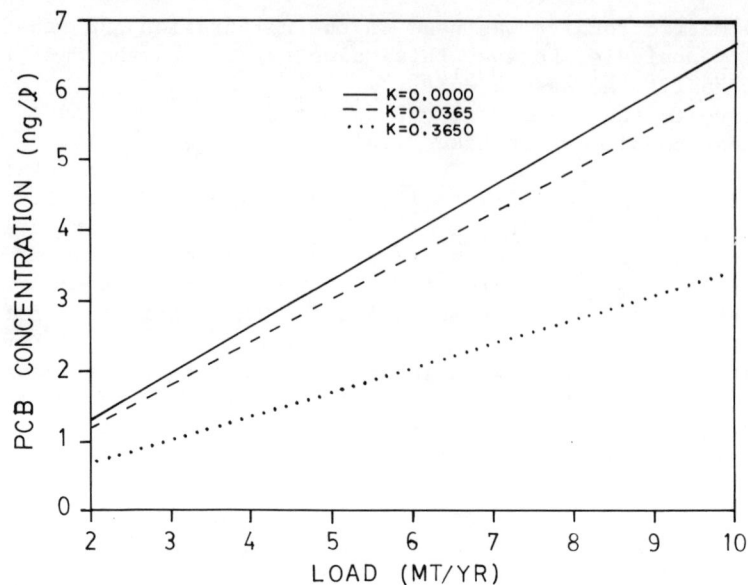

Figure 3. Steady-state response in total PCB concentration as a function of loading and the magnitude of the decay rate (yr^{-1}).

cent of the total PCB concentration, a result of the low solids concentration in Lake Michigan. This compares well to field observations of sub-surface samples from three cruises which indicated 31.1, 17.4, and 38.1 percent particulate PCB [6]. Since the particulate fraction of total PCB in the water column is less than 40%, the tracking of PCBs with solids dynamics is not as critical as it might be in systems with a higher solids concentration (e.g., the lower Great Lakes). Further illustrating the relative importance of the particulate fraction is the observation that while the overall net loss rate of solids was 68 m/yr the net loss of total PCB is only 25.5 m/yr.

Model output, using coefficients reported in Table I, indicated solids concentrations that are comparable to recent field observations. A steady-state water column concentration of 1.2 mg/ℓ is attained rapidly. Evaluation of whether this solids concentration is representative of a volume-weighted annual average is not straight forward. Samples collected in the open water during the last decade generally indicated 1-2 mg/ℓ suspended solids, but these concentrations vary considerably and no data is available for the same waters during the winter months. Sediment core data reported by Edgington and Robbins [11] indicate solids concentrations from $2 \cdot 10^5$ mg/ℓ to over $6 \cdot 10^5$

mg/ℓ, with the higher concentrations reflective of the high depositional zones. Further analysis of their data results in an area-weighted average of approximately $2.33 \cdot 10^5$ mg/ℓ. The steady-state model output for the sediment layer is $2.12 \cdot 10^5$ mg/ℓ.

Although partitioning of PCBs and their subsequent loss to the sediment is the predominant loss mechanism, there are several other processes which may affect the loss of soluble PCBs. These processes include chemical or biological decay and volatilization. Estimates for their rates in the aquatic ecosystem vary widely. Because of this uncertainty, k, the model coefficient representing a first order loss of soluble PCBs, cannot be estimated. Instead, as illustrated in Figure 3, k was varied and the effect on output was observed. The results of a sensitivity analysis, where selected coefficients were varied and model response observed, are reported in Table III. Note that k may have a value between 0.0 and 0.1% per day (equivalent to 0.0 and

Table III. Sensitivity Analysis
(Steady-State Values)

	Original Calibrated Value	$k(yr^{-1})$ (.0365)	$k(yr^{-1})$ (.365)	$k(yr^{-1})$ (3.65)	P_{cs} x 10	P_{cs} x 0.1
SSW (mg/ℓ)	1.17	–	–	–	–	–
SSSD (mg/ℓ)	2.12×10^5	–	–	–	–	–
PW (ng/ℓ)	3.71	3.39	1.91	0.37	3.71	3.71
PSD (μg/g)	1.18	1.08	0.57	0.09	1.18	1.23

0.365 yr^{-1}) and yield an output of total PCBs that can be compared to recent field observations of 1-5 ng/ℓ in the open waters of Lake Michigan [6,14]. However, a rate greater than this range would constitute a loss greater than presently can be reasoned from existing field data.

Even though there is evidence that the partitioning coefficient in the sediment layer (P_{cs}) should be less than in the water column because the solids concentration is higher, the sensitivity analysis, as reported in Table III, found that model response is not sensitive to the value of P_{cs} over those three orders of magnitude. This observation is due to the very high solids concentration typical in lake sediments. The effect of solids concentration on partitioning would be particularly noted in aquatic systems of highly variable solids, such as rivers and impacted near-shore zones.

HISTORICAL ANALYSIS

Completion of a historical simulation would not only add credence to the model calibration, but would also be an important tool for calibrating the components of the model which affect the rate of change, as opposed to the steady-state response. Regretably, recreating a legitimate historical loading scenario for PCBs involves speculation. A reasonable approximation of historical PCB loadings was to sum the known reservoirs of PCBs and an estimate of PCBs historically exported through the Straits of Mackinak. Table IV presents a sum where each component has a range in the mass estimate, based on their uncertainty.

Table IV. Estimate of Total Historical PCB Loading (Metric Tons)

Reservoirs:	
Water Column @ 2-7 ng/ℓ	10-35
Lake Michigan Sediment	17-32
Approximate of Outflow Loss:	
Mean water concentration (5-10 ng/ℓ) x Mean flow (4.51 · 10 ℓ/yr) x Yrs of activity (22 yrs)	5-10
Total:	32-77

The inventory of PCBs in Table IV should reflect an approximation of the total input of PCB mass to the open waters of Lake Michigan. An estimate of 32-77 metric tons for this total load has several important implications. If PCBs have been entering Lake Michigan since 1959, then the average annual load is between 1.5 and 3.5 metric tons per year. Another approach to estimating the average historical load was to vary this load until the model simulated present conditions. This reiterative process yielded an average historical load of between 1.5-3.0 metric tons/yr. These deductive approximations of historical loads seem to indicate the load reported in Table II, based on extrapolation of limited field measurements, may be an over estimate. Historical loading estimates by the above approaches yield only a total historical load and cannot suggest past loading trends, which would be expected to reflect a decline due to production cessation in the latter 1970s. Such indirect methods for estimating past loadings assume there are not significant decay or volatilization losses. Should these loss processes prove significant then the above conclusions may have to be reconsidered.

The introduction of plutonium as a "surrogate pollutant" for simulation has been an approach used by Thomann et al. [3]. The authors reasoned that the historical loads of plutonium are relatively well known [15], and that data for the mid-1970s are available for the water column and the sediments. Plutonium also partitions to solids at equilibrium levels similar to PCB. Inventory accounting of plutonium mass, similar to the PCB inventory presented in Table IV, succeeded in accounting for 90% of the historical plutonium load estimated by Edgington et al. [15] from field measurements.

An historical simulation using plutonium as a surrogate for PCB was run without changing any of the model coefficients. The results for the open waters of Lake Michigan are presented in Figure 4 as the dashed line. The solid line represents a refinement in model coefficients such that $V_{res} = 4 \cdot 10^{-4}$ (m/yr). To maintain the same overall loss rate, ω, V_{sett} is also set to 141 m/yr. These coefficients are also reported parenthetically in Table I. Note that both results simulate the observed field data reasonably well, although the model cannot simulate the seasonal oscillations. The model simulates the sediment layer in the mid-1970s as having a concentration of plutonium of about 0.14 pCi/g, which compares well with observed values ranging from 0.03 - 0.3 pCi/g [15]. To make judgements about the refinements in coefficient values it would be necessary to have data that might reflect lake response during the most dynamic period, between 1960-1967. However, this simulation of plutonium, as a surrogate pollutant, supports the overall loss rates and model framework previously developed for PCBs.

Although the modeling approaches presented offer methods for model calibration and estimation of the total PCB loading to Lake Michigan, these methods cannot suggest past

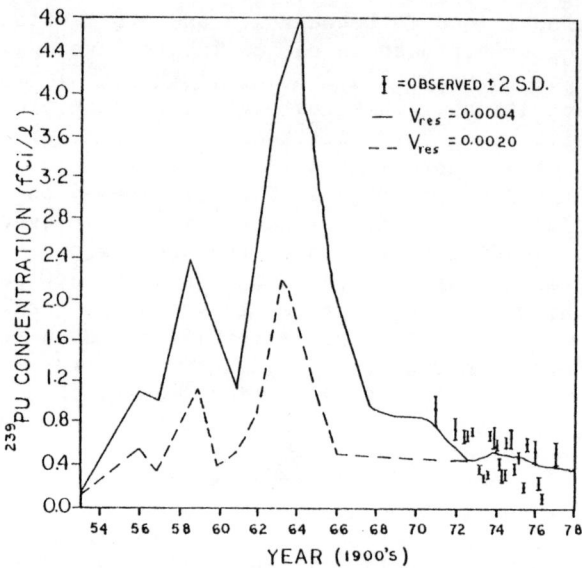

Figure 4. Simulation of plutonium by use of the PCB model and its calibrated coefficients (represented by the dashed line). The solid line reflects a change in model output due to adjustment of two coefficients, V_{res} and V_{sett}. (Data from Thomann [3]).

trends in the loading or in the lake response. For instance, the question of whether loads have decreased significantly is not answered by these modeling approaches. However, data describing fish body burdens for PCBs during the 1970s may offer an additional modeling approach for deriving these important trends. The approach of using toxic body burden of fish to infer concurrent water concentrations has already been implemented for the simulation of DDT in Lakes Michigan and Superior [16]. The combination of these modeling approaches (presently underway) should make possible the task of modeling PCBs in Lake Michigan, with the intent of assisting management decisions aimed at not exceeding fish threshold limits.

CONCLUSIONS

The model structure and calibration presented is capable of simulating long term lake responses to the loading of PCBs to Lake Michigan. The solids dynamics in Lake Michigan are not only important in determining the distribution of PCBs, but the processes of settling, resuspension, and sedimentation are influencial in fixing the rate at which the lake responds to a load or perturbation. The solids and PCB loading, along with the concurrent partitioning and overall loss rate, is sufficient for determining the long term response. These results indicate that the question of the ultimate response may be easier and more certain to answer than the question of response time.

PCB inventory analysis and historical modeling application indicated an average historical loading of PCBs to Lake Michigan of 1-3.5 metric tons/yr. However, trends in historical loading could not be discerned based on the application of the physicochemical PCB model. Knowledge of these trends would assist in quantitatively understanding the PCB loading - concentration response - biological response phenomena in Lake Michigan, and therefore remains an important research need.

The output for the Lake Michigan PCB model was designed to be flexible. This output format facilitated answering the questions of system response to various loading scenarios, losses, and coefficient values. The necessity for this kind of ouput underlines the fact that estimates used as model input are not fixed, and may be subject to refinement. It is also evident that very specific research still needs to be done in regard to the dynamics of PCBs in Lake Michigan. Most importantly, the loading of PCBs from all sources, the rates at which solids are vertically transported, and volatilization kinetics require continued scientific emphasis.

ACKNOWLEDGEMENTS

This research was supported by Environmental Protection Agency Grant No. 807412, Dr. Michael D. Mullin - Project Officer, Dr. Tom Murphy - Principal Investigator. Dr. Wayland R. Swain contributed valuable suggestions. Debra L. Caudill provided typing and editorial assistance.

REFERENCES

1. Environmental Protection Agency (Region V). 1981. The PCB contamination problem in Waukegan, Illinois. An agency progress report.

2. Schnoor, J.L., D.C. Avory, C.E. Ruiz-Calzada, and M.R. Wiesner. 1979. Verification of a pesticide transport and bioaccumulation model. EPA Ecological Research Series.

3. Thomann, R.V., et al. 1982. Preliminary model of the fate of toxic substances in the Great Lakes. Submitted to the J. of Great Lakes Res.

4. Horzempa, L.M. and D.M. DiToro. 1981. PCB partitioning in sediment-water systems: The effect of sediment concentration. A draft submitted for publication, Manhattan College, NY, NY.

5. O'Connor, D.J. and J.P. Connolly. 1980. The effect of concentration of adsorbing solids on the partition coefficient. Water Research, 14: 1517-1523.

6. Rice, C.P., P.A. Myers and G.S. Brown. 1981. Role of surface microlayers in the air-water exchange of PCBs. Presented at Physical Behavior of PCBs in the Great Lakes workshop, Toronto, Canada, Dec. 10-11.

7. Eisenreich, S.J., P. Capel and B. Looney. 1981. PCB dynamics in Lake Superior water. Presented at Physical Behavior of PCBs in the Great Lakes workshop, Toronto, Canada, Dec. 10-11.

8. Sonzogni, W.C., Monteith, T.J., Skimin, W.E. and S.C. Chapra. 1979. Critical assessment of U.S. land derived pollutant loadings to the Great Lakes, an IJC Report.

9. Murphy, T.J. and C.P. Rzeszutko. 1978. Polychlorinated biphenyls in the Lake Michigan Basin. EPA Ecological Research Series, EPA-600/3-78-071.

10. Murphy, T.J. 1982. DePaul University, personnel communication.

11. Edgington, D.N. and J.A. Robbins. 1976. Records of lead deposition in Lake Michigan sediments since 1800. Environ. Sci. and Technol., 10(3): 266-274.

12. Chambers, R.S. and B.J. Eadie. 1981. Nepheloid and suspended particulate matter in southeastern Lake Michigan. Sedimentology, 28: 439-447.

13. Rodgers, P.W. and D. Salisbury. 1981. Water quality modeling of Lake Michigan and consideration of the anomalous ice cover of 1976-1977. J. of Great Lakes Res., 7(4): 463-480.

14. Armstrong, D.E. 1981. University of Wisconsin, personnel communication.

15. Edgington, D.N. et al. 1976. The behavior of plutonium in aquatic ecosystems: A summary of studies on the Great Lakes, in Environmental and Toxicity of Aquatic Radioncludes: Models and Mechanisms. M.W. Miller and S.N. Stannard, Eds., Ann Arbor Science Publ., Inc., Ann Arbor, MI, pp. 45-75.

16. Bierman, V.J., Jr. and W.R. Swain. 1981. Mass balance modeling of DDT dynamics in Lakes Michigan and Superior. Submitted to Environ. Sci. and Technol. for publication.

CHAPTER 18

DYNAMIC MASS BALANCE OF PCB
AND SUSPENDED SOLIDS IN
SAGINAW BAY -- A CASE STUDY

William L. Richardson
U.S. E.P.A., ERL-Duluth,
Large Lakes Research
Station, 9311 Groh Rd.,
Grosse Ile, MI 48138

V. Elliott Smith
Cranbrook Institute of
Science, 500 Lone Pine
Rd., Bloomfield Hills,
MI 48013

Robert Wethington
Computer Sciences Corp.,
Large Lakes Research
Station, 9311 Groh Rd.,
Grosse Ile, MI 48138

INTRODUCTION

Although regulations have reduced their use and eliminated their manufacture in the U.S., polychlorinated biphenyls (PCBs) remain a serious environmental issue. Particularly in the Great Lakes from which large quantities of fish are caught for human consumption, the PCB question remains, "How long will PCB concentrations in some fish remain above the FDA action limit of 5 mg/kg?" Data collected by various agencies on PCB levels in Lake Michigan fish have revealed a decreasing trend since 1972 [1], but it is apparent that an extended period of time will be required for levels to decline below the FDA limits for some species [2]. Evidently, reservoirs of PCB remain in the environment which continue to act as sources of contamination to the Great Lakes. Continuing sources include landfills, road pavements previously sealed with PCB oil mixtures, contaminated sediments in harbors and bays and long-range atmospheric transport and deposition. Specific questions remain concerning the additional effort required to locate sources and decontaminate or seal these reservoirs. Problem areas include the contaminated sediments of Waukegan Harbor, the Kalamazoo River estuary on Lake Michigan, Raisin River, and the Saginaw River and its tributaries.

To assist in answering regulatory questions, research and surveillance data are needed to evaluate the relative importance of the existing inputs, sediment reservoirs, and the processes controlling transport and fate and exposure. In Saginaw Bay, Lake Huron, about 3.7 metric tons of PCB remain in the active sediment and inputs from the Saginaw River and atmospheric deposition contribute about 1.4 kg PCB per day. In 1977 the U.S. E.P.A. initiated a research effort on Saginaw Bay which was chosen because of the existing PCB contamination, its importance as a commercial and sports fishery, and because, within a relatively small area, many of the limnological processes occurring in the Great Lakes are represented. Therefore, findings from this work might be extrapolated to other parts of the Great Lakes or other similar water systems.

This paper presents an analysis of conditions in the bay during 1979 and a projection of future conditions using a dynamic mass balance model. The primary research questions addressed are: 1) whether simulation models of "total PCB" are sufficiently accurate or whether refined models considering at least mixtures are necessary, 2) whether volatilization of PCB is occurring, and 3) what is the expected longevity of PCB in the system.

DATA BASE DEVELOPMENT

The present research focuses on the 1979 surveys of Saginaw Bay. In 1979 ten cruises lasting about one week each were conducted and samples collected at 26 stations at one meter and one meter above the bottom (see Figure 1). Samples were serially filtered on-board to obtain subsamples for particulate fractions: 1) 0.7 to 37 μm; 2) 37 to 74 μm; 3) 74 to 210 μm; 4) 210 to 1000 μm. The dissolved fraction is operationally defined as the filtrate or portion of the sample containing solids $\leq$ 0.7 μm. On-board analyses included temperature, conductivity, pH, Secchi depth, and turbidity. Samples were analyzed later for suspended solids and PCB. PCB analyses were made using a dual column Hewlett-Packard gas chromatograph fitted with an auto sampler and electron capture detector and an Aroclor 1242-1260 standard. PCB results were reported in terms of total PCB, Aroclor 1242, Aroclor 1260, and in terms of six homo-

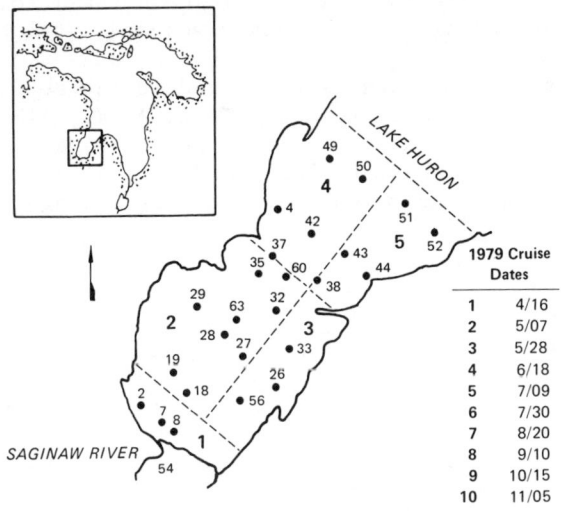

Figure 1. Saginaw Bay 1979 sampling network and segmentation.

logs (2 to 7 chlorines/biphenyl molecule) based on Sawyer [3]. Each sample collected was characterized by up to 45 PCB/solid fraction combinations. Details of the sampling procedures and PCB analyses are described by Smith et al. [4]. Although there continues to be much discussion about the accuracy of PCB analysis techniques, and fish samples analyzed by packed column GC have been as high as a factor of two higher than duplicate samples analyzed using capillary column GC [5], the data are relatively consistent. This is an important consideration in using the data for modeling. In the future the absolute levels of PCB reported herein may have to be adjusted as indicated by comparisons of packed column and capillary results.

DATA ANALYSIS

For the purpose of the data presentation, as well as for modeling, the data were stratified according to five segments shown in Figure 1 with Saginaw River represented by one station, SB0054. In addition, for this presentation the individual particulate fractions have been summed and the total particulate data (≥ 0.7 μm) presented. Only the data for total PCB (TPCB), Aroclor 1242 (A1242) and Aroclor 1260 (A1260) are presented. In all cases TPCB is the sum of A1242 and A1260. In addition only those samples where all data for dissolved and total particulate were available are used. Total PCB concentrations were calculated for each sample by the addition of the dissolved and total particulate.

A summary of the data is presented in Tables I-III. Averages for all data by segment, TPCB, A1242, A1260, (total, particulate, and dissolved) and suspended solids are presented. In these tables the distribution coefficients presented were calculated from the average data by Equation (1):

$$K_p = \frac{C_p/S}{C_D} \qquad (1)$$

	Units
K_p = distribution coefficient	(ℓ-mg^{-1})
C_p = particulate PCB concentration	(ng-ℓ^{-1})
C_D = dissolved PCB concentration	(ng-ℓ^{-1})
S = suspended solids concentration	(mg-ℓ^{-1})

As shown in Figure 2, concentrations of TPCB, A1242, and A1260 decrease from Saginaw River to Lake Huron. This could be due to simple dilution as well as losses via settling or volatilization. It is evident that A1242 is predominant in the river with an average concentration of about 130 ng/ℓ compared to 45 ng/ℓ for A1260; whereas, in the bay the concentrations of each mixture are almost identical, although A1260 is higher in the outer bay (Figure 3). This implies that whatever mechanisms are involved they are different for each of the mixtures.

For both mixtures the particulate fraction dominates in the river whereas the dissolved fraction becomes more dominant from the inner bay to Lake Huron, as indicated in Figure 3. This follows the pattern of higher solids concentrations in the river than in the bay and implies the importance of solids/PCB relationships. This is further explained in Figure 4 where K_p's as calculated from the data using Equation (1) are shown. Generally, K_p's for A1242 are lower than those for A1260 and, for both mixtures, K_p's are lower in the river and in the shallower segments, 1 and 3. This may be due to the higher proportion of sands and clays due to transport and resuspension than in the deeper segments where the lighter, organic and detrital matter predominate.

Finally, Figure 5 shows the temporal relationships of chloride, suspended solids, and fractions of mixtures of PCB for segment 2. Particulate PCBs track with suspended solids with lower concentrations in the summer. In contrast, the

Table I. Average Total PCB and Suspended Solids in Saginaw Bay, 1979

Segment		Total PCB Total ng/ℓ	Total PCB Dissolved ng/ℓ	Total PCB Particulate ng/ℓ	Sus. Sol. mg/ℓ	Mass PCB on solids ng/mg	Distribution Coefficient ℓ/kg
1	$\bar{x}$	43.1	27.0	16.2	15.2	1.07	39500
	S_d	20.2	22.0	6.04	3.96		
	S_e	6.09	6.62	1.82	1.06		
	N	11	11	11	14		
2	$\bar{x}$	26.4	14.8	11.6	9.68	1.20	81000
	S_d	11.5	9.47	7.95	5.73		
	S_e	1.69	1.40	1.17	0.906		
	N	46	46	46	30		
3	$\bar{x}$	25.6	15.7	9.91	12.2	0.81	51700
	S_d	24.4	21.4	4.43	5.72		
	S_e	6.09	5.36	1.11	1.65		
	N	16.0	16	16	12		
4	$\bar{x}$	18.1	14.1	3.98	3.03	1.31	93200
	S_d	10.2	10.2	2.24	1.94		
	S_e	1.96	1.97	0.469	0.389		
	N	27	27	27	25		
5	$\bar{x}$	16.2	13.7	2.57	2.65	0.970	70800
	S_d	14.7	15.2	1.32	2.57		
	S_e	3.47	3.59	0.312	0.624		
	N	18	18	18	17		

Table II. Average Aroclor 1242 and Suspended Solids Concentrations in Saginaw Bay, 1979

Segment		Aroclor 1242			Sus. Sol. mg/ℓ	Mass PCB on solids ng/mg	Distribution Coefficient ℓ/kg
		Total ng/ℓ	Dissolved ng/ℓ	Particulate ng/ℓ			
1	$\bar{x}$	23	15.67	7.45	15.2	0.490	31400
	S_d	12.6	12.3	2.36	3.96		
	S_e	3.80	3.72	0.712	1.06		
	N	11	11	11	14		
2	$\bar{x}$	13.4	8.09	5.31	9.68	0.549	67800
	S_d	7.81	6.51	3.79	5.73		
	S_e	1.15	9.959	0.558	0.906		
	N	46	46	46	40		
3	$\bar{x}$	12.7	8.13	4.52	12.2	0.370	45600
	S_d	13.2	11.3	2.48	5.72		
	S_e	3.31	2.83	0.621	1.65		
	N	16	16	16	12		
4	$\bar{x}$	7.66	5.95	1.71	3.03	0.564	94800
	S_d	3.87	3.79	1.20	1.94		
	S_e	0.745	0.730	0.231	0.389		
	N	27	27	27	25		
5	$\bar{x}$	6.87	5.83	1.04	2.65	0.392	67300
	S_d	7.07	7.23	0.515	2.57		
	S_e	1.67	1.70	0.121	0.624		
	N	18	18	18	17		

Table III. Average Aroclor 1260 and Suspended Solids Concentrations in Saginaw Bay, 1979

Segment		Aroclor 1260			Sus. Sol. mg/ℓ	Mass PCB on solids ng/mg	Distribution Coefficient ℓ/kg
		Total ng/ℓ	Dissolved ng/ℓ	Particulate ng/ℓ			
1	$\bar{x}$	20.1	11.4	8.70	15.2	0.572	50200
	S_d	8.59	10.4	4.12	3.96		
	S_e	2.59	3.14	1.24	1.06		
	N	11	11	11	14		
2	$\bar{x}$	13.0	6.68	6.34	9.68	0.655	98000
	S_d	11.5	5.49	4.33	5.73		
	S_e	1.69	0.81	0.64	0.906		
	N	46	46	46	40		
3	$\bar{x}$	12.8	7.45	5.39	12.2	0.442	59300
	S_d	11.5	10.5	2.3	5.72		
	S_e	2.88	2.64	0.57	1.65		
	N	16	16	16	12		
4	$\bar{x}$	10.4	8.13	2.27	3.03	0.749	92100
	S_d	7.54	7.62	1.29	1.94		
	S_e	1.45	1.47	0.249	1.389		
	N	27	27	27	25		
5	$\bar{x}$	9.36	7.83	1.53	2.65	0.577	73700
	S_d	8.26	8.58	0.896	2.57		
	S_e	1.95	2.02	0.211	0.624		
	N	18	18	18	17		

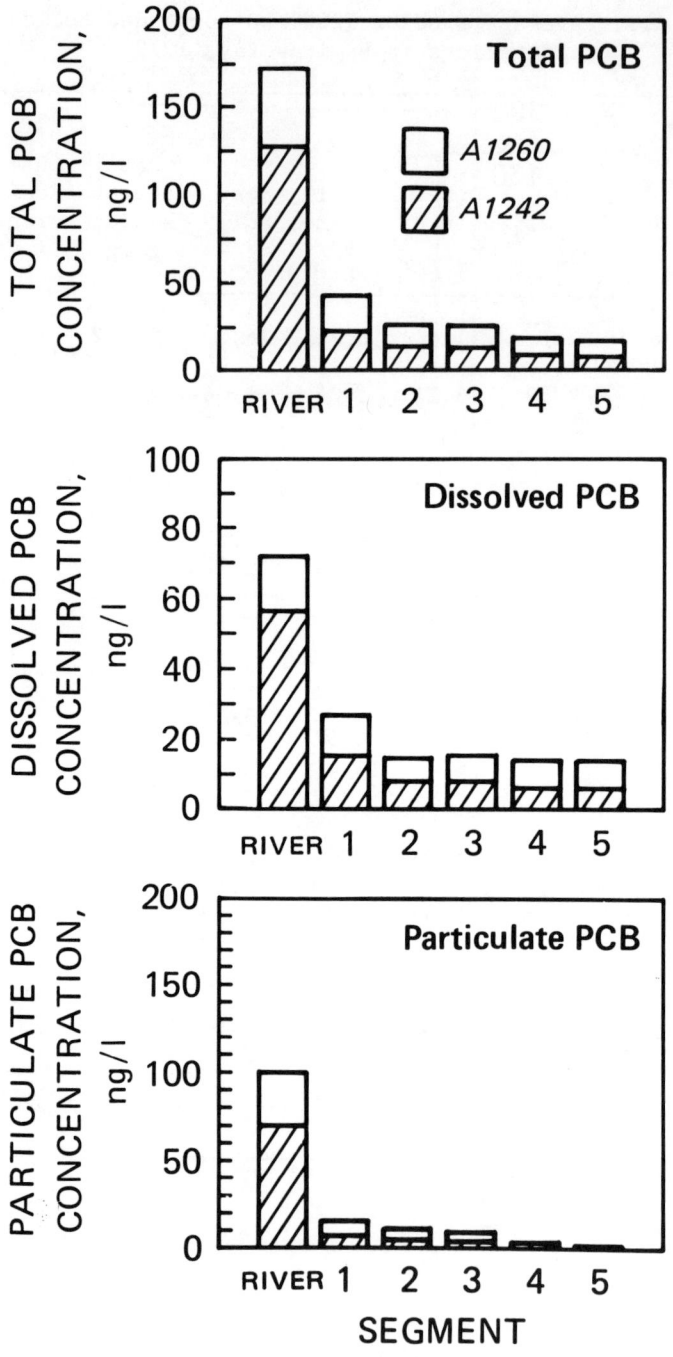

Figure 2. Average of total PCB concentrations by particulate and dissolved fractions, Saginaw River and 5 segments of Saginaw Bay, 1979.

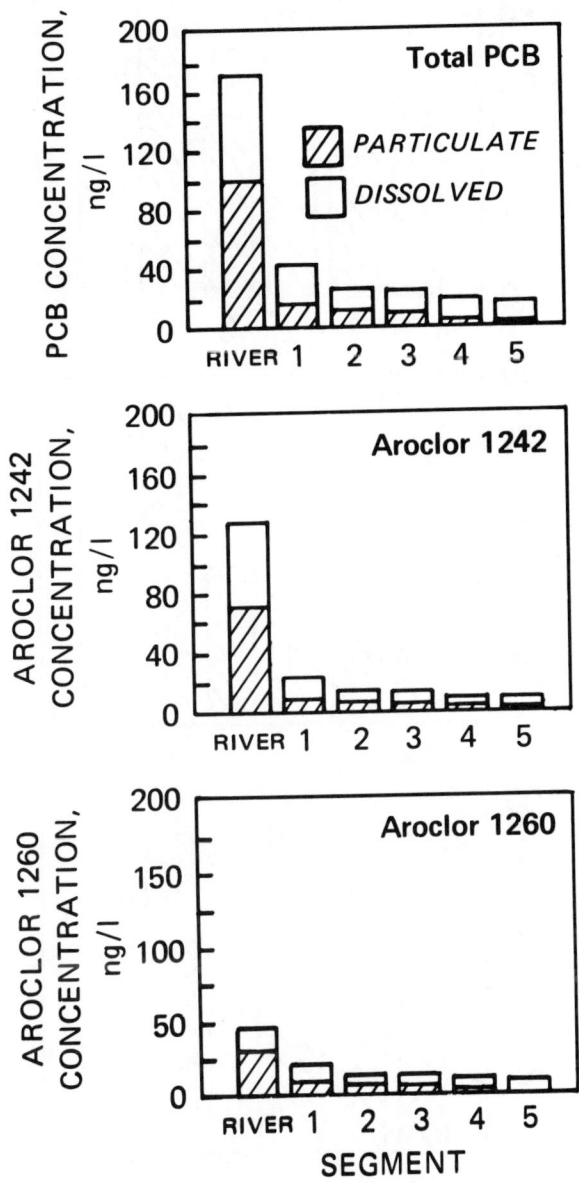

Figure 3. Average Aroclor 1242 and Aroclor 1260 concentrations in Saginaw River and 5 segments of Saginaw Bay.

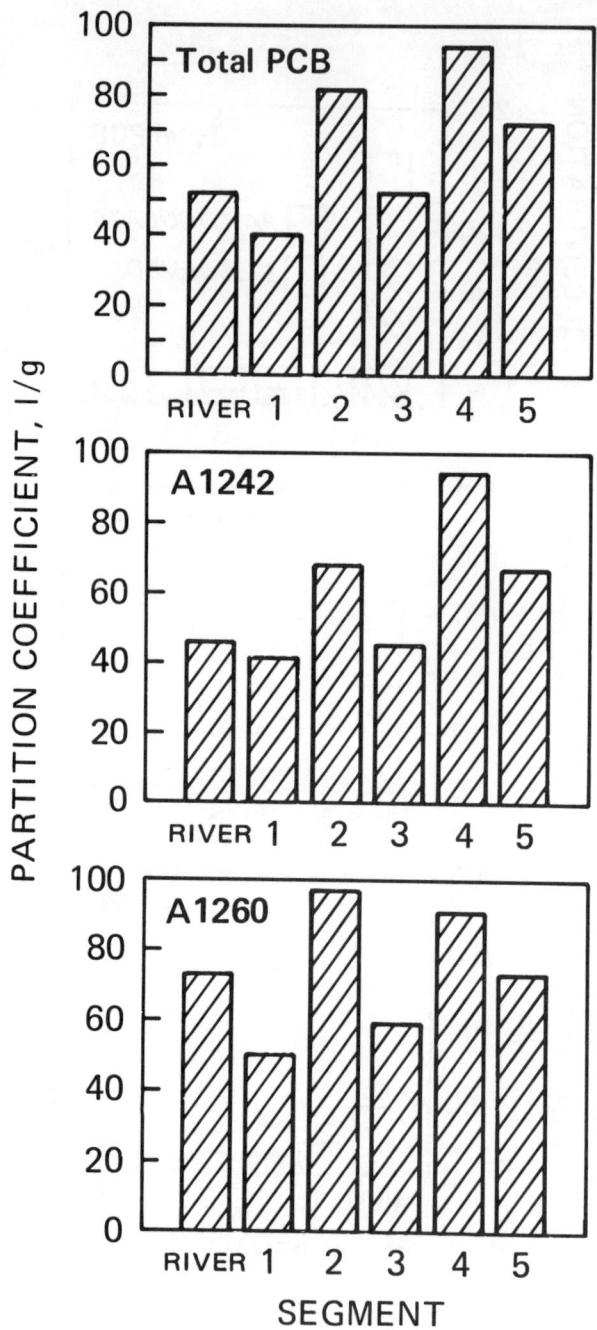

Figure 4. Distribution (partition) coefficients for total PCB, Aroclor 1242 and Aroclor 1260 in Saginaw Bay, 1979.

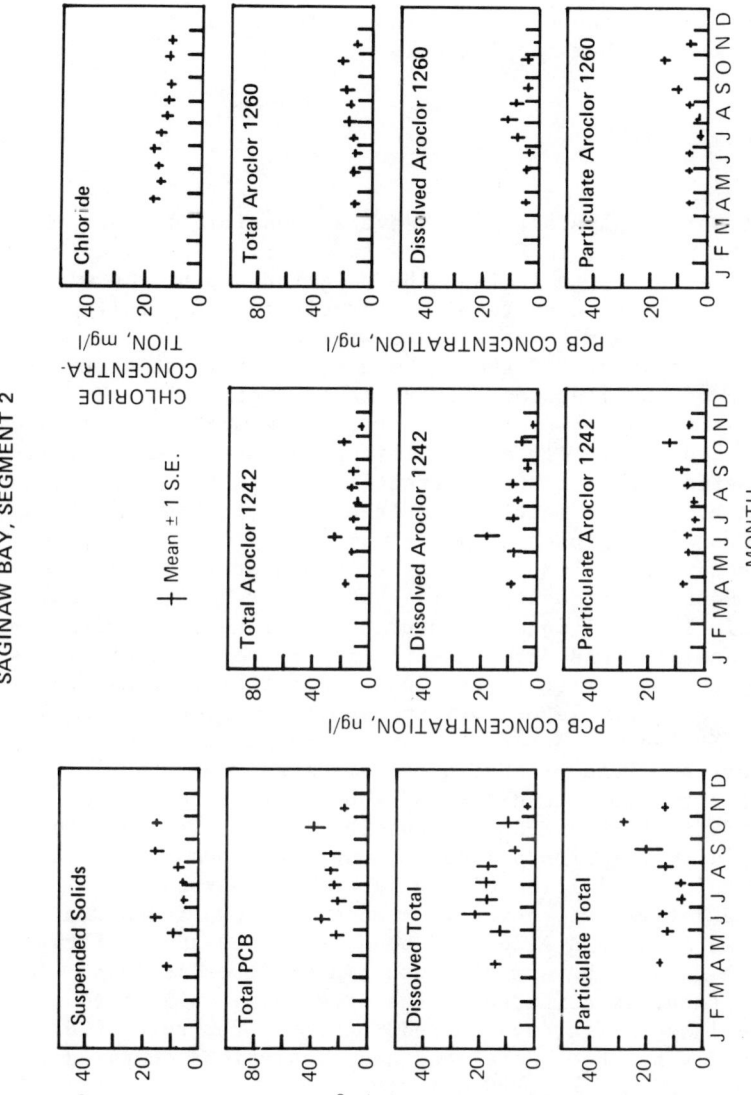

Figure 5. Cruise average PCB, suspended solids and chloride concentrations in Saginaw Bay, segment 2, 1979.

dissolved fraction is higher during the summer. This could be due to just the higher inputs in the spring and fall, or seasonality in some of the kinetic processes, or a combination of these.

Many inferences about cause and effect may be made by inspecting the data, however, because of the complex nature of PCB transport and kinetics it is difficult to discern the relative importance of each without the aid of a deterministic, dynamic mathematical model.

MODEL DEVELOPMENT

The data analysis reveals several possible factors that should be included in a modeling framework. For example, it appears necessary to include the relationship of PCB with suspended solids which requires formulation of settling-resuspension dynamics related to wind and waves. The data also indicate that PCB should be modeled at least at the resolution of PCB mixtures rather than as total PCB. In addition, particulate and dissolved fractions must be calculated and tracked separately as specific processes like volatilization affect the dissolved fraction and settling and resuspension affect the particulate fraction. It is well known that PCBs are mixtures of large numbers of compounds [6], and further that TPCB and particular mixtures do not have a "true" Henrys law constants (H) but rather have "apparent" H's. Similarly, TPCB and mixtures do not have true distribution coefficients, bioconcentration factors, solubilities, boiling points, or other physical properties that can be determined in the laboratory and applied directly in a model. It becomes apparent that to accurately predict the transport and fate of PCB, single isomers would have to be modeled and summed to determine longevity of total PCB. Since 209 isomers of PCB are possible [7], this would be an overwhelming task for a modeler, and would push the chemical analytical capabilities of any laboratory beyond its limit.

As a refinement, however, a model for two industrial mixtures, A1242 and A1260 has been developed and calibrated to field data. The sum of these two mixtures is TPCB. This derived total is compared to the originally calibrated total PCB model and differences noted. A model for total PCB in Saginaw Bay has been developed and simulations compared to 1977 data [8] and to 1979 data [9].

The mass balance framework shown in Figure 6 is similar to that developed by Thomann [10] with the addition that PCB is represented by the two, noncoupled compartments for A1242 and A1260. The framework includes:

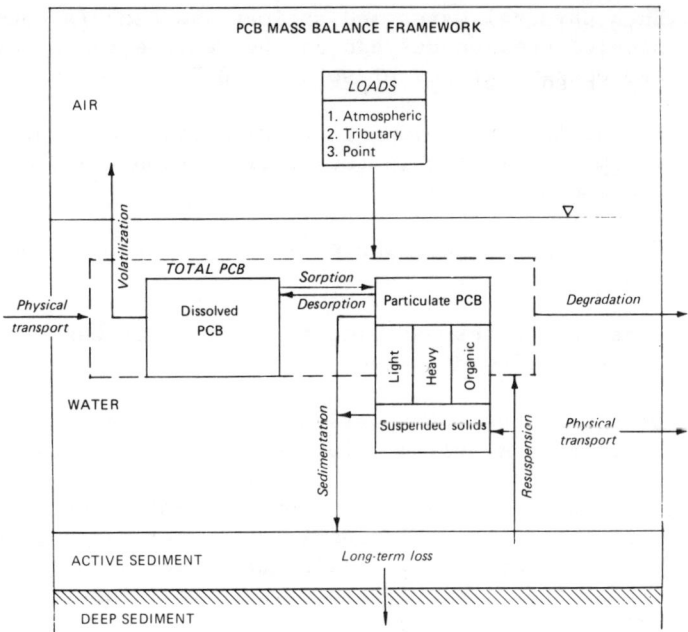

Figure 6. PCB mass balance framework.

1. Input of suspended solids and Aroclor mixtures from tributaries and the atmosphere. Solids also originate from shoreline erosion and primary production.

2. Transport of solids, PCB adsorbed to solids and dissolved PCBs within the water column and to and from Lake Huron (boundary conditions specified at Lake Huron).

3. Settling and resuspension of solids and associated particulate PCBs to and from active sediments as a function of wind, depth of water, and sediment type.

4. Long-term burial of PCB to deep sediments.

5. Volatilization of dissolved PCBs.

The underlying assumptions of the model include:

1. Adsorption-desorption of PCBs associated with solids occurs at a rapid rate relative to other processes in the system, i.e., transport and settling-resuspension so a local equilibrium assumption can be applied [10]. This allows the calculation of the particulate and dissolved fractions from total concentrations using a partition coefficient.

2. Each physical water and sediment segment is considered homogeneous and can be represented by average concentrations of solids and PCBs.

3. The horizontal movement of the active sediment layer is small relative to the water transport and can be assumed to be zero.

4. No biodegradation, photolysis, or hydrolysis of PCBs.

The modeling process applied to Saginaw Bay has progressed in the following sequence:

1. Determine the transport characateristics of the bay in 1979 by modeling chloride as a tracer substance.

2. Model solids relating bottom sediment to water concentration through an empirically derived and calibrated formulation. Solids are modeled as three types--heavy, light, and organic (originating from primary production).

3. Model the two mixtures of PCB coupled to suspended solids. Calibrate the model by adjusting the distribution coefficients until a reasonable comparison is obtained between calculated and measured total, dissolved, and particulate PCB.

4. Simulate fate and longevity of PCB by simulating solids and PCB mass balance over an extended period into the future (ie., about fifty years).

External Material Loadings

The key driving force is the material inputs for the state variables, chloride, solids, and the PCB mixtures. The annual loading for 1979 are summarized in Table IV. Saginaw River loadings were estimated by the product of flow and concentration. The phytoplankton contribution to the solids load was obtained by estimates of the gross phytoplankton biomass production in the bay. Atmospheric loadings of PCB were provided by Murphy [11]. For a more detailed explanation of loading estimates, see Richardson et al. [9]. The loadings are input to the model as a function of time with primary loading occurring in the spring with high runoff. The estimated monthly average PCB loadings from Saginaw River are shown in Figure 7.

Table IV. Average 1979 Loadings to Saginaw Bay (Kilograms/day) for Chloride, Suspended Solids and PCB

CHLORIDE		520000	
SUSPENDED SOLIDS			
SAGINAW RIVER		351500	
OTHER TRIBUTARIES		89700	
SHORELINE EROSION		129000	
ATMOSPHERIC		37800	
PHYTOPLANKTON		473300	
TOTAL		1081300	
PCB	TOTAL PCB	AROCLOR 1242	AROCLOR 1260
ATMOSPHERIC [11]	0.620	0.258	0.361
TRIBUTARY	0.758	0.494	0.264
TOTAL	1.378	0.752	0.626

The model was implemented using a computer program, "Water Analysis Simulation Program (WASP)" [12], and applied to Saginaw Bay. The bay was subdivided into 19 segments (5 water and 14 sediment) as shown in Figure 8.

Physical Transport

Chloride was used as a transport tracer using the mass balance equation

$$V_k \frac{dC_k}{dt} = W_{ck} + \sum_j Q_{jk}C_j - \sum_j Q_{kj}C_k + \sum_j E'_{jk} (C_j - C_k) \qquad (2)$$

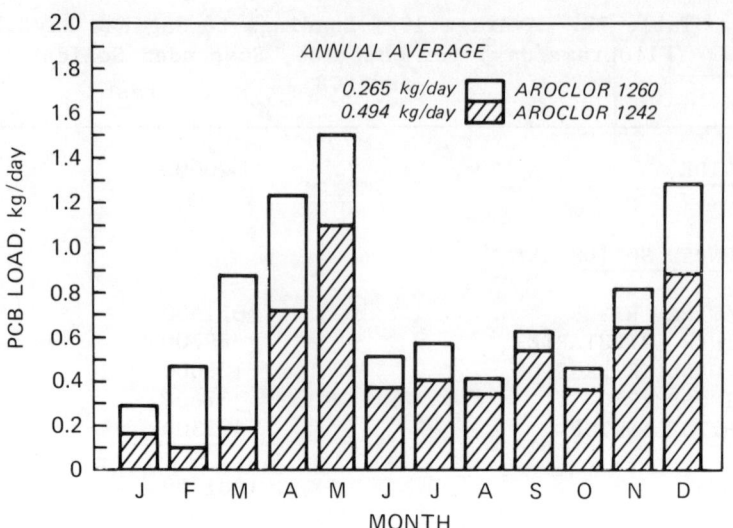

Figure 7. Monthly average PCB loads by mixture from Saginaw River, 1979.

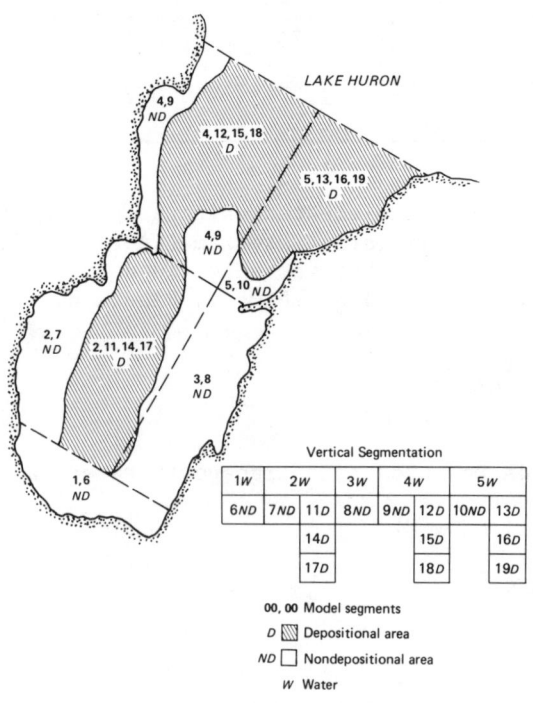

Figure 8. Segmentation for Saginaw Bay solids/PCB model.

```
      | j |
——————+———+——————
  j   | k |  j
——————+———+——————
      | j |
```

k = segment of interest

j = adjacent, interacting segments

where: Units

    k   = water segment considered
    j   = adjacent water segments
    V_k = volume of segments (L^3)
    C_k = concentration of chloride $(M\ L^{-3})$
    t   = time   (T)
    W_{ck} = total input load of substance,
            chloride $(M\ T^{-1})$
    Q   = advective transport coefficient $(L^3\ T^{-1})$
    E'_{jk} = dispersive transport coefficient $(L^3\ T^{-1})$

The water transport parameters Q and E' (Equation 2) were obtained through an iterative process described by Richardson [13] whereby chloride is used as a tracer substance and values of E' and Q were selected so that the calculated concentration "fits" those measured. The final calibration of computed versus measured concentrations is shown in Figure 9.

Solids

The solids submodel includes three categories of solids: 1) light; 2) heavy; 3) organic, following the formulation of Thomann (10). Light solids include inorganic, clay fractions and detrital materials of terrestrial origin; heavy solids include non-cohesive particles, particularly sand originating from runoff and shoreline erosion; and organic solids are materials of aquatic origin including gross phytoplankton and zooplankton production. For a more detailed description of the solids model and associated data, see Richardson [9]. A summary of the solids-related model coefficients is given in Table V.

For solids the mass balance equation is

$$V_k \frac{dS_{ki}}{dt} = W_{ski} + \sum_j Q_{jk} S_{ji} - \sum_j Q_{kj} S_{ki}$$

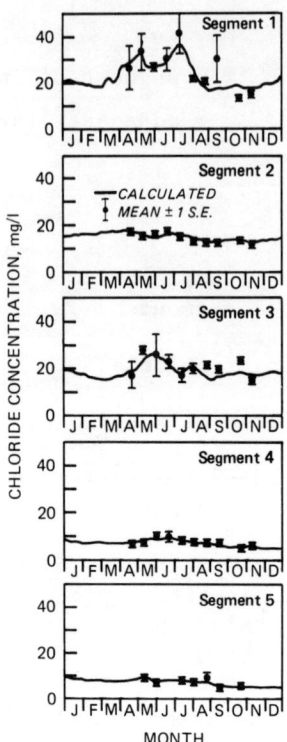

Figure 9. Calibrated chloride concentrations in Saginaw Bay, 1979.

$$+ \sum_j E'_{kj} (S_{ji} - S_{ki}) \pm \frac{S_{vi} V_k S_{ki}}{D_k} \pm R_{ski} \qquad (3)$$

where: Units

S_{ki} = concentration of solids type i   ($M\ L^{-3}$)
D_k = average depth of segment k   (L)
S_{vi} = gross turbulent settling velocity   ($L\ T^{-1}$)
   for solids type i

$$R_{ski} = \frac{R_{vi} S_{ki} V_{sk}}{D_{sk}} \times W_f \times WRS_k \qquad (4)$$

R_{ski} = mass resuspended for solids type i   ($M\ T^{-1}$)
R_{vi} = resuspension velocity for solids   ($L\ T^{-1}$)
   type i

Table V. Solids Submodel Coefficients and Variables

Description	Value	Units
Load for "light" solids (average annual)	469,000	kg-day^{-1}
Load for "heavy" solids (average annual)	122,000	kg-day^{-1}
Load for "organic" solids (average annual)	474,260	kg-day^{-1}
Settling velocity of "light" solids in water column	0.2	m-day^{-1}
Settling velocity of "light" solids in sediment layers	0.24×10^{-4}	m-day^{-1}
Settling velocity of "heavy" solids in water column	1.5	m-day^{-1}
Settling velocity of "heavy" solids in sediment layers	0.88×10^{-5}	m-day^{-1}
Settling velocity of "organic" solids in water column	0.1	m-day^{-1}
Settling velocity of "organic" solids in sediment layers	0.82×10^{-6}	m-day^{-1}
Resuspension velocity for "light" solids from active sediment	0.005	m-day^{-1}
Resuspension velocity for "heavy" solids from active sediment	0.000001	m-day^{-1}
Resuspension velocity for "organic" solids from active sediment	0.008	m-day^{-1}
Julian day on which ice cover disappears	74	day
Julian day on which ice cover appears appears	345	day
Wind speed below which no resuspension occurs	13	knots
Constant in resuspension/wind, Equation (5)	0.02	-
Resuspension suppression factor, Seg. 6	3	-
Resuspension suppression factor, Seg. 7	1	-
Resuspension suppression factor, Seg. 8	3	-
Resuspension suppression factor, Seg. 9	.5	-
Resuspension suppression factor, Seg. 10	.6	-
Resuspension suppression factor, Seg. 11	.35	-
Resuspension suppression factor, Seg. 12	.005	-
Resuspension suppression factor, Seg. 13	.005	-

[1]Loads input on a daily basis.
[2]No resuspension during ice cover.

D_k = depth active sediment layer (L)
WRS_k = resuspension, morphometric suppression factor

$$W_f = A V_w$$

$$W_f = 0 \text{ if } V_w \leq W_{min} \tag{5}$$

W_f = wind function
W_{min} = minimum wind velocity (L/T)
V_w = wind velocity (L/T)
A = constant

Total suspended solids is computed by the sum of the three sizes.

The particular empirical formulation for the solids resuspension was derived after reviewing research by Sheng [14] and Onishi [15]. These investigators have formulated detailed hydrodynamic models for sediment transport that relate wind speed and morphometry to shear stress at the sediment/water interface and subsequent resuspension. Time and space scales for these formulations are relatively small--seconds and meters--and are difficult to apply to large-scale situations such as PCB transport in Saginaw Bay. In addition, the detailed models usually consider the sediment as a boundary layer providing an infinite source of material. To answer the longevity question, a model must track mass of the sediment and sorbed PCB and simulate responses over time scales of decades.

The empirically derived formulation considers the physical factors described in the hydrodynamic models and attempts to simplify them so long-term calculations can be made. The model coefficients are determined by calibration to the suspended solids data both in the water column and sediment. The primary constraint in the long-term calculation is to maintain steady-state solids concentration in the sediment layers for each of the sediment types. The final water column calibration is given in Figure 10. Sediment solids concentrations over a 50 year period are maintained within 10% of the initial solids concentrations using 1979 wind conditions.

PCB

Two kinetic processes are assumed to predominate in the fate of PCB, adsorption-desorption and volatilization.

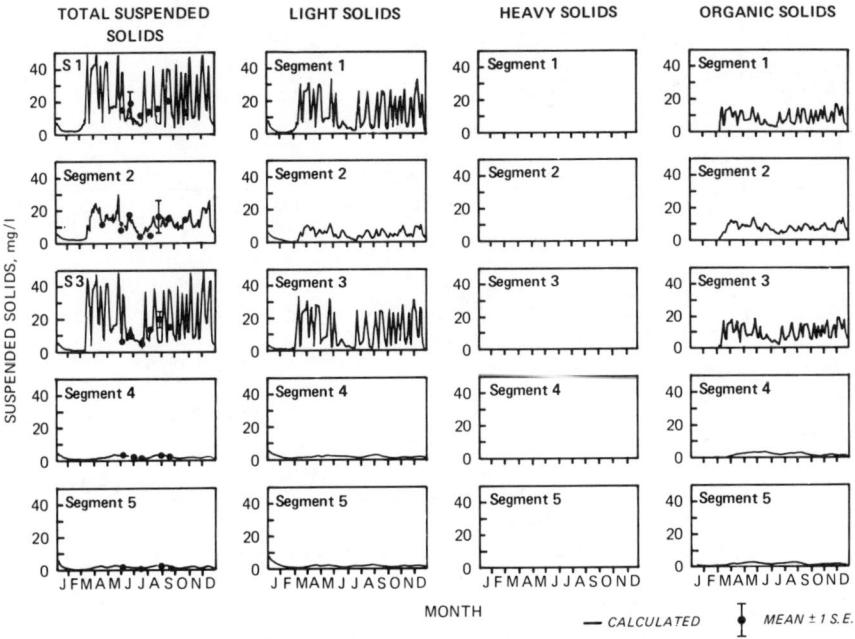

Figure 10. Calibrated suspended solids in Saginaw Bay, 1979.

Other processes affecting fate are physical transport and settling/resuspension. These processes act simultaneosly on particulate and/or dissolved fractions of each of the PCB mixtures to determine the fate of the PCB loads, attenuation in the sediments, and longevity of PCB in the system. A record of historical loadings is retained in the sediment. The sediment attenuates PCBs if they subsequently become available to the water column. If sedimentation rates are sufficient some of the PCB may become buried and permanently lost.

The PCB mass balance equation which follows the formulation of Schnoor [16] for the water column was modified by Thomann [10] and Richardson [9] to track PCB in the sediment. Further refinements to the formulation is to consider total PCBs as the sum of two mixtures, Aroclor 1242 and Aroclor 1260, which are modeled separately and added to represent total PCB.

For PCB the mass balance equation is

$$V_k \frac{dC_{Tmk}}{dt} = W_{Tmk} \pm \sum_j T_{mjk} \qquad (6)$$

$$- \sum_i \left[\frac{S_{vi} K_{pim} S_{ki}}{(1 + \sum_i K_{pim} S_{ki})} C_{Tmk} \frac{V_k}{D_k}\right] - \sum_i K_{Lm} \frac{C_{Tmk} V_k}{(1 + \sum_i K_{pim} S_{ki}) D_k}$$

$$- \sum_i K_{\ell m} \frac{V_k}{D_k} \frac{(C_{Tmk} K_{pim} S_{ki})}{(1 + \sum_i K_{pim} S_{ki})} + R_{kp}$$

where

C_{Tmk} = total Aroclor mixture, (m) concentration ($M\ L^{-3}$)
W_{Tmk} = total input load from all sources ($M\ T^{-1}$)
T_{mjk} = net physical transport ($M\ T^{-1}$)
K_{Lm} = volatilization rate ($L\ T^{-1}$)
$K_{\ell m}$ = average loss rate for all other loss processes ($L\ T^{-1}$)

$$R_{kp} = \sum_i \frac{(R_{vi} K_{pim} S_{ki})}{(1 + K_{pim} S_{ki})} C_{Tmk} \frac{V_k}{D_k} \qquad (7)$$

$$C_{dmk} = \frac{C_{Tmk}}{(1 + \sum_i K_{pim} S_{ki})} \qquad (8)$$

$$C_{pmk} = \frac{(C_{Tmk} \sum_i K_{pmi} S_{ki})}{(1 + \sum_i K_{pmi} S_{ki})} \qquad (9)$$

C_{dmk} = dissolved concentration for PCB mixture m ($M\ L^{-3}$)
C_p = particulate concentration for PCB mixture m ($M\ L^{-3}$)

$$C_T = \sum_m C_{Tm} \qquad (10)$$

C_T = total PCB (sum of all mixtures)

Great Lakes models have not yet included volatilization even though reseach by MacKay [17] and Murphy [11] have indicated its potential as a primary loss mechanism. The effective volatilization rates for A1242 and A1260 can be calculated by a method described by O'Connor [18]:

$$\frac{1}{K_L} = \frac{1}{K_\ell} + \frac{1}{K_g H} \quad (11)$$

K_L = volatilization rate (L T^{-1})
K_ℓ = liquid transfer coefficient (L T^{-1})
K_g = gas transfer coefficient (L T^{-1})
H = Henry's law constant (dimensionless)

Henry's law constant is calculated by

$$H = \frac{p}{C_s} \frac{M}{T} \times 16 \quad (12)$$

p = vapor pressure (mmHg)
C_s = saturation concentration (mg-ℓ^{-1})
M = molecular weight (g/mol)
T = absolute temperature (K)

further,

$$K_g = \frac{D_g}{\nu_g} C_D W \quad (13)$$

$$K_\ell = \frac{D_\ell}{\nu_\ell} C_D W \quad (14)$$

D_g = diffusivity of PCB in air (L^2 T^{-1})
D_ℓ = diffusivity of PCB in water (L^2 T^{-1})
ν_g = kinematic viscosity of air (L^2 T^{-1})
ν_ℓ = kinematic viscosity of water (L^2 T^{-1})
C_D = drag coefficient (dimensionless)
W = wind speed (L T^{-1})

From these relationships, theoretical K_L's of 0.21 m/day meters per day and 0.17 m/day were calculated for Aroclor 1242 and Aroclor 1260, respectively. (See Table VI for values of parameters used to develop K_L). These compare

Table VI. Volatilization Parameter Values

Parameter			Aroclor 1242	Aroclor 1260
D_g	Diffusivity of PCB in air	cm^2/sec	0.04652	0.03673
D_ℓ	Diffusivity of PCB in water	cm^2/sec	5.387×10^{-6}	4.253×10^{-6}
ν_g	Kinmatic Viscosity of air	cm^2/sec	0.15	0.15
ν_ℓ	Kinmatic Viscosity of water	cm^2/sec	0.01	0.01
C_D	Drag Coefficient	–	0.001	0.001
W	Wind Speed	M/sec	5.0	5.0
p	Vapor Pressure	mm Hg	4.06×10^{-4}	4.06×10^{-5}
C_s	Saturation Concentration	mg/ℓ	0.35	0.027
M	Molecular Weight	g/mol	261	372
T	Temperature	°K	289	289
H	Henry's Law Constant (calculated Eq. 12)	–	0.01676	0.0309

to 1.37 and 1.6 m/day estimated by MacKay [17]. As will be described below, these K_L's were used as initial estimates for volatilization but were changed during the calibrations.

RESULTS OF PCB MODELING

Model results for four basic simulations are compared in Figure 11, where the average field data are compared to the average calculated concentrations over the same period of time (April-November). Simulation A represents the total PCB model; B, the total PCB model with volatilization; C,

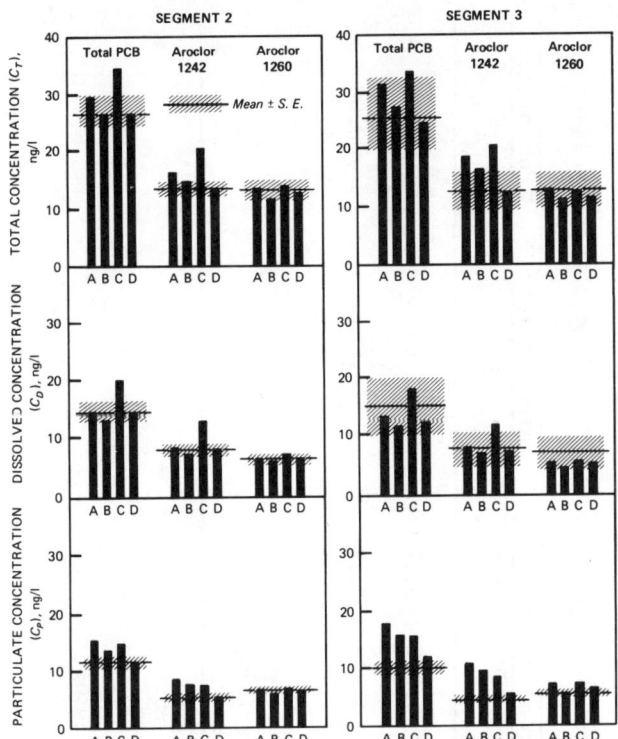

Figure 11. Comparison of average model simulations with average field data: A) Total model, no volatilization; B) Total model with volatilization; C) Two mixture model, no volatilization; D) Two mixture model with volatilization.

the two mixture model with different K_p's; and D, the two mixture model with volatilization. The K_p's and K_L's used for each simulation are summarized in Table VII.

Simulation A (Figure 11) represents the best fit possible of the total model after adjusting K_p. K_p's are constant for all water and sediment segments and for both mixtures KPH = .002 ℓ/mg, KPL = .050 ℓ/mg and KPP = .1 ℓ/mg. These values correspond approximately to those reported by Hiraizumi et al. [19]. The total PCB simulation is somewhat high for both fractions. Note that the problem appears to be with the particulate fraction of Aroclor 1242. Similarly, in segment 3 the particulate fraction for Aroclor 1242 is well above the data. Values of Aroclor 1260 in both segments 2 and 3 are within one standard error of the mean.

The time variable results for Simulation A are shown for segments 2 and 3 in Figure 12. PCBs over the year can be simulated reasonably well, however, it is not possible with the present formulation to simulate some of the temporal de-

Table VII. Calibration Coefficients for PCB Model Simulations

Simulation	Distribution Coefficients (ℓ/mg)						Volatilization Rates (m/day)	
	Aroclor 1242			Aroclor 1260			Aroclor 1242	Aroclor 1260
	KPL	KPH	KPP	KPL	KPH	KPP	K_L	K_L
A	0.05	0.002	0.100	0.05	0.002	0.100	0	0
B	0.05	0.002	0.100	0.05	0.002	0.100	0.08	0.08
C	0.02	0.002	0.06	0.04	0.002	0.10	0	0
D	0.02	0.002	0.06	0.04	0.002	0.10	0.20	0.05

KPL = Distribution coefficient for light solids.
KPH = Distribution coefficient for heavy solids.
KPP = Distribution coefficient for organic solids.

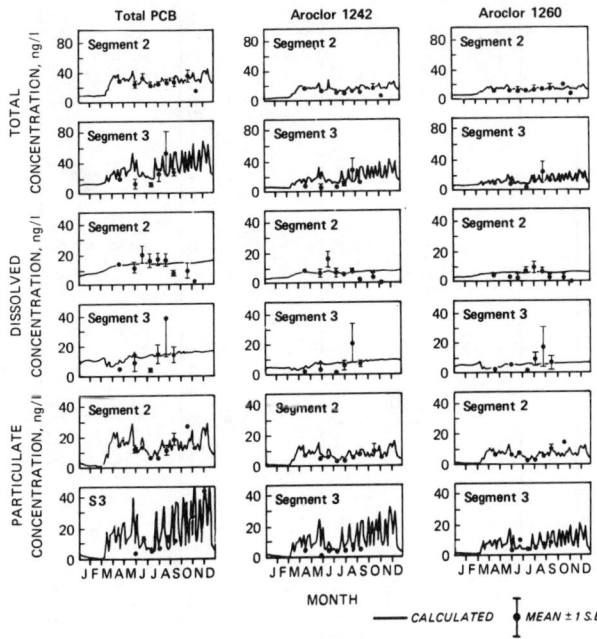

Figure 12. PCB simulation in Saginaw Bay, segment 2 and 3.

tail represented by the cruise average data. A comparison of the average of all cruise data by segment to the simulated concentrations averaged over the same period (April through November) is presented in Table VIII.

The table indicates that total PCB is simulated within 25% of the data for all segments with the best comparison in segments 1 and 2 and the worst in segments 4 and 5. Also, A1260 compares better (21%) than A1242 (38%). Also, for TPCB, A1242, and A1260 the dissolved fraction compares better than the particulate. It is also obvious that segment 2 simulations are generally more accurate than the other segments. Therefore, most of the following discussion focuses on segment 2, although segment 3 results will be shown as representative of a non-depositional segment. Segment 4 and 5 are driven primarily by the boundary concentrations and the simulations are not as accurate. This could be a result of missing loads from the outer bay tributaries, inaccurate boundary concentrations, or other phenomena not represented by this modeling framework. However, segment 4 and 5 simulations are sufficiently accurate to provide good transition concentrations for the inner bay to Lake Huron.

Table VIII. Comparison of Computed PCB Concentrations to Average Field Data in Saginaw Bay, 1979 (Simulation A)

PCB	Segment	Concentrations (ng/ℓ)				
		1	2	3	4	5
Total PCB Total	Data $\bar{x}$	43.1	26.4	25.6	18.1	16.2
	S_e	(6.09)	(1.69)	(6.09)	(1.96)	(3.47)
	Model $\bar{x}$	41.7	29.6	31.1	10.8	8.4
	% Error	3.2	12.1	21.5	40.3	48.0
Dissolved	Data $\bar{x}$	27.0	14.8	15.7	14.1	13.7
	S_e	(6.62)	(1.4)	(5.36)	(1.97)	(3.59)
	Model $\bar{x}$	17.9	14.5	13.5	9.6	7.3
	% Error	33.7	2.0	14.0	31.9	96.7
Particulate	Data $\bar{x}$	16.2	11.6	9.91	3.98	2.57
	S_e	(1.82)	(1.17)	(1.11)	(.47)	(.31)
	Model $\bar{x}$	23.8	15.0	12.2	1.2	1.1
	% Error	46.9	29.3	23.2	69.8	57.0
Aroclor 1242 Total	Data $\bar{x}$	23	13.4	12.7	7.7	6.87
	S_e	(3.8)	(1.15)	(3.31)	(.75)	(1.67)
	Model $\bar{x}$	26.9	16.4	18.6	3.7	3.3
	% Error	17.0	22.3	46.4	51.9	52.0

Dissolved	Data $\bar{x}$	15.6	8.09	8.13	5.95	5.83
	Se	(3.72)	(.96)	(2.83)	(.73)	(1.70)
	Model $\bar{x}$	11.2	8.1	8.1	3.3	2.9
	% Error	28.2	0.12	0.37	44.5	50.3
Particulate	Data $\bar{x}$	7.45	5.31	4.52	1.71	1.04
	Se	(.712)	(.56)	(.62)	(.23)	(.121)
	Model $\bar{x}$	14.8	8.4	10.5	0.4	0.4
	% Error	98.7	58.1	132.0	76.6	61.5
Aroclor 1260 Total	Data $\bar{x}$	20.1	13.02	12.8	10.4	9.36
	Se		(1.69)	(2.88)	(1.45)	
	Model $\bar{x}$	15.7	13.2	12.5	7.0	5.0
	% Error	21.9	1.38	2.34	32.7	46.6
Dissolved	Data $\bar{x}$	11.4	6.68	7.45	8.13	7.83
	Se		(.81)	(2.64)	(1.47)	
	Model $\bar{x}$	6.7	6.5	5.4	6.2	4.4
	% Error	41.2	2.69	27.5	23.7	43.8
Particulate	Data $\bar{x}$	8.7	6.34	5.39	2.27	1.53
	Se		(.64)	(.57)	(.25)	
	Model $\bar{x}$	9.0	7.7	7.1	0.8	0.6
	% Error	3.44	5.67	31.7	64.8	60.8

The next simulation (B), shown in Figure 11, resulted from inclusion of a volatilization rate which was calibrated until a reasonable fit was obtained with a rate of 0.08 m/day. This results in a better comparison but with particulate A1242 remaining above the data for both segments.

Simulation C represents an attempt to reduce the particulate fraction for A1242 by adjusting the K_p's (See Table VII). The difference between this simulation and simulation A is a slight improvement in the particulate fraction, especially for A1242. The difficulty is that the dissolved fraction is very sensitive in this K_p range and increases dramatically as shown in Figure 13. This results in a slightly improved particulate simulation, but with a concurrent increase in the dissolved fraction, far beyond the data, in segment 2.

It becomes clear that to calibrate A1242 the correct combination of distribution coefficients and loss rate would have to be found. To accomplish this a series of sensitivity calculations were made over a range of K_p's and K_L's. The results are shown in Figure 13, where calculated average TPCB, A1242, and A1260 for segment 2 are plotted as a function of K_L and K_p. A separate calibration for each mixture is obtained by that K_p curve which intersects the data (represented by the horizontal, dashed line) for both dissolved and particulate concentrations, at the same K_L. For A1260, that correct curve is near the curve for the K_p scale factor of 1.0 for a K_L at or near zero. For A1242 the correct combination lies between the the curves for K_p scale factors of 0.5 and 1.0, or a K_L of about 0.2 m/day and a K_p scale factor of about 0.6 (K_p's for a scale factor of 1 are those in Table VII, for Simulation A). So K_p for A1242 is about 0.6 that of A1260. This final calibration with reduced K_p's for A1242 and K_L's of 0.2 and 0.05 m/day for A1242 and A1260 are indicated by Simulation D in Figure 11. This final calibration results in an average error in segment 2 of 2.4% for TPCB, A1242, and A1260 for total, dissolved, and particulate compared to 14.9% for Simulation A (Table IX).

The mass balance for this one year simulation (D) is presented for the inner bay (segments 1, 2 and 3) in Figure 14. The magnitudes of the mass transfers provide insight into the relative importance of each process. Volatilization represents the primary loss process (787 kg/yr). A mass of 312 kg/yr is transported to the outer bay (Lake Huron). Because only 398 kg are input, the difference results in a loss from the active sediment layer (717 kg/yr) and the water column (12.2 kg/yr).

If the simulation is carried out over a long period, the concentrations in the water and sediment eventually reach a new equilibrium in about 15 years. This is shown by Simula-

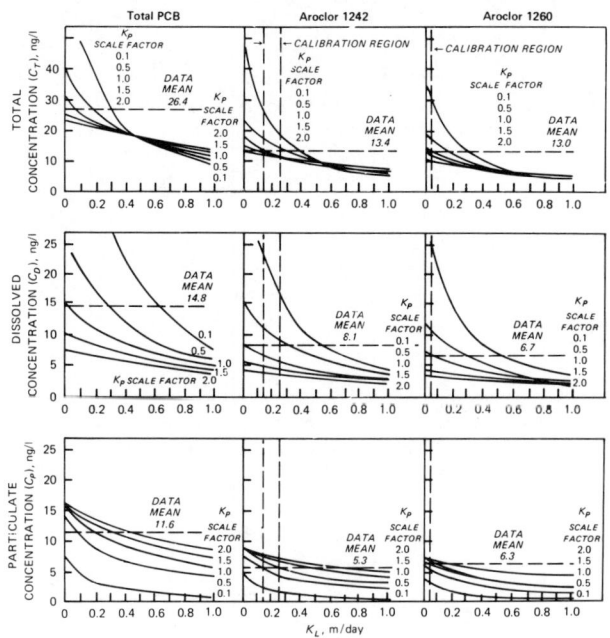

Figure 13. Sensitivity of average calculated PCB in segment 2 to volatilization rate (K_L m/day) and distribution coefficient (K_p). (See Table VII, Simulation A, for K_p values at scale factor = 1).

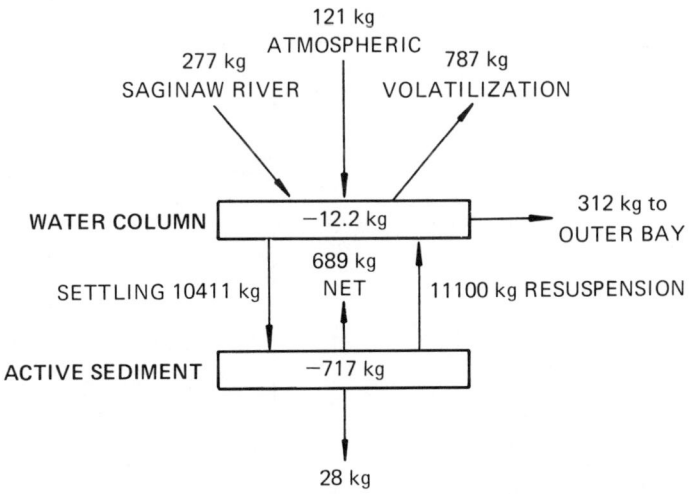

Figure 14. Mass balance for total PCB (kilograms) in inner Saginaw Bay, segments 1-3, 1979, for Simulation D.

Table IX. Comparison of Computed PCB Concentrations to Average Field Data in Saginaw Bay, 1979 (Simulation D)

PCB	Segment	Concentrations (ng/ℓ)				
		1	2	3	4	5
Total PCB Total	Data $\bar{x}$	43.1	26.4	25.6	18.1	16.2
	S_e	(6.09)	(1.69)	(6.09)	(1.96)	(3.47)
	Model $\bar{x}$	45.2	26.2	24.3	12.4	10.3
	% Error	4.9	1.5	5.1	31.5	36.4
Dissolved	Data $\bar{x}$	27.0	14.8	15.7	14.1	13.7
	S_e	(6.62)	(1.4)	(5.36)	(1.97)	(3.59)
	Model $\bar{x}$	22.5	14.8	12.6	10.5	8.9
	% Error	16.7	.7	19.7	25.5	35.0
Particulate	Data $\bar{x}$	16.2	11.6	9.91	3.98	2.57
	S_e	(1.82)	(1.17)	(1.11)	(.47)	(.31)
	Model $\bar{x}$	22.7	11.3	11.8	1.9	1.5
	% Error	40.1	2.6	19.1	52.3	41.6
Aroclor 1242 Total	Data $\bar{x}$	23.0	13.4	12.7	7.7	6.87
	S_e	(3.8)	(1.15)	(3.31)	(.75)	(1.67)
	Model $\bar{x}$	25.9	13.4	12.6	4.4	4.3
	% Error	12.6	0	.8	42.9	37.4

Dissolved	Data x̄	15.6	8.09	8.13	5.95	5.83
	S_e	(3.72)	(.96)	(2.83)	(.73)	(1.70)
	Model x̄	14.3	8.4	7.3	4.0	3.8
	% Error	9.1	3.8	10.2	32.8	34.8
Particulate	Data x̄	7.45	5.31	4.52	1.71	1.04
	S_e	(.71)	(.56)	(.62)	(.23)	(.121)
	Model x̄	11.6	5.0	5.3	.4	.43
	% Error	55.7	5.8	17.3	76.6	58.7
Aroclor 1260 Total	Data x̄	20.1	12.8	12.8	10.4	9.36
	S_e		(1.69)	(2.88)	(1.45)	
	Model x̄	19.3	12.2	11.7	7.9	6.1
	% Error	4.0	2.3	8.6	24.0	58.7
Dissolved	Data x̄	11.4	6.68	7.45	8.13	7.83
	S_e		(.81)	(2.64)	(1.47)	
	Model x̄	8.2	6.4	5.3	6.5	6.1
	% Error	28.1	4.2	28.9	20.0	34.8
Particulate	Data x̄	8.7	6.34	5.39	2.27	1.53
	S_e		(.64)	(.57)	(.25)	
	Model x̄	11.1	6.4	6.4	1.4	1.0
	% Error	27.6	.6	18.7	38.3	34.6

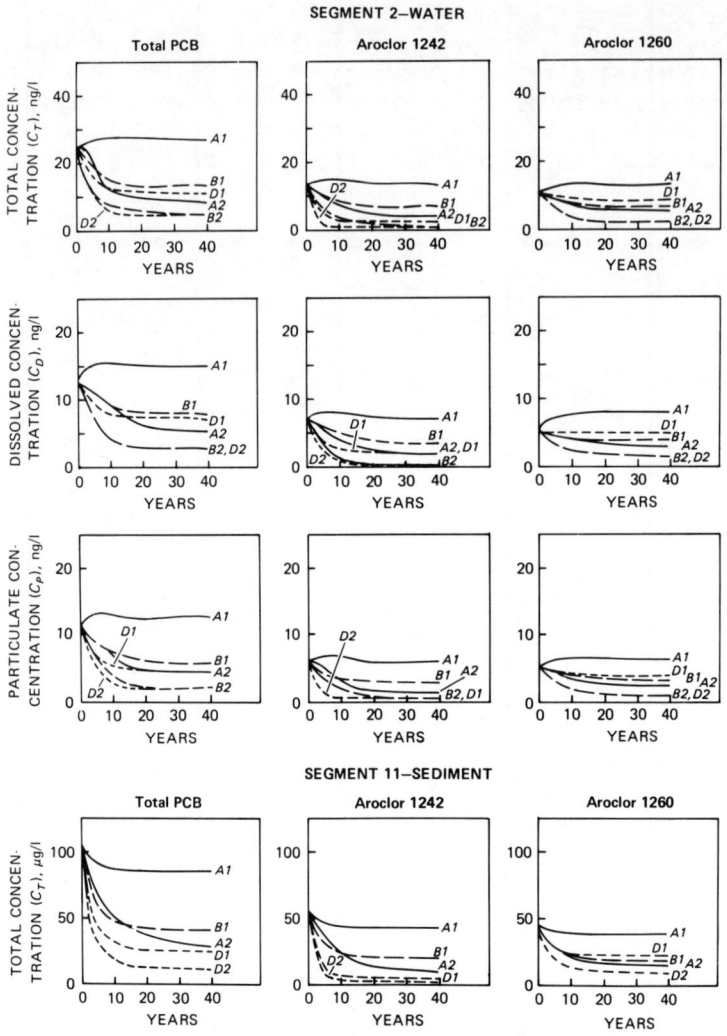

Figure 15. Long term simulation of average PCB concentrations in Saginaw Bay, segment 2. (See text for labels).

tion D-1 in Figure 15. Simulation D-2 represents the projected concentrations in segment 2 if the loads (Saginaw River and atmospheric) were reduced to zero. For comparison Simulations A and B with and without loads (1 and 2, respectively), are also shown in Figure 15. The difference between the total PCB model and the two-mixture model both calibrated with volatilization is indicated by B-1 and D-1. Therefore, it might be concluded that little was gained by the refinement of the model to include two mixtures. However, the two-mixture model has revealed the problem of calibrating A1242 and required the inclusion of a loss mechanism. Thus, the real difference between the modeling results would be represented by Simulation A-1, without volatilization calculated where concentrations actually increase slightly, and B-1, with volatilization.

CONCLUSIONS

PCB data collected in 1979 from Saginaw River and Bay have revealed differences in inputs, transport, and fate of two industrial mixtures, Aroclor 1242 and Aroclor 1260. Although the A1242 composes over 65% of the tributary load, its concentration is about equal to that of A1260 in the bay. This is partly due to the fact that atmospheric loads of A1242 are only 42% of the total but also due to the different partitioning with solids and different volatilization rates as revealed by use of a dynamic, mass balance model. The model included PCB-solids interaction, water-sediment interaction, and volatilization and was used to simulate 1979 PCB and solids concentrations. The best comparison of model simulations to the measured concentrations was made with the distribution coefficients for A1242 of about 60% of those for A1260 and with a volatilization rate for Aroclor 1242 of 0.2 m/day. The volatilization rate for A1260 was determined to be no more than 0.05 m/day. The calibrated volatilization rate for Aroclor 1242 compares well with that calculated from the theory described by O'Connor (0.21 m/day). In contrast, the calibrated volatilization rate for 1260 is much smaller than the theoretical (0.17). The reason for this disparity is not known.

The simulations of PCB also reveal the relative importance of each transport mechanism. In the inner bay for total PCB (the sum of A1242 and A1260), Saginaw River contributes 277 kg/yr whereas 121 kg/year in input from the atmosphere. A net mass of 689 kg is resuspended from the active sediment layer, 787 kg loss from volatilization and 312 kg transported to the outer bay. Twenty-eight kilograms are lost to the deep sediment. This all results in an annual loss of 12.2 kg from the water column and 717 kg from the active sediment during one year (1979).

Long-term simulation indicate that if volatilization is occurring at the rates of 0.2 and 0.05 m/day for Aroclor 1242 and Aroclor 1260, respectively, total PCB in segment 2 will decrease about 50% within about 10 years. Most of this reduction, however, would be for Aroclor 1242 (80%).

Aroclor 1260 would only decrease by about 20%. If all loads were eliminated (Simulation D2, Figure 15), total PCB would decrease by 90% within 10 years. These conclusions depend on the validity of the volatilization process and the rationale for its inclusion. Confirmation should be done by conducting a similar monitoring effort of water and sediments in about five years to obtain verification data.

ACKNOWLEDGEMENTS

The authors have been privileged to work with many talented people at the EPA, Large Lakes Research Station. Modeling research requires an infrastructure of computers, software, data bases, and research budgets which represent the efforts of numerous individuals. We are appreciative of the support of Dr. Norbert Jaworski during a difficult transition period. We acknowledge the assistance and encouragement received from Wayland Swain, Nelson Thomas, Michael Mullin, David Dolan and Victor Bierman. It should also be recognized that much of the modeling approach and framework was first developed by Robert V. Thomann and Dominic DiToro and John Mueller, of Manhattan College. We thank all of these colleagues for their advice, encouragement, and review of the manuscript. The Cranbrook Institute of Science staff under the supervision of John Filkins, collected and analyzed all of the samples. Kevin McGunagle of Computer Science Corporation should be commended for his role in data base development for the PCB data. Finally, we thank Debra Caudill for the excellent typing and editing of the manuscript.

REFERENCES

1. Wilford, W. "Results of Fish Contaminant Monitoring," Great Lakes Fishery Laboratory, Ann Arbor, Michigan, Unpublished Report (1981).

2. Swain, W.R. "What is the Scientific Basis for the Present Concern [for PCBs] in the Great Lakes," International Symposium on PCBs in the Great Lakes, In preparation (1982).

3. Sawyer, L.D. "Quantitation of Polychlorinated Biphenyl Residues by Electron Capture Gas-Liquid Chromatography: Reference Material and Characterization and Preliminary Study," J. Assoc. Off. Analytical Chem, 61:2 (1978).

4. Smith, E.V. "Surveys of PCB in Saginaw Bay," Cranbrook Institute of Science, Bloomfield Hills, MI, In preparation.

5. Mullin, M. Personal Communication, U.S. E.P.A., ERL-D, Large Lakes Research Station, Grosse Ile, MI (February, 1982).

6. Murphy, T.J. Personal Communication, DePaul University (1981).

7. Hutzinger, O., S. Safe and V. Zitko. "The Chemistry of PCB's," CRC Press, Cleveland, Ohio (1974).

8. Richardson, W.L., John C. Filkins and Robert V. Thomann. "Preliminary Analysis of PCB in Saginaw Bay-1977," U.S. E.P.A., Environmental Research Lab-Duluth, Large Lakes Research Station, Grosse Ile, MI, Unpublished Report (1981).

9. Richardson, W.L., R.V. Thomann and E.V. Smith. "Model of PCB in Saginaw Bay, 1979," In preparation.

10. Thomann, R.V. and D.M. DiToro. "Preliminary Model of the Fate of Toxic Substances in the Great Lakes," submitted to Journal of Great Lakes Research, International Association of Great Lakes Research (1982).

11. Murphy, T.J., J.C. Pokojowczyk and Guiseppi Paolucc. "Henry's Law Constants From Equilibrium Measurements; Large Particulates in PCB Deposition," Presented at Workshop for Physical Behavior of PCBs in the Great Lakes, Toronto, Canada, (December, 1981).

12. DiToro, D.M., J.J. Fitzpatrick and R.V. Thomann. "Documentation for Water Quality Analysis Simulation Program (WASP) and Model Verification Program (MVP)," U.S. E.P.A., ORD, Environmental Research Lab-Duluth, Large Lakes Research Station, Grosse Ile, MI (1982).

13. Richardson, W.L. "An Evaluation of the Transport Characteristics of Saginaw Bay Using a Mathematical Model of Chloride," In: Modeling Biochemical Processes, ed., R.P. Canale, pp. 113-132, Ann Arbor Science Publishers, Ann Arbor, MI (1976).

14. Sheng, Y.P. and W. Lick. "The Transport and Resuspension of Sediments in a Shallow Lake," Journal of Geophysical Research, 84: 1809-1826 (1979).

15. Onishi, Y. "Sediment-Contaminant Transport Model," Journal of the Hydraulics Division, Proceedings of the American Society of Civil Engineers, 107: 9 (September, 1981).

16. Schnoor, J.L., D.C. McAvory, C.E. Ruiz-Calzada and M.R. Wiesner. "Verification of a Pesticide Transport and Bioaccumulation Model," EPA Ecological Research Series (1979).

17. MacKay, D. and P.J. Leinonen. "Rate of Evaporation of Low-Solubility Contaminants From Water Bodies to Atmosphere," ES&T, Vol. 9, pp. 1178 (December, 1975).

18. O'Connor, D.J. "Physical Transport Processes," Modeling of Toxic Substances in Natural Water Systems, Manhattan College, Bronx, NY (1980).

19. Hiraizumi, Y., M. Takahashi and H. Nishimura. Adsorption of Polychlorinated Biphenyl onto Sea Bed Sediment, Marine Plankton, and Other Adsorbing Agents. Environmental Science and Technology, pp. 580 (1979).

CHAPTER 19

SURVEY OF POLYCHLORINATED
BIPHENYLS IN AMBIENT AIR ACROSS
THE PROVINCE OF ONTARIO

Eugen Singer, Thomas Jarv* and Michael Sage
Ministry of Environment of Ontario
Air Resources Branch
Toronto, Ontario

INTRODUCTION:

The Ministry of the Environment has performed or initiated seven studies of polychlorinated biphenyls (PCB's) in ambient air in Ontario since 1970. Three studies were of a preliminary nature and involved a limited number of samples. The first one, conducted in 1972 (1) measured PCB's in Hamilton and at Sheridan Park in Mississauga using 1,2-Ethanediol (ethyleneglycol) as a collection medium. Only a few samples were collected and analyzed. The reported concentrations ranged from approximately 12 to 99 ng/m^3. The identity of PCB's in some samples was confirmed by GC/MS. A similar study was done in 1975 (2) with results of approximately 1 to 8 ng/m^3. Again, the presence of PCB's was confirmed by GC/MS in some of the samples.

A survey carried out in 1976 (3) reported PCB's levels at several selected sites in Ontario with values ranging from 1 to 29 ng/m^3. The presence of PCB's was confirmed by analyzing each sample on two different packed columns.

The first intensive survey was performed in February - March 1977 (4). Air samples were collected in the vicinity of the St. Lawrence Cement Plant, Mississauga, and in the adjacent residential districts before and during incineration of PCB's in the cement kiln. A total of 54 samples were collected and the reported concentrations ranged from 0.1 to 6 ng/m^3 with no significant difference

*present address: Ontario Hydro Research Div.; Atmospheric Research Section - Toronto, Ontario.

between the periods before and during the incineration. An additional 17 samples were collected on the plant property, downwind of the storage and pumping area for PCB's containing oil. Values between 3 and 2300 ng/m^3 were measured, with the higher concentrations recorded during the unloading operation.

A second survey was performed in May 1978 (5) at locations in residential districts of Mississauga identical with those used in the above survey. A total of 53 samples were collected. The results ranged from 2 to 40 ng/m^3 with no appreciable difference between the locations. No attempt was made to eliminate the possibility of interferences by using different analytical techniques.

A later survey in Mississauga was carried out (6) during June - July, 1978. The residential sites of the previous survey were included in the seven residential sampling sites of this survey and in addition, four industrial sites were selected.

A total of 533 samples were collected. The range of observed concentrations was from 0.1 to 120 ng/m^3, with no appreciable difference between most industrial and residential sites when possible interferences were removed.

In order to improve the comparative data base of typical concentrations of airborne PCB's in areas of Ontario, one province-wide survey was carried out for a period of 28 days in September and October of 1979 and a second for a period of 14 days in June 1980 (7,8,9,10). A total of 25 sites were selected. The sites were typical of industrial, urban, suburban and rural land use in Ontario.

SAMPLING EQUIPMENT AND PROCEDURE

The sampling equipment used were low volume gas samplers "Nutech" Model 221 AC/DC (Nutech Corp., Durham, N.C.). The dry gas meters were checked before each survey with a wet test meter and found to measure the volume within $\pm$5%.

Twenty-four hour integrated samples were collected on a daily basis; each sample corresponding to a total volume of 10 to 15 m^3 of ambient air.

CARTRIDGE PREPARATION

Cartridges used for the field collection of PCB's were manufactured in accordance with a design previously published (11) and prepared and tested following the procedures outlined in (7). On average, the tested cartridges had a blank assay equivalent to less than 0.3 ng PCB's/m^3 air.

CARTRIDGE ELUTION AND CLEAN UP

Ambient air samples, which were collected onto "Florisil" cartridges, were extracted and processed for analysis, utilizing procedures described in (7). All processed samples were sealed in glass ampoules and labelled for subsequent analysis.

GAS CHROMATOGRAPHIC CONDITIONS:

Gas-chromatograph - Hewlett - Packard 5840A and 5880A with split/splitless injection ports and EC detectors.

column - Hewlett - Packard, fused silica capillary columns, 0.2 mm ID, 25 m long, coated with SP-2100, OV-1 and SE-54.

carrier gas - Helium, linear velocity $\bar{u}$ =, 23 cm/sec at $120°C$

make-up gas - Nitrogen, 25 ml/min

ECD Temp. - $250°C$

Inj. Port Temp. - $250°C$

Injection - Splitless, 2 µl, valve closed from 0 to 30 sec., with H-P Auto-sampler 7672A, alternate vials with clean solvent (2,2,4-trimethylpentane) as wash after each injection.

Temp. Programming - $70°C$, hold for 1 min, then temperature programe rate $10°C$/min to $130°C$, then temperature program rate $3°C$/min. to $220°C$ and hold for 13 min.

DETECTION AND QUANTIFICATION OF POLYCHLORINATED BIPHENYLS

During the previous work on monitoring and detection of PCB's in ambient air (1,2,3,4,5,6), the limitations of the widely used gas-chromatographic methods with packed columns and quantification based on the characteristic fingerprints of commercial Aroclors were soon realized. These methods (12,13,14) assumed that:

a) The ratio of individual PCB's in the vapour phase was identical or at least very close to the ratio in the liquid phase.

b) The ratio of the PCB's did not change after exposure to biological or ambient environment.

c) All interfering compounds could be removed from the sample during the clean-up procedure and, if any stayed in the processed sample, they were resolved and, thus, the PCB peaks could be identified and quantified with a high degree of confidence.

d) If the fingerprint resembled a mixture of two or more commercial Aroclors in an unknown ratio, a new standard could be easily synthesized (by mixing commercial Aroclors) to match the fingerprint and enable quantification.

However, the distribution patterns of PCB's in ambient air can be altered because of the differences in volatilities, solubilities, and chemical and physico-chemical reactivities of the individual components of the original mixture. The fingerprint of a commercial Aroclor may also change after exposure to the environment (15,16). The fact that PCB's may be photolytically degraded has been shown (17,18). During the previous work on PCB's in air, samples were found rarely where the fingerprint would resemble the fingerprint of a commercial Aroclor mixture unless the sample was collected close to a known source and the concentrations of PCB's were significantly higher than the concentrations usually encountered in ambient air.

The efficiency of the clean-up procedures of ambient air samples could be questioned as well. It has been estimated that in the concentration range of PCB's in ambient air, at least 10^5 organic compounds could be expected to be present in similar level in a sample of "clean" air (19). Only a fraction of these low level impurities were identified and it was a reasonable assumption that a significant number may not be removed by the clean-up procedure. Some may have retention times close to the retention times of the PCB's and also may be detected by the EC detector.

This analysis was further complicated by the fact that we were not looking for a single or small number of species. There are 209 possible isomers and homologues of

PCB's; 102 of which have been identified in commercial Aroclors and likely to be encountered in environmental samples (20,21,22).

Most analytical methods to date accomplished identification and quantification by comparison of the total peak area or height of all matching peaks in the sample with the peaks of standard commercial Aroclors (13,14). For most ambient air samples, the source of contamination was usually unknown and any similarity was likely to be purely fortuitous.

The problem of identification and accuracy in PCB analysis is not easily solved. Because of the complexity of commercial mixtures, identification of individual isomers and homologues was not practical until recently. One possible solution to the problem was to convert all of the PCB components to one entity by perchlorination (23). The total PCB content was then measured as decachlorobiphenyl. Analytical methodology for this technique, however, has not been established and accepted.

Until recently, high resolution gas chromatography, $(GC)^2$, was limited to research and development laboratories only. In the last few years, coated glass-capillary columns of good quality have become commercially available as well as instruments designed for work with capillary columns. The advent of fused silica capillary columns together with micro-processor controlled gas-chromatographs made the use of $(GC)^2$ even more attractive. Almost at the same time, a number of individual PCB's were synthesized and the $(GC)^2$ was being slowly introduced into routine PCB analysis. The advantages were obvious,

a) High resolution power of capillary columns as compared to packed columns enable better resolution of the majority of individual PCB's from each other and from impurities.

b) Because of the high resolution power of $(GC)^2$ and commercial availability of many individual PCB's, the possibility of identifying and quantifying PCB's is improved. PCB's may be identified and quantified with higher degree of confidence even when the fingerprint does not match a commercial Arochlor at all.

However some questions were not resolved at the time the analyses of the 1979 survey were performed. These were:

a) not all the individual PCB's, as they were identified and present in commercial Aroclors were commercially available.

b) not even the $(GC)^2$ was able to resolve all individual PCB's.

c) There still remained the possibility that some impurities, present in ambient air sample and not removed by the clean-up procedure, may be identified as PCB's because of the close retention times under the used conditions.

Calibration of the HP 5840A GC for the 1979 survey was performed with a mixture of 44 individual PCB's and column coated with SP-2100. All samples were spiked with 2,2-bis-(p-chlorophenyl)-1,1-dichloroethylene (p,p'-DDE) which served as an internal reference standard to correct for any changes in the retention times and was also used as a reference peak for the calculation of relative retention times. The GC was automatically recalibrated after every five samples.

As mentioned before, one drawback of the procedure was that not all the PCB's identified in commercial Aroclors were commercially available. Because only peaks which were in the standard mixture were automatically detected, quantified and reported, the analysis may eventually underestimate the concentration of PCB's in ambient air.

To overcome, at least temporarily, this disadvantage, the following procedure was adopted.

a) retention times, relative to pp'-DDE, of all the PCB's in the standard mixture and all the major peaks in commercial Aroclors were calculated. (Table I).

b) Because the relative response factors of the higher chlorinated biphenyls are fairly close (Table II), the response factors of the major Aroclor peaks were calculated as the average of the response factors of the two neighbouring PCB's of the standard mixture.

c) Relative retention times for all major peaks, which were not automatically quantified by the gas-chromatograph, were manually calculated and compared

Table I Relative Retention Times of PCB's in the Standard Mixture and Major Peaks in Commercial "Aroclors"

$$(RRT_{pp'-DDE}=1)$$

Std.	RRT	PCB Peak
2	0.352	2-
3	0.412	3-
4	0.417	4-
5	0.452	2,2'-
6	0.500	2,4-
	0.518	1016,1221,1232,1242
7	0.528	2,3-
8	0.558	3,5-
	0.563	1016,1232,1242,1248
9	0.586	2,46-
10	0.592	3,3'
11	0.602	3,4-
12	0.609	2,2',5-
	0.639	1221,1232,1260
	0.643	1016,1221,1232,1242,1248
13	0.666	2,2'6,6'-
14	0.683	2,3'5-
15	0.697	2,4',5-
16	0.717	1,3',4'-
	0.729	1016,1221,1232,1242,1248
	0.739	1016,1221,1232,1242,1248,1254,1260
	0.754	1016,1232,1242,1248
17	0.770	2.2'.5.5'-
18	0.778	2,2',4'5-
19	0.783	2,2'4,4'-
20	0.790	2,3,5,6
	0.796	1016,1221,1232,1242,1248,1254,1260
21	0.800	2,2',4,6,6'-
22	0.806	2,2'3'5-
	0.811	1016,1232,1242,1248
23	0.831	2,3',5,5'
24	0.842	2,2',3,3'-
25	0.853	2,2',4,5',6-
26	0.863	2,2'4,4',6-
	0.869	1242,1248
27	0.878	2,3,4,5-
28	0.885	2,3',4',5-
	0.890	1016,1232,1242,1248
	0.893	1016,1254,1260

Table 1 Continued

Std. Peak	RRT	PCB
29	0.907	2,3',4,5'6-
	0.911	1242,1249,1254
30	0.928	2,2',4,4',6.6'-
	0.936	1242,1248,1254,1260
31	0.944	2,2',4,5,5'-
	0.951	1232,1242,1248,1254
32	0.965	2,3',4,4',6-
	0.970	1248,1254
	0.980	1232,1242,1248,1254
33	0.989	2,2', 3,4,5'
	0.995	1232,1242,1248,1254
34	1.005	3,3',4,4'
35	1.019	2,2',4,4',5',6
	1.024	1242,1248,1254
36	1.038	2,2',3,5,5',6-
	1.051	1248,1254,1260
	1.061	1254
37	1.064	2,2',3,4,4'6-
38	1.075	2,2',3.4,5,6'-
	1.083	1254
	1.098	1254,1260
39	1.110	2,2',4,4'5,5'-
40	1.131	2,2',3,4,5,5'-
41	1.141	2,2',3,4,4',5_
	1.145	1254,1260
	1.156	1248,1254
	1.162	1221,1254,1260
42	1.171	2,2',3,3',4,5-
	1.199	1254,1260
43	1.207	2,2',3,3',4,4'-
	1.222	1254,1260
44	1.234	2,2',3,4,5',6-
	1.247	1254,1260
	1.257	1254,1260
	1.269	1254,1260
45	1.273	2,3,3',4,4',5-

Table II RELATIVE RESPONSE FACTORS OF SOME PCB'S

MONO'S	REL. RESPONSE	PENTA'S	REL. RESPONSE
2-	21	2,2',4,6,6'-	313
3-	1	2,2',4,5',6-	340
4-	11	2,2',4,4',6-	355
	AVG.=11	2,2',3,4,6-	351

DI'S	REL. RESPONSE		
2,2'-	48	2,2',3,5,6-	364
2,5-	206	2,3'4,5',6-	381
2,3-	224	2,2',4,5,5'	395
3,5'-	150	2,3',4,4',6-	369
3,3'-	82	2,2',3,4',5-	427
3,4'-	166		AVG.=366
	AVG.=146	HEXA'S	REL. RESPONSE

TRI'S	REL. RESPONSE	2,2',4,4',6,6'-	367
2,4,6,-	289	2,2',4,4',5',6-	351
2,2',5-	211	2,2',3,5,5',6-	378
2,4,5-	326	2,2',3,4,4',6-	341
2,3',5-	319	2,2',3,4,5,6'-	399
2,4',5-	308	2,2',4,4',5,5'-	401
	AVG.=290	2,2',3,4,5,5'-	400

TETRA's	REL. RESPONSE	2,2',3,4,4',5-	440
2,2',6,6'-	224	2,2',3,3',4,5-	370
2,2',4,6-	352	2,2',3,3',4,4'-	344
2,2',6'-	248	2,3,3',4,4',5-	291
2,3',4,6-	320		AVG.=371
2,2',5,5'-	320	HEPTA'S	REL RESPONSE
2,2',4',5-	335	2,2',3,4,5,5',6-	402
2,4,4',6-	527		
2,3,5,6-	361	OCTA'S	REL. RESPONSE
2,2',3',5-	388	2,2',3,3,5,5',6,6'-	378
2,3',5,5'-	351	2,2',3,3',4,5',6,6'-	290
2,2',3,3'-	371	2,2',3,4,4',5,6,6'-	366
2,3,4,5-	296		AVG-345
2,3',4'5-	358		
3,3',4,4'-	287		
	AVG.=338		

with the relative retention times of the major peaks
identified in commercial Aroclors. If the relative
retention time was within ± 0.002 units, the peak was
considered identified as a PCB and the amount
calculated from the average response factor.

d) All peaks which were automatially detected and
quantified as PCB's by the programme but the measured
retention time did not agree with the expected
retention time to within ± 0.02 min. were rejected.

e) The sum of PCB's as reported by the gas-chromatographs
microprocessor and corrected as in c) and d) was
reported as total PCB's in the sample.

The calibration and quantification procedure for the
1980 survey was improved. This was possible by the use of
a GC combined with a more powerful microprocessor -the H-P
5880 A - and availability of superior fused silica
capillary columns with OV-1 and SE-54 liquid phase.

Because of it, we were able to inject the sample
simultaneously on two columns of slightly different
polarity, record both chromatograms and perform the
necessary data reduction by a single instrument.

In the time period between the two surveys we were
able to develop a method of identification of PCB's in
commercial Aroclors based on retention indices. This
enabled us to synthesise a standard from commercially
available Aroclors, which was used for the detection and
quantification. To cover the majority of PCBs, which might
be present in the sample, the standard was a mixture of
Aroclors 1016, 1221 and 1254. 1,2,3,4-tetrachlorobenzene
(1234-TCB) and p,p'-DDE were used as internal standards, to
correct for any changes in retention times during the
analysis. The programme, written for the microprocessor,
correlated both chromatograms and identified and quantified
the PCB's applying the following criteria.

1) The retention time of the peak must have been within
limits (usually $\pm 0.1\%$) of the expected retention time,
otherwise the peak was rejected.

2) If the peak for a given PCB was identified on both
columns and the quantities were within limits (usually
$\pm 20\%$), the average was calculated and printed followed
by the word "confirmed".

3) If the peak for a given PCB was identified on both columns, but the quantities were not within given limits, than the lower value was taken as the result, followed by word "interference" (we assumed that the higher value was caused by an impurity coeluting with the PCB).

4) If the peak for a given PCB was identified on one column only, the value was rejected and identified by the words "not PCB".

5) Although the capillary columns have very high resolving power some PCB's could not be resolved at all or could be resolved on one column only. If this happened, then the determined quantities of unresolved PCB's on one or both columns were summed up and evaluated under identical criteria.

In Figure 1 and 2 is an example of a typical analysis. Comparing the chromatograms it is obvious that only few from the large number of peaks on each column were identified as PCB's. The correlation programme rejected additional significant number of peaks, because they were not identified on both columns. A good example is the 3-monochlorobiphenyl which was identified in significant quantity on OV-1 column but could not be identified on the SE-54 column.

In Table III, is a brief summary of results of some samples collected in different areas of Ontario and analysed by the three mentioned techniques. The packed column was a 9 feet long 1/8" O.D. SS column, packed with 1% Dexsil 400 on Anachrom Q. The results of single capillary columns are in fair agreement; however they are at least one order of magnitude lower than data obtained on the packed column. The results for dual capillary column analysis are another order of magnitude below the results of single capillary column. Similar conclusions can be drawn from Table IV, where are briefly summarized the results of the 1979 and 1980 province wide surveys for PCB's across the Province of Ontario.

CONCLUSION

The study confirms that the resolution of a packed column is insufficient to resolve the PCB's from interfering impurities remaining in the ambient air sample after the clean-up procedure.

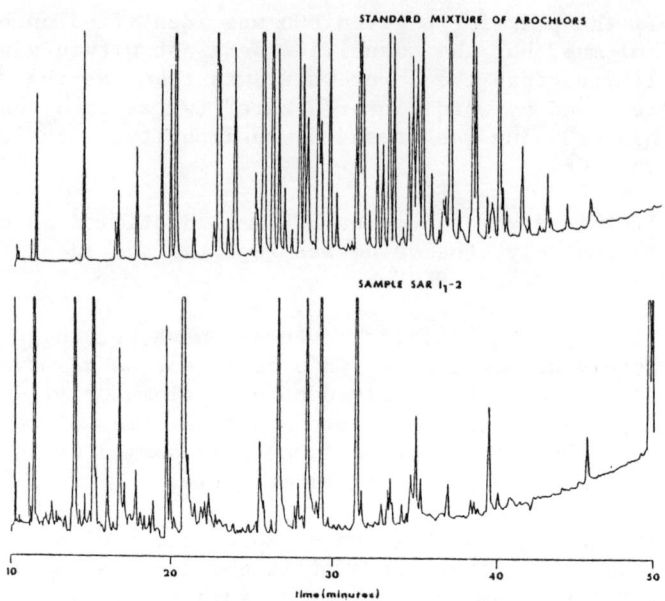

Figure 1. Standard Aroclor mixture and air sample SAR-I_1-2 on OV-1 capillary column.

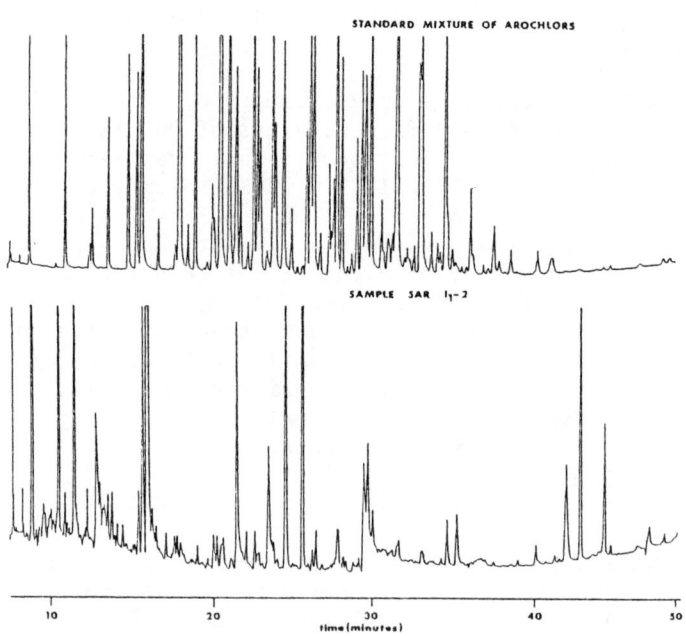

Figure 2. Standard Aroclor mixture and air sample SAR-I_1-2 on SE-54 capillary column.

Table III. Analysis of Ambient Air Samples for PCB's with Packed Column. $(GC)^2$ and $(GC)^2$ with Two Columns of Different Polarity

No.	Sample Code	Packed	Single Capillary Column			Dual Column
			SP-2100	OV-1	SE-54	OV-1/SE-54
1	HAM-S1-12-16	66	4.2	5.3	1.4	0.4
2	HAM-S1-12-17	44	1.0	2.7	3.0	0.7
3	HAM-S1-12-18	7.1	2.2	2.6	1.2	0.0
4	SAR-S1-U1-16	58	4.6*	6.1	4.3	0.6
5	SAR-S1-U1-18	6.2	1.6*	3.2	1.0	0.02
6	SUD-S1-S1-18	52	3.8*	2.3	2.1	0.0
7	SUD-S1-S1-17	53	1.7*	2.9	2.0	0.0
8	SUD-S1-S1-4	47	4.1*	3.9	2.8	0.3
9	SUD-S1-S1-1	66	3.9*	3.9	3.1	0.8
10	SAR-S1-U1-15	115	3.5	3.7	14	0.6
11	SAR-S1-U1-17	19	3.7*	3.3	2.3	0.2
12	HAM-S1-U-4	173	4.8*	5.2	4.1	1.4
13	HAM-S1-U-1	317	11*	11	7.0	3.3
14	TOR-S1-U1-1	28	12*	10	10	5.4
15	TOR-S1-U1-2	238	23*	16	15	6.9
16	TOR-S1-U1-18	6.4	4.1*	5.5	4.7	1.6
17	HAM-S1-U2-16	31	3.6*	2.3	1.7	0.2
18	MIS-S1-U4-15	6.7	2.4*	3.4	3.3	0.2
19	MIS-S1-U4-19	13	1.9*	2.2	2.4	0.2
20	MIS-S1-U4-17	1.5	4.0	2.4	1.8	0.4
21	HAM-S1-U2-4	128	2.4*	3.8	3.0	0.9
22	MIS-S1-U4-16	19	1.8*	2.5	2.0	0.1
23	HAM-S1-U2-18	25	4.4	2.2	1.6	0.2
24	HAM-S1-U-5	52	8.7*	0.7	1.6	0.2
25	LON-S1-S1-18	35	11	9.4	9.1	6.3
26	SAR-S1-U-2	1615	6.8*	4.3	6.5	2.1
27	HAM-S1-U-15	68	20*	5.9	9.7	1.9
28	MIS-S1-U1-3	–	11*	2.9	3.1	0.6

*A not well resolved peak is present around RT=26 min, which was originally considered as PCB. In this set of data the peak was excluded.

Concentrations are expressed as ng PCB's/m^3 air.

Table IV Summary of the 1979 and 1980 PCB's
Levels across the Province of Ontario

Location and Sample Code	1979 Averg. conc.	Std. Dev.	1980 Averg. Conc.	Std. dev.
Th. Bay-U1	5.7	4.2	0.01	0.02
Th. Bay-R1	3.8	2.6	0.01	0.26
Burlington-U1	5.6	5.2	0.10	0.13
St.Catherine-R1	6.6	5.6	0.22	0.55
Nanticoke-R1	5.2	3.2	0.03	0.04
Nanticoke-R2	5.1	4.3	0.80	2.8
Sarnia-I1	9.8	9.4	0.28	0.31
Sarnia-U1	7.3	5.6	0.10	0.03
Moor T.-R1	4.9	3.6	0.03	0.05
Windsor-U1	5.6	3.6	0.06	0.09
London-S1	8.1	4.8	1.4	1.5
Hamilton-I1	7.0	5.6	0.06	0.07
Hamilton-I2	5.9	4.6	0.08	0.10
Hamilton-U1	5.7	5.0	0.06	0.10
Hamilton-U2	9.2	14	0.08	0.26
Oshawa-U1	5.7	4.4	0.11	0.23
Toronto-S1	5.3	3.3	0.20	0.30
Toronto-U1	9.6	7.3	0.64	0.89
Mississauga-U1	5.3	3.9	0.13	0.18
Mississauga-U2	5.0	3.9	0.05	0.06
Mississauga-U3	6.3	5.7	0.14	0.12
Mississauga-U4	5.4	3.6	0.19	0.20
Kingston-U4	4.7	3.0	-	-
Sudbury-R1	6.8	4.7	0.03	0.04
Sudbury-S1	4.4	2.9	0.02	0.03

U=ruban   R=rural   all results are in ng/m^3

S=subarban I=industrial

The data from the dual capillary column analysis with the quantification method based on individual PCB's are probably better than the results from the other procedures but further development remains to be done on the method.

It is possible, that most of the data for PCB concentrations in ambient air as indicated in reports (1 to 7) and literature and which have been obtained with packed columns or single capillary column are biased high due to the relative inefficiency of the methods unless the PCB's were completely separated from all interfering compounds.

REFERENCES:

1) Collection and Analysis of PCBs in The Atmosphere. Ontario Research Foundation Report ORF-72-1

2) Collection and Analysis of Polychlorinatedbiphenyls (PCBs) in Ambient Air. Ontario Research Foundation Report ORF-75-1

3) An Ambient Air Survey for Polychlorinated Biphenyls in Ontario -1976. Ministry of Environment of Ontario Report ARB-TDA-49-78 (1978).

4) Report on PCB's Monitoring in Mississauga St. Lawerence Cement Plant, February to March 1977. Ministry of Environment of Ontario Report ARB-TDA-40-77 (1977).

5) Don Ogner, Ontario Ministry of Environment, Central Region: private communication.

6) Sampling and Analysis of Emissions from a Wet Process Kiln at St. Lawrence Cement Company in Mississauga. Ontario Research Foundation Report No. P-2818/G (Revised) -01

7) A Survey of Polychlorinated Biphenyls in Ambient Air in Ontario, Phase I., Ministry of Environment of Ontario Report ARB-08-80-ARSP (1981)

8) A Survey of Polychlorinated Biphenyls in Ambient Air in Ontario, Phase II, Volume I, Ministry of Environment of Ontario Report ARB-11-81-ARSP (1981)

9) A Survey of Polychlorinated Biphenyls in Ambient Air in Ontario, Phase II., Volume II., Ministry of Environment of Ontario Report ARB-25-81-ARSP (1981)

10) A Survey of Polychorinated Biphenyls in Ambient Air in Ontario, Phase III., Ministry of Environment Report ARB-26-81 (1981)

11) Sampling Procedures for Airborne Polychlorinated Biphenyls (PCB's). Ministry of Environment of Ontario Report ARB-TDA-51-78 (1978)

12) Assoc. Off. Anal. Chem. Methods 29.018

13) Sawyer L.D.: Quantitation of Polychlorinated Biphenyls Residues by Electron Capture Gas-Liquid Chromatography: Reference Material Characterization and Preliminary Study. J. Assoc. Off. Anal. Chem. $\underline{61}$,272 (1978)

14) Sawyer L.D.: Quantitation of Polychlorinated Biphenyl Residues by Electron Capture Gas-Liquid Chromatography: Collaborative Study J. Assoc. Off. Anal. Chem. $\underline{61}$,282 (1978).

15) Fries G.F.: Polychlorinated Biphenyl Residues in Milk of Environmentally and Experimentally Contaminated Cows. Environmental Health Perspectives $\underline{1}$, 55 (1972). Published by National Institute of Environmental Health Science, N.C., USA.

16) Cook J.W.: Some Chemical Aspects of Polychlorinated Biphenyls. Ibid. p.10.

17) Hutzinger O., Safe S. and Zitko V.: Photochemical Degradation of Chlorobiphenyls (PCB's). Ibid.p..15.

18) Hutzinger O., Safe S. and Zitko V.: The Chemistry of PCB's, CRC-Press, 1974 pp. 113-148.

19) Lewis R.G.: Accuracy and Trace Organic Analysis. National Bureau of Standards Special Publication 422, pp 17-20.

20) Weidmark, G. The OECD Study of the Analysis of PCB, Report from the Institute of Analytical Chemistry, Stockholm, Sweden 1968.

21) Albro P.W., Haseman J.K., Clemmer T.A., and Corbett B.J., Identification of the Individual Polychlorinated Biphenyls in a Mixture by Gas-Liquid Chromatography. J. Chromatogr. $\underline{136}$,147 (1977)

22) Albro P.W., and Parker C.E., Comparison of the Composition of Arochlor 1242 and Arochlor 1016., J. Chromatogr. <u>169</u>, 161 (1979)

23) Berg, O.W., Diosady P.L., and Rees G.A.V., Column Chromatagraphic Separation of Polychlorinated Biphenyls from Chlorinated Pesticides, and their Subsequent Gas Chromatographic Quantitation in Terms of Derivatives. Bull. Environ. Cont. Toxicol. <u>7</u>, 338 (1972).

CHAPTER 20

MONITORING OF PCBs IN
WATER, SEDIMENTS AND BIOTA
OF THE GREAT LAKES - SOME
RECENT EXAMPLES

P. B. Kauss

K. Suns

A. F. Johnson

    Ontario Ministry of the
      Environment
    Water Resources Branch

INTRODUCTION

This paper indicates the types of monitoring and surveillance efforts being conducted by the Water Resources Branch and West-Central Region of the Ontario Ministry of the Environment (MOE) in the study of the impact, fate and temporal trends of PCBs in the Great Lakes nearshore environment. Some recent data on PCBs analyses of both abiotic and biotic samples are presented as examples of this work.

Examples from the MOE's fish contaminants monitoring programs in the Great Lakes (both sports and forage fish) are discussed, with respect to regional differences as well as temporal trends. Although the Niagara River is not the only source of PCBs to Lake Ontario, it is the major tributary, and trace contaminants carried by the river have been shown to impact Lake Ontario sediments and biota (1,2,3). Examples of water, sediments and biota studies being carried out in an attempt to delineate source areas of various trace contaminants in the river, including PCBs, are therefore included as well.

METHODS

All chemical analyses were performed by the MOE laboratories in Rexdale. Analytical methods employed were as described in the "Handbook of Analytical Methods for Environmental Samples" (4). For field methodologies, the reader is referred to the references cited as well as "A Guide to the Collection and Submission of Samples for Laboratory Analysis" (5). Concentrations of PCBs discussed in the text refer to "total PCBs".

RESULTS

1. Water, Sediments

Data on PCBs in samples taken in the Niagara River as part of the intensive surveillance and monitoring efforts under the Canada-Ontario Agreement on Great Lakes Water Quality are presented here. These, as well as data on other trace contaminants, have been reported in the Canada-Ontario Review Board reports "Environmental Baseline Report of the Niagara River" (1,2).

As a background to the following results, figure 1 shows the locations of intakes, outfalls and disposal sites near the Niagara River or its tributaries in both Ontario and New York State. These indicate the density of urbanization and industrialization (not PCBs inputs per se) on the U.S. side of the river, particularly in the section from Buffalo to Niagara Falls, New York. Although industrial and combined sewer discharges have not been differentiated or identified because of space limitations, wastewater/sewage treatment plant discharges, water treatment plant intakes and major disposal sites have been labelled. Some of the New York State disposal sites, such as Hooker's Hyde Park, S & N, 102nd Street and Love Canal and Dupont's Necco Park, have been identified as having received large quantities of hazardous wastes (6).

(a) Water

(i) Municipal Water Supplies

Drinking water supplies on the Ontario side of the Niagara River have been monitored by the MOE's West-Central Regional staff for PCBs as well as other parameters since late 1974 at the Municipality of

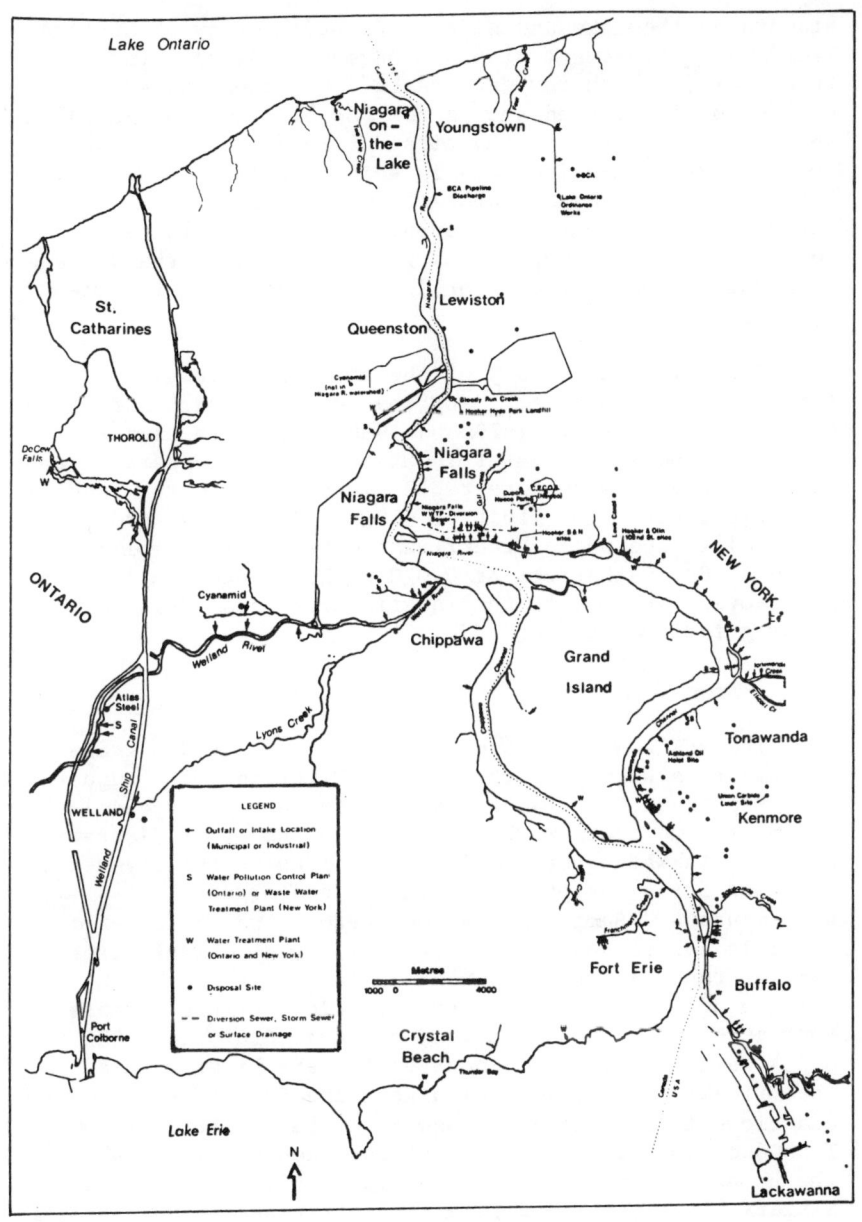

FIGURE 1. INTAKES, OUTFALLS AND DISPOSAL SITES IN THE NIAGARA RIVER (Source: Niagara River Toxics Control Committee, July, 1981).

Niagara-on-the-Lake and since 1976 at the City of Niagara Falls water treatment plants. During 1979, sampling frequency for both raw and treated (finished) water samples was increased and is now (as of 1980) conducted on about a monthly basis. Sampling of river water at Fort Erie has also been included in this program. In July, 1980, this program was expanded to include the St. Catharines municipal water supply at DeCew Falls (as a control) which is supplied by water drawn from the Welland Ship Canal (see Figure 2 for locations of water treatment plants).

Table I is a summary of the analytical results of this monitoring program. For the years in which more than one sample was taken (1979 and 1980) PCBs were usually only detected in raw water in 8% to 16% of the samples and concentrations, when detectable, were close to the routine analytical detection limit of 0.020 ug/L. PCBs were detected in one out of the five treated water samples taken at St. Catharines at 0.020 ug/L. Concentrations in all samples were below the interim provincial drinking water guideline of 3 ug/L.

(ii) River Water

The MOE's Water Resources Branch conducted a trace contaminants survey of the Niagara River in 1979 (August 25 - September 5). Water, bottom sediments, suspended sediments and biota were sampled at a number of discrete stations in the upper and lower sections of the river and analyzed for a variety of inorganic and organic contaminants. Sampling stations were situated to reflect possible source and/or depositional areas. River water samples (one per station) were collected at mid-depth using a solvent-rinsed Kemmerer bottle. Figure 3 shows that detectable concentrations of PCBs (0.020 to 0.080 ug/L) were mainly found in upper Niagara River waters, specifically in samples taken near the Buffalo River mouth and close to the mainland (New York State) shore of the Tonawanda Channel. Four out of fourteen samples (29%) in the upper and one out of nine samples (11%) in the lower Niagara River contained PCBs concentrations above the detection limit and exceeding the 1978 provincial objective for the protection of aquatic life (0.001 ug/L). If only samples taken near the Buffalo River mouth and adjacent to the mainland in the Tonawanda Channel are considered, the percentage of detections increases to 40%.

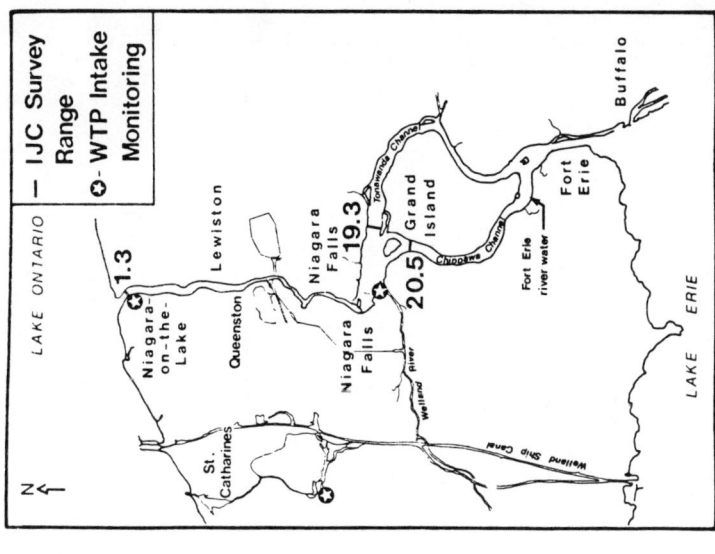

FIGURE 3. PCBs CONCENTRATIONS (ug/L=ppb) IN NIAGARA RIVER WATERS (Aug. 25 - Sept. 5/79). ND = Not detected (below detection limit of 0.020 ug/L)

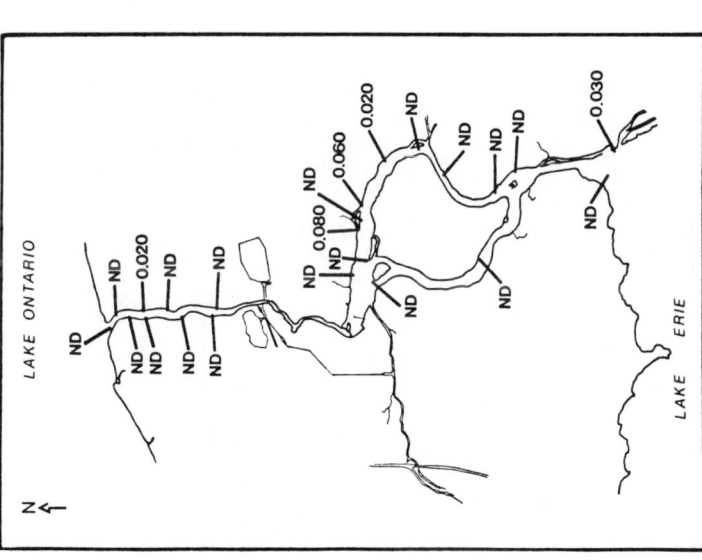

FIGURE 2. NIAGARA RIVER SURFACE (1.5 m depth) WATER QUALITY AND WATER TREATMENT PLANT INTAKE MONITORING LOCATIONS FOR TRACE CONTAMINANTS - 1980.

TABLE 1. CONCENTRATION OF PCBs* (ug/L=ppb) IN RAW AND TREATED WATER SAMPLES FROM THE NIAGARA RIVER AND WELLAND CANAL, 1974-1980.

Year	Niagara-on-the-Lake Raw			Niagara-on-the-Lake Treated			Niagara Falls Raw			Niagara Falls Treated			Fort Erie River Water			St. Catharines (DeCew Falls) Raw			St. Catharines (DeCew Falls) Treated		
	N	n	Range	N	n	Range	N	n	Range	N	n	Range	N	n	Range	N	n	Range	N	n	Range
1974	1	0	ND	-	-	-	-	-	-	-	-	-	-	-	-	-	-	-	-	-	-
1975	1	0	ND	-	-	-	-	-	-	-	-	-	-	-	-	-	-	-	-	-	-
1976	1	0	ND	-	-	-	1	0	ND	-	-	-	1	0	ND	-	-	-	-	-	-
1978	1	0	ND	-	-	-	1	0	ND	-	-	-	-	-	-	-	-	-	-	-	-
1979	7	1	ND-0.020	5	0	ND	7	0	ND	4	0	ND	7	0	ND	-	-	-	-	-	-
1980	13	2	ND-0.032	9	0	ND	12	2	ND-0.030	8	0	ND	12	1	ND-0.030	5	0	ND	5	1	ND-0.020

*Detection Limit = 0.020 μg/L.
ND = Not detected (below detection limit).
N = Number of samples.
n = Number of samples with detectable concentration.
- = No samples taken.

Note: interim provincial guideline for PCBs in drinking water is 3 μg/L.

Routine river monitoring for PCBs as well as other trace contaminants during 1980 consisted of four surveys (June, August, September, and October) along three International Joint Commission (IJC) river ranges (see Figure 2): one each at the downstream (northern) ends of the Chippawa and the Tonawanda Channels (river ranges 19.3 and 20.5, respectively) to reflect inputs from these channels to the lower Niagara River, and one in the lower Niagara River at Niagara-on-the-Lake (range 1.3) to reflect inputs from the Niagara River to Lake Ontario. During each survey, surface water (1.5 m depth) samples were taken by Water Resources Branch personnel on three consecutive days at each of five discrete stations along each river range.

TABLE II. CONCENTRATION OF PCBs (ug/L=ppb) IN NIAGARA RIVER SURFACE WATER DURING 1980.

IJC River Range												
1.3				19.3				20.5				
$\bar{x}$	N	n	Range	$\bar{x}$	N	n	Range	$\bar{x}$	N	n	Range	
<0.020	60	0	ND	<0.020	45	4	ND-0.040	<0.020	42	3	ND-0.080	

$\bar{x}$ = Mean annual concentration, based on data of four surveys (June, August, September and October).
N = Number of samples.
n = Number of samples at or above detection limit (0.020 µg/L) and above 1978 provincial aquatic life objective of 0.001 µg/L.

As shown by Table II, mean yearly concentrations were non-detectable (below the routine laboratory detection limit of 0.020 ug/L) for all three ranges. However, in contrast to the lower Niagara River, PCBs levels in the upper Niagara River were above the detection limit and the provincial objective for the protection of aquatic life in some (7% to 9%) of the samples taken in both the Tonawanda and Chippawa Channels. Concentrations, when detectable, ranged from 0.020 to 0.080 ug/L.

(iii) Industrial and Municipal Inputs

Monitoring is continuing on both sides of the Niagara River to gather information on trace contaminants inputs. When complete, this information will enable a comparison of loadings of PCBs as well as other organic and inorganic trace contaminants from point and non-point sources (see Figure 1) with estimates of inputs from Lake Erie to the Niagara River and the loadings carried by the river to Lake Ontario [see (2)].

(b) Sediments

(i) Suspended Sediments

Figure 4 indicates the results of analyses for total PCBs in suspended sediments collected from different areas of the Niagara River during the 1979 trace contaminants survey. These samples (one per station) were collected using a ship-based continuous-flow centrifuge (Alfa-Laval, Model 103B) taking water from mid-depth. Collection plates and bowl of the centrifuge were solvent-rinsed between samples.

All seven samples contained detectable concentrations of PCBs, but concentrations were generally higher in the upper than in the lower Niagara River. The high concentrations observed in samples collected downstream of the Buffalo River mouth and in the Tonawanda Channel (1200 and 660 ug/kg = ppb, respectively) are probably indicative of the importance of these areas to PCBs inputs/loadings to the river, although levels at the former location may be at least partially due to resuspension of bottom sediments into the water column as a result of storm events in Lake Erie. PCBs in the Chippawa Channel sample (160 ug/kg) were within the range of levels found in lower Niagara River samples (69 to 230 ug/kg). A sample from the control station at Thunder Bay, Lake Erie contained a surprisingly high concentration of PCBs (450 ug/kg) as well. This may also be a result of resuspension of bottom material.

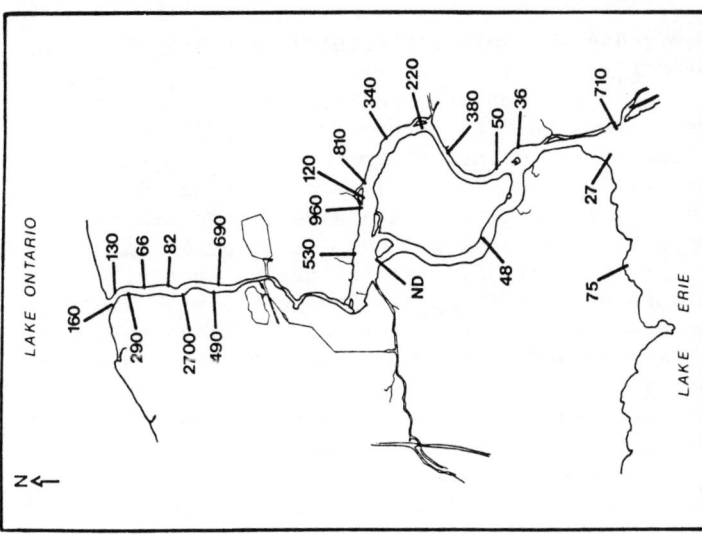

FIGURE 5. PCBs CONCENTRATIONS (ug/kg=ppb, dry weight) IN NIAGARA RIVER BOTTOM SEDIMENTS (Aug. 25 - Sept./79). ND = Not detected (below detection limit of 10 ug/kg). Adapted from (2).

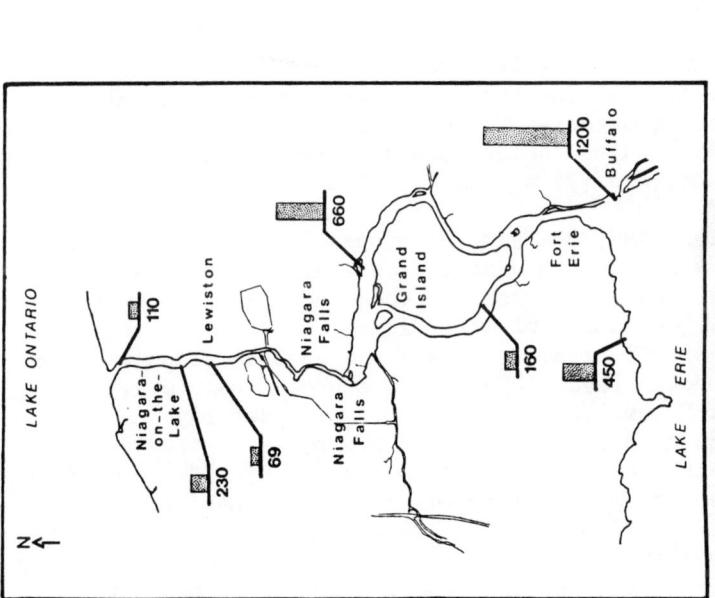

FIGURE 4. PCBs CONCENTRATIONS (ug/kg=ppb, dry weight) IN NIAGARA RIVER SUSPENDED SEDIMENTS (Aug. 25 - Sept. 5/79).

(ii) Bottom Sediments

Bottom sediments were collected in the Niagara River as part of MOE's 1979 trace contaminants survey. At each station, the top 3 cm of at least three Shipek or Ponar grabs were composited. Concentrations of PCBs in these bottom sediment samples are shown in Figure 5. All eight of the samples from the lower Niagara River and 9 out of 13 samples from the upper Niagara River contained total PCB concentrations exceeding the MOE's 50 ug/kg (ppb) guideline for open-lake disposal of dredge spoils. Concentrations of PCBs in upper Niagara River sediments were generally higher just downstream of the Buffalo River mouth (710 ug/kg) and in the Tonawanda Channel (50 to 960 ug/kg) than in the Chippawa Channel (not detected to 48 ug/kg) and these were also the areas of exceedence of the 50 ug/kg guideline. The highest concentration of PCBs detected (2700 ug/kg) was in a sample from the lower Niagara River downstream of Queenston in a large back-eddy area, suggesting that this may be an at least temporary depositional area for trace contaminants associated with particulates.

2. Biotic Samples

(a) Freshwater Clams

Freshwater clams (<u>Elliptio complanatus</u>) placed in cages anchored at the bottom have proven to be a valuable biomonitoring organism for detecting organic trace contaminants in the Humber and Moira River systems in southern Ontario (7).

During 1980, the same clam species was utilized in a study of trace contaminant source areas in the Niagara River and the uptake rates of these contaminants by the clams (8). A number of sites in the upper and lower sections of the river were selected so that clams would be situated downstream of potential source areas (e.g. Tonawanda Channel, Buffalo River). Thunder Bay, Lake Erie was chosen as a control site on the basis of the 1979 young-of-the-year spottail shiner data (9). Clams (6.5 to 7.2 cm length) were collected from Balsam Lake in southern Ontario, placed in galvanized wire cages and then exposed to Niagara River water for 1 to 101 days before their removal, shucking and subsequent analysis of the soft tissues.

Significant concentrations of PCBs were accumulated by clams within two days of exposure to water at the northern end of Tonawanda Channel and in the lower Niagara River at Niagara-on-the-Lake, but maximum concentrations were not observed until 8 to 16 days' exposure had elapsed (Figure 6).

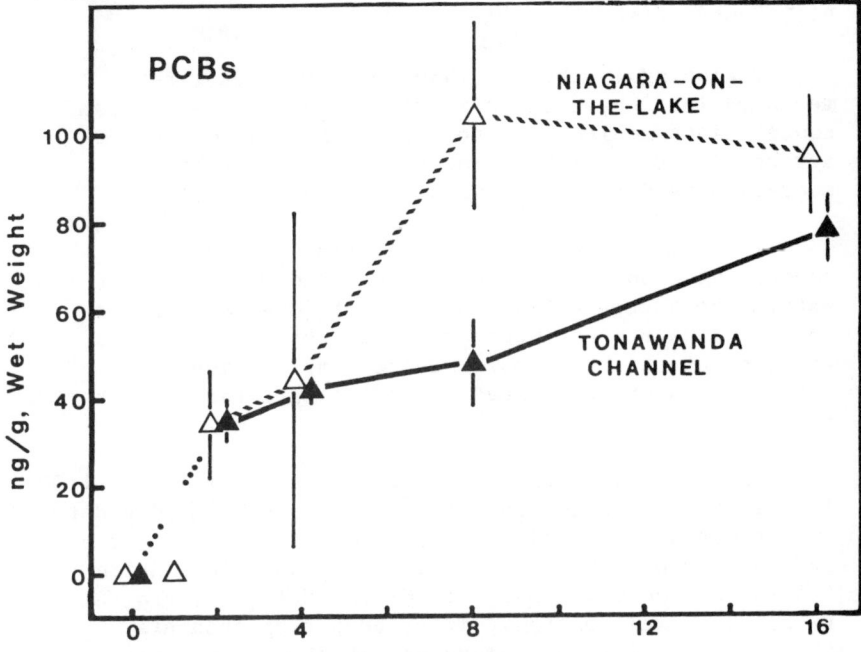

FIGURE 6. EFFECT OF EXPOSURE TIME ON PCBs ACCUMULATION BY FRESHWATER CLAMS (Elliptio complanatus) PLACED IN THE NIAGARA RIVER (Aug. 12 – Aug. 28/80). Values (ng/g=ppb, wet weight) are mean and standard deviation of three whole clam soft tissue analyses. Detection limit = 20 ng/g. From (8).

As shown by Figure 7, the highest mean concentrations of PCBs after 16 days' exposure were detected in clams left at these two sites (79 ppb and 95 ppb, respectively). Mean concentrations in clams left in the Chippawa Channel were, at 23 ppb, just above the detection limit (20 ppb) for this class of compound. Only trace quantities were accumulated by clams situated near the Buffalo River mouth, and clams exposed to water at the other three sites (Thunder Bay in Lake Erie, Welland River and below Queenston in the lower Niagara River) did not accumulate detectable quantities.

(b) Fish

(i) Sports Fish

Beginning in the mid-1960s when DDT was first measured in fish, Ontario instituted an extensive fish contaminants monitoring program which came to include mercury in 1969 and PCBs, mirex and other trace contaminants during the 1970s.

This fish contaminants monitoring program is a co-ordinated undertaking of the Ontario Ministries of Natural Resources (MNR), Environment (MOE) and Labour (MOL). Fish are collected primarily by MNR staff and analyzed at the MOE labs. Medical implications of contaminants are evaluated by medical specialists with the MOL.

Data on contaminants in sports fish from nearly 1100 locations in Ontario are available [see (10)]. PCB data on various species and sizes of sports fish are currently available from 119 locations in lakes Superior, Huron (including Georgian Bay), St. Clair, Erie, Ontario, St. Francis, as well as their interconnecting channels (Detroit, St. Clair, Niagara and St. Lawrence rivers). Data on PCBs in fish are also available for 247 inland lakes. Results of analyses are reported annually in the Ontario Government publication "Guide to Eating Ontario Sport Fish" and in interim news releases as new data becomes available. These indicate consumption guidelines for different sizes of fish species sampled in a given location. As an example, 1980 data for several species of sports fish collected in the upper and lower sections of the Niagara River have been extracted from the 1981 edition of the publication and are shown in Table III. This indicates that fish fillets from the larger sizes of coho salmon and American eel (>55 cm in length) collected in the lower Niagara River contained PCBs and mirex concentrations in excess of the National Health and Welfare Canada guidelines for human consumption (2 ppm and 0.1 ppm respectively) and are suitable for "occasional meals only". However, yellow perch (15 to 30 cm in length) and white sucker (less than 45 cm in length) from

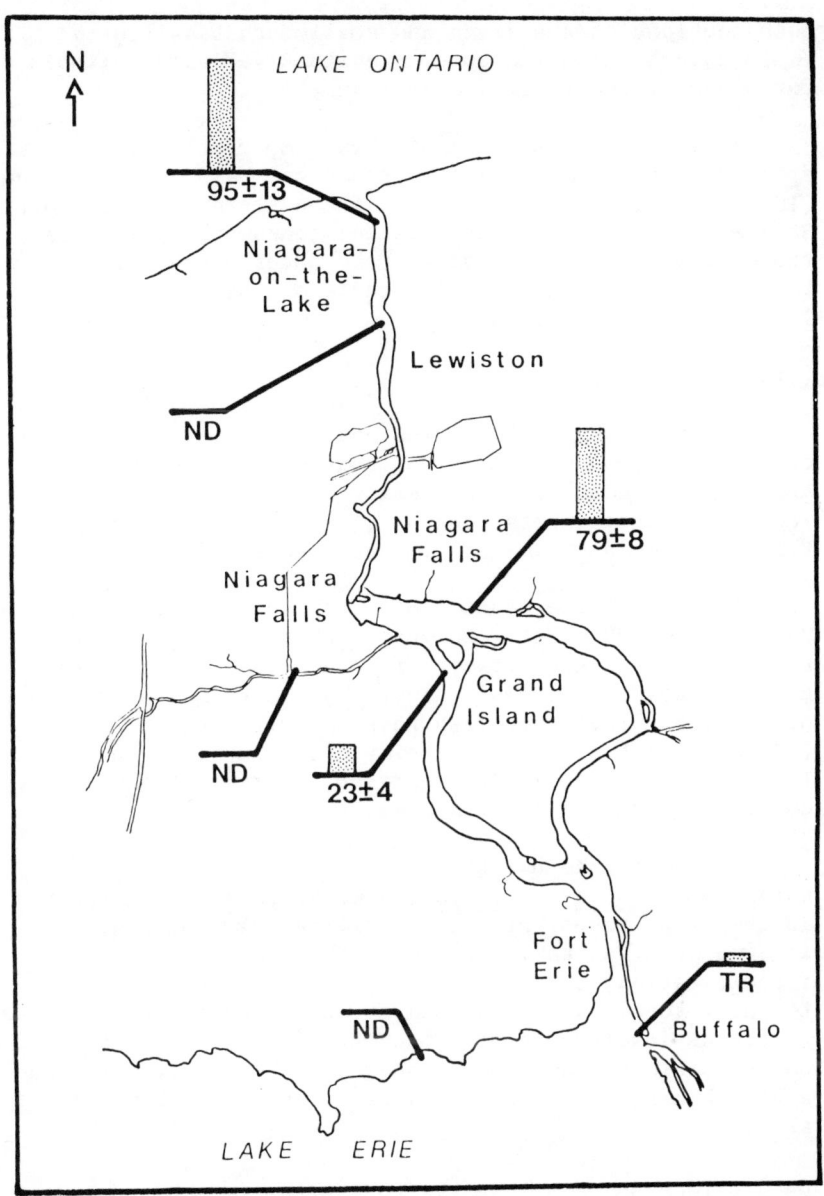

FIGURE 7. MEAN AND STANDARD DEVIATION OF PCBs CONCENTRATION (ng/g=ppb, wet weight) IN FRESHWATER CLAMS (Elliptio complanatus) AFTER 16 DAYS' EXPOSURE TO NIAGARA RIVER WATER AT VARIOUS LOCATIONS. Values are mean of three whole clam soft tissue analyses. Tr = Trace. ND = Not detected (below detection limit of 20 ng/g). From (8).

both sections of the river and smallmouth bass (20 to 45 cm in length) from the upper Niagara River were suitable for unrestricted (unlimited) consumption.

A regional comparison of three important fish species, coho salmon, rainbow trout and rainbow smelt [see (10)] shows that fish caught in the eastern basin of Lake Erie are suitable for unrestricted consumption, whereas those from the western end of Lake Ontario are suitable for "occasional meals only" because of PCBs and/or mirex levels in their tissues.

(iii) Forage Fish

Due to aging difficulties and the wide-ranging habit of many species of adult fish, contaminant residue data are not very helpful in answering a number of questions regarding contaminants control (e.g. location of sources, temporal trends and mechanisms of uptake).

To augment the adult sport fish surveys, the Toxicity Unit of the MOE's Water Resources Branch began in 1975 to conduct contaminant surveys of young fish. The program concentrates on young-of-the year or yearlings of a number of species. The young fish have a limited range (~ 2 km^2) and their age is easily determined, so that tissue contaminant levels represent current uptake conditions in a restricted area.

Young-of-the-year (3- to 4- months old) spottail shiners (Notropis hudsonius), a major forage fish species in the Great Lakes, are being utilized as a biological integrator of contaminants for problem area identification, the development of a data base for temporal trend assessment and as an indicator of hazardous subtances loadings to the Great Lakes (9). Fish are collected in the nearshore zones of lakes or rivers using a bag-seine net and processed as described by Suns et al (9).

TABLE III. CONSUMPTION GUIDELINES FOR SPORTS FISH FROM THE NIAGARA RIVER – 1980. Adapted from (10).

Legend:
- 🐟 (outline) = Unlimited Consumption / Consommation sans limites
- 🐟 (with dot) = Limited Consumption / Consommation limitée
- 🐟 (with bar) = Limited Consumption / Consommation limitée
- 🐟 (filled) = Don't Eat At All / Ne pas manger
- 🐟 (small) = Occasional Meals Only / Repas occasionnels seulement
- * = Data Not Available / Données non disponibles

Waterbody / Cours d'eau — Fish Species / Espèces de poisson	<15 (<6)	15-20 (6-8)	20-25 (8-10)	25-30 (10-12)	30-35 (12-14)	35-45 (14-18)	45-55 (18-22)	55-65 (22-26)	65-75 (26-30)	>76 (>30)
(Upper) Niagara River / Rivière (Supérieur) Niagara (Miller Creek) (Ruisseau Miller) 4258/7857 (Niagara R.M.) (M.R. de Niagara)										
Smallmouth Bass[2] / Achigan à petite bouche[2]	*	*	Limited	Limited	Limited	Limited	*	*	*	*
Yellow Perch[2] / Perchaude[2]	*	Limited	Limited	Limited	*	*	*	*	*	*
White Sucker[2] / Meunier noir[2]	*	*	Limited	Limited	Limited	Limited	Occasional	*	*	*
(Lower) Niagara River / Rivière (Bas) Niagara (Queenston) 4310/7903 (Niagara R.M.) (M.R. de Niagara)										
Coho[3] / Saumon coho[3]	*	*	*	*	*	*	*	Unlimited	Unlimited	Unlimited
Yellow Perch[2] / Perchaude[2]	*	Limited	Limited	Limited	*	*	*	*	*	*
American Eel[2] / Anguille d'Amérique[2]	*	*	*	*	*	*	Unlimited	Unlimited	Unlimited	Unlimited
White Sucker[2] / Meunier noir[2]	*	*	*	Limited	Limited	Limited	Occasional	*	*	*

Contaminant identification

The type of laboratory tests carried out on each species listed in the tables can be determined by noting the small number (1,2,3,4,5 or 6) that appears after each species name and then checking this number against the following:

Key to Analysis:

1 – Mercury
2 – Mercury, PCB, mirex and pesticides.
3 – PCB, mirex and pesticides
4 – Mercury, PCB and mirex.
5 – Mercury, other metals, PCB, mirex and pesticides.
6 – Mercury, other metals.

Example:

A species name followed by a 2 has been analysed for mercury, PCB, mirex and pesticides.

The fact that testing for a particular contaminant was carried out does not necessarily mean that the fish will contain this substance.

Fish with less than 0.5 parts per million mercury are identified in the tables of this book by 🐟.

Fish containing between 0.5 and 1.0 parts per million, identified by 🐟, and fish with 1.0 to 1.5 parts per million, identified by 🐟, can be eaten by adults (except women of childbearing age) in restricted amounts as indicated in the consumption guidelines. Fish with mercury levels in excess of 1.5 parts per million, identified by 🐟, should not be consumed.

Fish exceeding 2.0 ppm PCB, 0.1 ppm mirex, or 5.0 ppm DDT are identified in the tables of this book by 🐟.

The 1979 collections included a total of 37 sites on Lake Superior, Georgian Bay, Lakes Huron, St. Clair, Erie, Ontario and the Niagara and St. Lawrence rivers. PCBs levels in fish collected in 1979 from each of the above-listed areas were averaged on a regional basis to permit a lake-by-lake comparison.

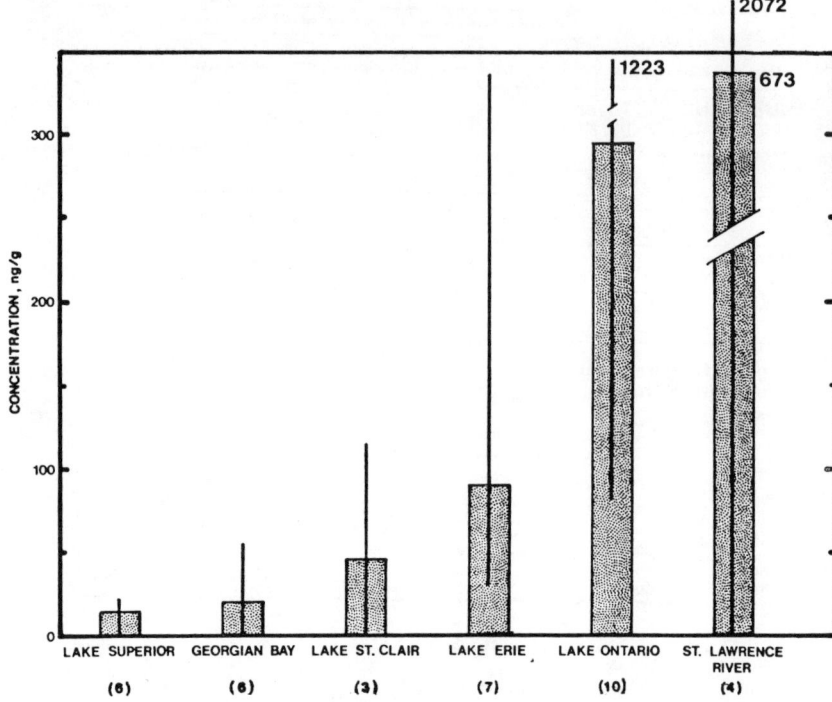

FIGURE 8. PCB CONCENTRATIONS (ng/g=ppb, wet weight) IN YOUNG-OF-THE-YEAR SPOTTAIL SHINERS (Notropis hudsonius) FROM THE GREAT LAKES - REGIONAL AVERAGES, 1979. Stippled bars denote regional averages while vertical lines and figures in brackets indicate, respectively, the range of means and number of collection sites for each region. Adapted from (9).

As shown in Figure 8, there is a pronounced increase in the average PCB concentrations in spottail shiners as one progresses from Lake Superior to the lower Great Lakes (Erie and Ontario) and the St. Lawrence River. PCBs concentrations in these fish exceeded the IJC specific

objective for the protection of birds and animals which
consume fish (100 ng/g, whole fish, wet weight) at 43% of
all sites sampled. The considerable range of means for
PCBs levels in fish of some regions (e.g. Lake Erie, Lake
Ontario and St. Lawrence River) is believed to reflect the
efficacy of this species as a source area biomonitor.

The data base for trend evaluations of contaminants
in spottail shiner fish is at present limited to nine
sites on Lakes St. Clair, Erie and Ontario and the Niagara
River (Figure 9).

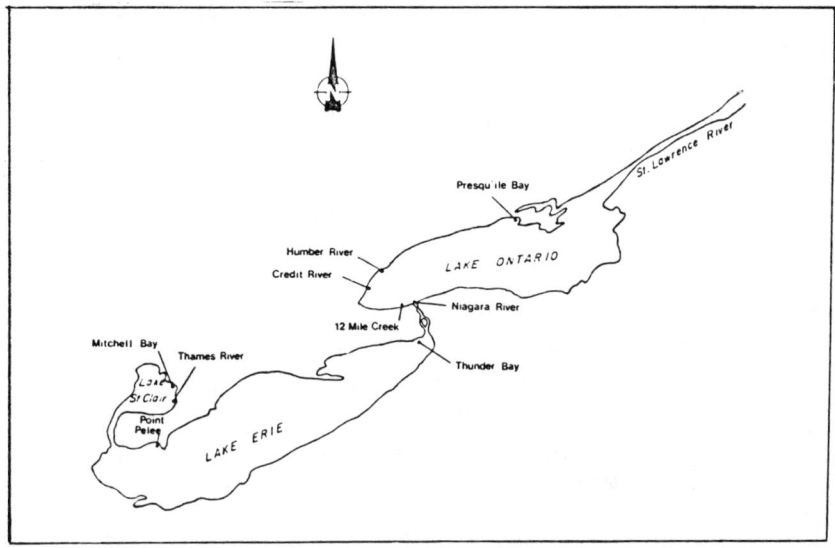

FIGURE 9. SPOTTAIL SHINER (Notropis hudsonius)
COLLECTION SITES ON LAKES ST. CLAIR, ERIE
AND ONTARIO USED IN PCBs, DDT and MIREX
RESIDUES TREND EVALUATION (1975-1979).
Adapted from (9).

The 1979 contaminant residue concentrations in
young-of-the-year spottail shiners for several of the
principal organics (PCBs, total DDT, mirex) are lower than
those in the earliest lower Great Lakes collections of
1975. The most pronounced reductions in terms of absolute
values were noted for PCBs, with decreases at the nine
sites ranging from 22% to 89% (Table IV).

TABLE IV. ORGANOCHLORINE RESIDUE CHANGES IN YOUNG-OF-THE-YEAR SPOTTAIL SHINERS IN THE LOWER GREAT LAKES, 1975-1979*.

Location	Time Span, Years	Percent Reduction		
		PCBs	ΣDDT	Mirex
Mitchell Bay, Lake St. Clair	1978-79	89	76	-
Thames River, Lake St. Clair	1977-79	66	31	-
Point Pelee, Lake Erie	1975-79	60	80	-
Thunder Bay, Lake Erie	1978-79	80	73	-
Niagara River, Lake Ontario	1975-79	78	89	>62
12 Mile Creek, Lake Ontario	1978-79	22	26	>75
Credit River, Lake Ontario	1976-79	86	75	>84
Humber River, Lake Ontario	1977-79	45	72	0**
Presqu'ile Bay, Lake Ontario	1975-79	77	58	-

- = Data not available from previous years to enable comparison.
* = Adapted from: Suns et al. (1981)
** = Not indicative of a problem, since mirex levels in Humber River fish were non-detectable in 1979 (below detection limit of 5 ppb) and at the detection limit (5 $\pm$ 2 ppb) in 1977.

While the overall downward trend was common to all sites, the reductions were site-specific and probably reflect PCB inputs and control measures in individual watersheds. As an example, trends for Point Pelee (Lake Erie), Niagara-on-the-Lake (lower Niagara River), and the Credit and Humber Rivers (Lake Ontario) are shown in Figure 10. These indicate a significant ($p<0.05$) decline in PCBs in 1978 and 1979 collections from Niagara-on-the-Lake, while over the same period, levels did not decrease significantly in the Point Pelee collections. PCBs levels decreased steadily and significantly in collections from the Credit River between 1976 and 1979. A decline in PCBs was evident in the Humber River collections between 1978 and 1979, but levels were still more than ten times the IJC specific objective of 100 ng/g.

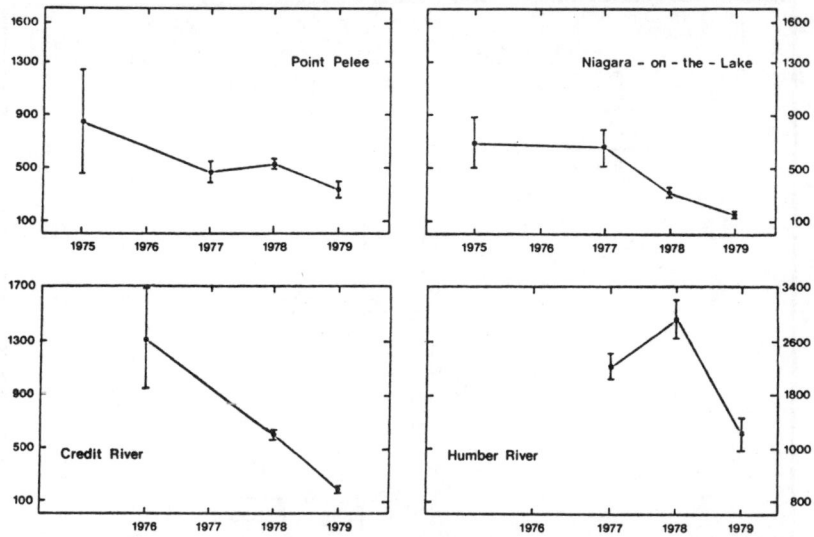

FIGURE 10. PCBs RESIDUE TRENDS IN YOUNG-OF-THE-YEAR SPOTTAIL SHINERS (Notropis hudsonius) COLLECTED AT POINT PELEE (LAKE ERIE) AND NIAGARA-ON-THE-LAKE, CREDIT RIVER AND HUMBER RIVER (Lake Ontario) FROM 1975-1979. Mean concentrations (ng/g=ppb, wet weight) and 95% confidence limits for four to ten analyses (each a composite of ten whole fish) are given. Detection limit = 20 ng/g. Adapted from (9).

The observed PCB residue declines in nearshore juvenile spottail shiners suggest that PCB loadings to the lower Great Lakes are being reduced by the restricted use and discharge regulations implemented. However, fish collected from Niagara-on-the-Lake in 1980 contained significantly (p 0.05) higher PCBs levels (266±51 ppb) than those collected in 1979 (153±23 ppb) (2). Data from the 1981 collection at this site will aid in evaluating this change with respect to the downward trend observed between 1975 and 1979.

Additional collection sites are added to the program as needed and as resources allow, and Figure 11 indicates the 1980 Niagara River collection sites (seven as opposed to two in 1979) as well as the mean PCBs levels in

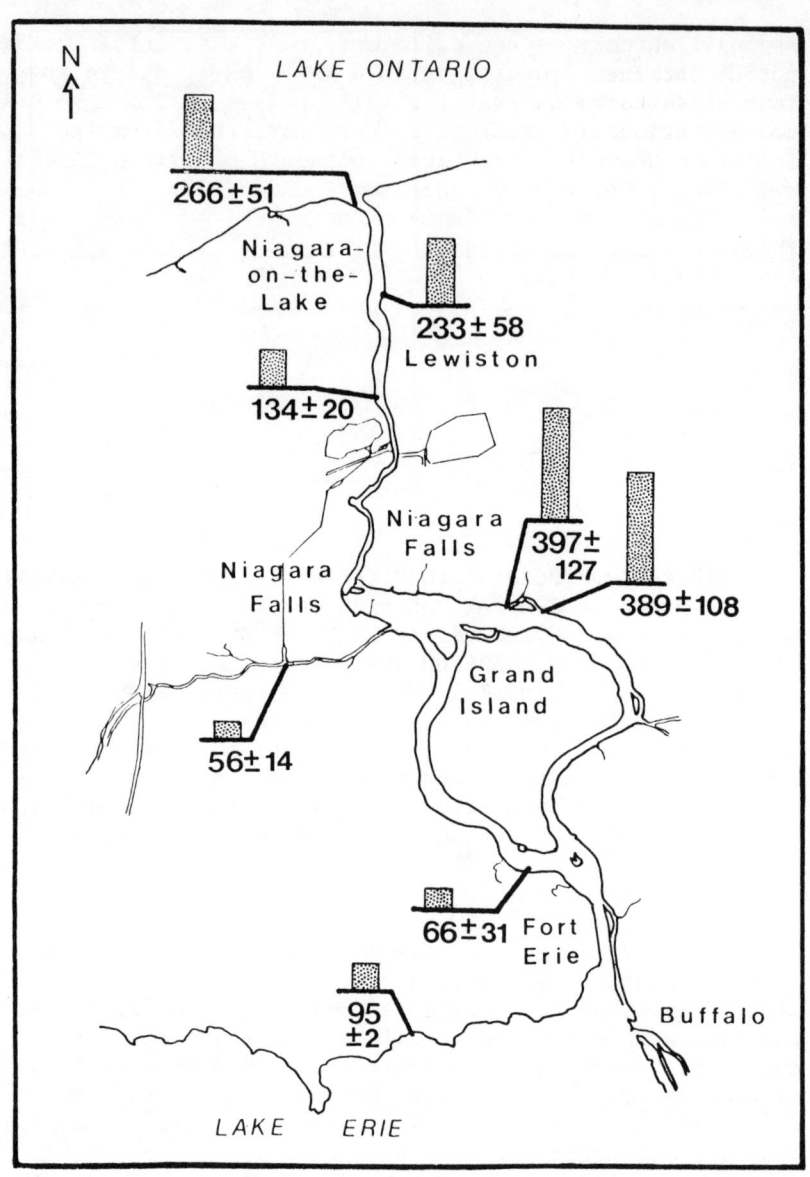

FIGURE 11. MEAN AND STANDARD DEVIATION OF PCBs CONCENTRATION (ng/g=ppb, wet weight) IN YOUNG-OF-THE-YEAR SPOTTAIL SHINERS (Notropis hudsonius) COLLECTED IN THE NIAGARA RIVER (1980). Means are of five to ten analyses (each a composite of ten whole fish). Detection limit = 20 ng/g. Adapted from (2).

spottail shiners at the different locations. (Fish could not be obtained from the Buffalo River mouth due to the lack of suitable habitat for this species.) The high concentrations in spottail shiners collected from the Tonawanda Channel (Love Canal and mouth of Little River) relative to Chippawa Channel and Thunder Bay, Lake Erie fish indicate the importance of point and non-point source inputs from the U.S. side of the river to loadings in the lower Niagara River.

SUMMARY AND CONCLUSIONS

Data collected by the Ontario Ministry of the Environment's Water Resources Branch in the Niagara River illustrates the advantages of using a variety of approaches in studying the impact, fate and temporal trends of PCB inputs to the river system.

PCB concentrations in bottom sediments of the Niagara River sampled in 1979 ranged from non-detectable to 2700 ug/kg. Although these values have not been corrected for station-to-station differences in particle size composition or organic carbon content, it is evident that samples containing high concentrations in the upper Niagara River, such as those collected at the mouth of the Buffalo River and those from the Tonawanda Channel (Tonawanda to Niagara Falls, N.Y.), represent areas close to or downstream of either past or existing PCB sources.

High PCB concentrations in some samples from the lower river, in the absence of obvious shore-based sources, reflect (at least temporary) depositional areas for particulate-associated contaminants from the upper river. However, the fact that these concentrations were detected in the top 3 cm (surface) of the bottom sediments suggests the PCBs are of recent origin and/or deposition. This is further substantiated by the data for suspended sediments collected in 1979: PCBs were detected in all samples at concentrations ranging from 69 to 1200 ug/kg. Although the data base, in terms of spatial distribution and numbers of sample sites is much more limited than that for bottom sediments, a similar picture with respect to regional differences in the upper Niagara River is evident: higher concentrations of PCBs were detected at the mouth of the Buffalo River and in the Tonawanda Channel downstream of the Love Canal/102nd Street disposal sites than in the Chippawa Channel and the lower Niagara River.

Because of the small size of the water-borne particles collected by centrifugation, and hence a higher adsorptive surface area per unit volume than a given bottom sediment sample, one could expect higher concentrations of contaminants in suspended sediments. This was the case in samples from the upper Niagara River, with ratios of concentrations in suspended to bottom sediments from 1.7:1 to 5.5:1. In the lower river, concentrations were generally higher in bottom sediments, resulting in ratios ranging from 0.09:1 to 0.9:1.

Data on raw water samples collected at Ontario water treatment plants along the Niagara River in 1979 and 1980, as well as at stations along ranges at the downstream ends of the Tonawanda and Chippawa Channels in 1980, indicate a relatively low percentage of samples (7-17%) containing detectable concentrations of PCBs (0.020-0.080 ug/L). Of the latter samples, those from the Tonawanda Channel were collected downstream of a number of suspected source areas of PCBs and other trace as well as more conventional contaminants. Apart from the station closest to the U.S. mainland, these data do not reflect shore-based contributions (point and non-point) so much as they do the general overall PCB levels in the channel, although data on other trace contaminants (mercury, iron, BHC isomers) as well as conventional parameters (total phosphorus, bacteria) exhibited a pronounced (decreasing) gradient away from the U.S. mainland towards Grand Island (11). When samples were taken closer to shore and near suspected source areas, as in the 1979 survey, PCBs were detected in 29% of upper Niagara River samples at levels from 0.020 to 0.080 ug/L. Virtually all of the samples with detectable concentrations were from stations with high bottom and suspended sediment PCB levels (i.e. Buffalo River mouth, Love Canal/102nd Street and mouth of Little River).

Since the present MOE routine analytical detection limit of PCBs in a one liter water sample is 0.020 ug/L (when analyzed as part of the PCBs/organochlorine pesticides scan, as the Niagara River samples were), the above percent detections in river water are only a conservative estimate of the actual percentage of samples exceeding the provincial objective for the protection of aquatic life (0.001 ug/L) which was set to prevent bioaccumulation in fish to levels above those deemed safe for human consumption. Further work is planned (or underway) by various agencies, including MOE, using field-concentration or large volume water sampling and extraction to determine the actual concentrations of PCBs in various sections of the Niagara River.

PCBs in migratory or far-ranging sports fish species (e.g. coho salmon, lake trout, walleye) result in part from other sources to Lake Ontario besides the Niagara River area, and hence reflect a lake-wide problem. However, mirex levels (and, more recently, 2,3,7,8-tetrachlorodibenzo-p-dioxin in lake trout) can be attributed to past or present sources in the Niagara River (2). In 1980, sport fish species more or less resident in the river, such as smallmouth bass, yellow perch and white sucker did not contain PCBs or mirex concentrations in excess of Canadian government guidelines. Such species can still move throughout the upper or lower sections of the river and therefore are not useful in pinpointing source areas of a given contaminant.

Data for young-of-the-year spottail shiners from other locations in the Great Lakes and Lake St. Clair as well as the Niagara River, indicate their usefulness in the identification of problem (source) areas, in temporal trend assessments, and in the assessment of hazardous substances loadings to the Great Lakes. For example, the higher PCB concentrations in young spottail shiners collected from the Love Canal/102nd Street and Little River sites show that PCBs are coming from these areas in large enough quantities to impact this component of the local aquatic fauna. Trend data on PCBs in fish from various Ontario nearshore sites on Lakes St. Clair, Erie and Ontario as well as in the lower Niagara River (Niagara-on-the-Lake) do, however, indicate that restrictions on the use and discharge of these chemicals are successful in reducing loadings to the lower Great Lakes. Differences in the rate of decrease in fish at the various sites likely reflect PCB inputs and control measures in individual watersheds.

While concentrations of PCBs in freshwater clam tissues were lower than those in young-of-the year spottail shiners taken from the same or nearby locations in the Niagara River, both species contained higher levels when exposed to or taken from water in the Tonawanda Channel and lower Niagara River than from Thunder Bay (Lake Erie), the Chippawa Channel or the Welland River. Differences in tissue PCB concentrations are likely due to species-related differences in age, food sources, habitat, metabolic rate, and, of course, differences in exposure time. However, the data indicate that the species of clams used can accumulate PCBs to detectable levels over a relatively short time period (2-4 days) and be in equilibrium with ambient river levels within 8 days. Also, ease of handling permits placing of clams in locations with unsuitable habitats for endemic species.

ACKNOWLEDGEMENTS

We would like to thank J. Bourns for typing the manuscript and J. Barnes for preparing the graphics.

We are also grateful to F. C. Fleischer, M. Griffiths, D. N. Jeffs, J. D. Kinkead and R. R. Weiler for their constructive review of the manuscript.

The authors also gratefully acknowledge the assistance of field personnel in the collection of samples and the efforts of the Pesticides Section of the Laboratory Services Branch in the analyses of samples.

REFERENCES

1. Canada-Ontario Review Board. "Environmental Baseline Report of the Niagara River", 32 pp. (June, 1980).

2. Canada-Ontario Review Board. "Environmental Baseline Report of the Niagara River", 31 pp + figures and tables (November, 1981).

3. Frank, R., R.L. Thomas, M. Holdrinet, A.L.W. Kemp and H.E. Braun. "Organochlorine insecticides and PCB in surficial sediments (1968) and sediment cores (1976) from Lake Ontario". J. Great Lakes Res., $\underline{5}$:18-27 (1979).

4. "Handbook of Analytical Methods for Environmental Samples". Ont. Min. of the Env., Lab. Services Br.

5. "A Guide to the Collection and Submission of Samples for Laboratory Analysis". Ont. Min. of the Env., Lab. Services Br., 46 pp., July, 1979.

6. "Interagency Task Force on Hazardous Wastes". Draft Report on Hazardous Waste Disposal in Erie and Niagara Counties, New York, March, 1979.

7. Curry, C.A. "The freshwater clam (Elliptio complanatus), a practical tool for monitoring water quality", Water Poll. Res. Canada, $\underline{13}$:45-52 (1977/78).

8. Kauss, P.B., M. Griffiths and A. Melkic. "Use of freshwater clams in monitoring trace contaminants source areas", paper presented at Technology Transfer Conference No. 2, Toronto, Ontario, November 24, 1981 (in press).

9. Suns, K., C. Curry, G.A. Rees and G. Crawford. "Organochlorine Contaminant Declines and Their Present Geographic Distribution in Great Lakes Spottail Shiners (Notropis hudsonius)". Ont. Min. of the Env. Report, 18 pp. (April 1981).

10. "Guide to Eating Ontario Sport Fish - 1981, Northern Ontario, Southern Ontario and Great Lakes". Government of Ontario, 1981.

11. Kauss, P.B. "Studies of trace contaminants, nutrients and bacteria levels in the Niagara River. Paper presented at XXV IAGLR Conf., Sault Ste. Marie, Ont., May 4-6, 1982 (to be published in J. Great Lakes Res.).

CHAPTER 21

VOLATILIZATION OF PCB FROM
SEDIMENT AND WATER: EXPERI-
MENTAL AND FIELD DATA

T. J. Tofflemire and T. T. Shen
New York State Department of
Environmental Conservation

E. H. Buckley
Boyce Thompson Institute

INTRODUCTION

Studies of the Hudson River PCB problem have indicated that the volatilization losses of PCB are substantial in spite of the fact that the chemical has a very low vapor pressure. As noted in another report (1) this is also true for a number of other toxic chlorocarbons. For a number of uncapped PCB dump sites and dredge spoil sites along the Hudson River the calculations indicated that for PCB the volatilization loss (3205lb/yr) was greater than the erosional (590lb/yr) or groundwater losses ~4lb/yr losses from the sites. Approximate calculations of the water to air volatilization also gave substantial PCB losses as noted in prior studies (2). Thus, additional studies were done on PCB volatilization, air monitoring was conducted and vegetation sampled for PCB analysis. This paper summarizes the initial data from these studies.

THEORY: WATER:AIR TRANSFER

The transfer of substances such as oxygen and PCB can be described by the two resistance model for mass transfer in which the liquid phase resistance controls (3,4,5) which yields the equation

$$\text{Mass flux} = dw/dt = -K_L (C - P_i/H)$$

where w is the mass of substance
    t is time
    K_L is the liquid phase mass transfer coefficient

C is the aqueous concentration
P is the atmospheric partial pressure
H is the Henry's Law constant

Compounds with a low vapor pressure and a very low solubility may thus have a high Henry's Law constant and will volatilize rapidly from water. Values of K_L for rivers have been discussed by O'Connor (3), Southworth (6) and for oceans by Liss and Slater (7). Variation of vapor pressure with temperature can be expressed by the Clapeyron equation (8) and used to estimate the effect of temperature on H.

THEORY: SOIL-AIR TRANSFER

For this process, the effective vapor pressure of the hydrocarbon is reduced in accordance with its strength of adsorption to the solid adsorbent or soil (8). However, the effective surface area for desorption is increased on fine particles over what it would be for liquid only. Also, much higher concentrations of hydrocarbons can be adsorbed on soils than are soluble in water. Studies by General Electric (GE) demonstrated that the rate of volatilization of PCB 1242 is much less for topsoil than it is for coarse sand (9). This was also confirmed in a study by Hague et al. (10). Several additional studies have confirmed that PCB adsorption to substrates is proportional to their organic carbon content (11,12). One of the common substrates high in organic carbon is the A horizon of an organic top soil. In a surface layer of soil, the mass flux rates for chlorinated hydrocarbons increased with soil moisture content (8).

The rate of diffusion of the PCB vapors through the soil must also be considered. Finer soils and an increased moisture content slows this diffusion (13, 14). Thus, the calculation of PCB volatilization from a particular soil is complicated. Experiments on the particular contaminated soil are needed to accurately predict the volatilization rate.

Equations cited by Shen (15) can be used to give an estimate of volatilization rates of various chemicals from capped or exposed soils. For open dumps the volatilization rate was a function of the square root of the wind speed. For capped sites the volatilization rate was directly proportional to the cap thickness.

WATER TO AIR PCB VOLATILIZATION TANK STUDY

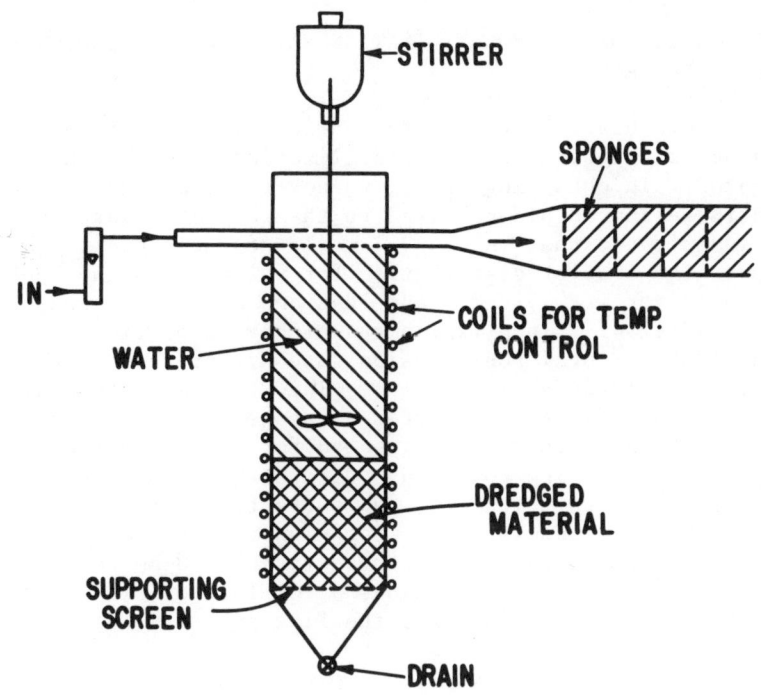

Figure 1 System for Environmental Loss Tests

Figure 1 shows the tank GE used for a series of volatilization tests. It was 0.2 m (8 in) in diameter and .78 m (31 in) high from the supporting screen to the water surface. The results of the water to air tests indicated that wind speed, turbulence and temperature were important variables. Detailed results were given in other reports (2,16).

To allow comparison of the oxygen reaeration rate constant k_2 with the PCB volatilization rate, a study of the k_2 rate in the GE tank was made. The methods and detailed data are given elsewhere (16).

It was found that stirring at 60 to 80 rpm increased the k_2 rate. The ratios of k_{PCB}/k_2 were about .13 to .31 for this study. This is in the same range as the ratios reported by Paris et al. (17) which ranged from .15 to .25 for Aroclors 1016 and 1242 at k_2 values of 1.2 to 5.0/hr.

These turbulence levels are higher than those used in the GE tank study. The k_2 values and turbulence levels used in the GE tank were similar to those in the Hudson River.

CALCULATION OF PCB VOLATILIZATION RATES FOR THE HUDSON RIVER INCLUDING DAMS

If accurate k_2 values for various reaches of the Hudson can be obtained, then PCB volatilization rates can be calculated. In obtaining the k_2 values for the Hudson, special consideration should be given to the reaeration at the dams in addition to that in the 3.7 m (12 ft) deep pools between dams. The Aroclor distribution for sediment in the Upper Hudson River was 75% 1016 and 1242 and about 12% each of 1221 and 1254. Since 1221 has a higher volatilization rate than 1342, a total PCB emission rate higher than that predicted for 1242 is likely (16). The PCB loss rate mass transfer coefficient for water saturated with 1242, obtained in the GE tank tests and corrected for average Hudson River temperature and ice cover was .01 m/hr (2). A K_L value of several times this results when correction for average Hudson River winds is made. For higher wind speeds or PCB concentrations the volatilization rate would be correspondingly higher. At normal winds a PCB loss of 910 kg/yr (2000 lb/yr) is possible for the flat pools.

US Geological Survey did two gas and dye tracer studies to determine the gas transfer coefficients and oxygen reaeration k_2 values for the Hudson River near the Ft. Miller dam. The theory, the procedure used and results are described elsewhere (18,19,20,21,22). The first reach was an upstream pooled area of a mean depth of about 2.3 m, while others included dams. It is found that the k_2 values or gas transfer rates increase for the dam sections. If a section with a 20 min travel time over the dam is used, the k_2 rate doubles to 19.6 day^{-1}. Thus, it is possible that PCB air concentrations are 10-20 times higher over a dam than over a quiet river pool. It was found that the PCB concentration change over a one km river reach was only 2.3%, while the loss across the 3.5 m dam was 5.6%. The Upper Hudson (Lock 7 to Troy Dam) has 8 dams thus approximately 45% of the PCB may volatilize. To use this approximate calculation one would have to assume that PCB is resupplied to the water by downstream bed sediments. It is possible that 450 kg of PCB were lost from the Upper Hudson dams which have an average head loss of 4.5 m. Thus, a total annual Upper Hudson PCB loss of 450 to 900 kg/yr for the quiet pools and 450 kg for the dams was possible at an assumed .2 µg/L PCB concentration. Rathburn (19) also noted that the presence of an oil film on water reduced the oxygen, ethylene and propane gas transfer rates.

SEDIMENT TO AIR PCB VOLATILIZATION STUDY AND AIR DATA

In the earlier studies, described elsewhere (2, 16) the tank in Figure 1 and sandy sediment from the Buoy 212 area of the Hudson River were used. The results indicated a substantial increase in volatilization with temperature over the 10 to 39°C range.

In May 1979, a second series of runs were made with the GE tank, employing sediments higher in PCB, to determine the effect of sediment PCB concentration on volatilization rates and the effectiveness of various capping materials. Three sediments were used. The Aroclor distribution was 75-80% 1016 and 1242 with lesser percentages of 1254 and 1221.

Using the GE tank and two sediments, various cover materials were tested for the degree to which they reduced PCB volatilization. The materials included manure (3"), muck soil (6"), clay (3.5") a 50/50 sand, paper mill sludge mixture (6"), Becket soil (6") and a chlorinated polyethylene liner (30 mil). The percent PCB emission reduction varied from 90% to 99.85%. The manure, muck soil and clay shrank as they dried and allowed PCB vapor to escape. The silty soil (Becket B horizon), the polyethylene liner and a layer of water performed best in the tank study. It was theorized that the materials high in organic content would be effective in adsorbing PCB vapors. This may have occurred, but shrinking masked the effect. In field situations, shrinking and cracking of these capping materials may also be a problem, unless they are kept moist.

PCB emission concentrations were measured using Florisil cartridges (23). Air samples were also taken 1 m above the ground at dump or dredge sites with 3 to 5 replicates. Background air concentrations in the Fort Edward area were less than 20 ng/m³ which is about the detection limit for a 24 hour sample. For comparison, air samples for PCBs near Lake Michigan averaged 7 ng/m³ (24), in Maryland 3-9 ng/m³, and in Providence, R.I. as high as 9.4 ng/m³ (25). The results showed that the air concentration (in µg/m³) is 0.002 to 0.011 of the sediment concentration (in µg/g) in both the tank and at the dredge sites with a trend for the lower ratios to apply at the higher sediment concentrations (26).

PCB IN PLANTS

Standard procedures were used to collect, prepare and analyze plant samples for PCBs. Field samples were collected in plastic bags, frozen in coolers containing dry ice, freeze dried, ground and stored in glass bottles. Ground tissues were extracted with petroleum ether, concentrated, passed through Florisil columns, concentrated to measured volumes and analyzed by gas chromatography using an electron capture detector. Both Aroclor 1242 and 1254 were measured but only total PCBs are reported here since the quantities of Aroclor 1254 (20% of total) were too variable. In field surveys, the standard deviation for total PCB content in foilage was $\pm$14% (27).

PCB contamination in plants around PCB dumps and dredge spoil sites of the upper Hudson River was first documented by Weston, Inc. in 1977 (11). A variety of plants were analyzed including trees, grasses, weeds and a few agricultural crops. All contained PCBs in the environs of PCB sources. Extremely high levels of PCBs (up to 2800 ppm in leaves on a dry weight basis) were found in plant species growing on PCB dumps, while in areas removed from known PCB sources, concentrations were usually below the Weston limits of detection (0.2 ppm or 0.1 ppm in some cases). Subsequently, staff of Boyce Thompson Institute documented PCB gradients along transects radiating out from PCB dumps and showed that measurable PCB accumulations in foilage extended as far as 700 and 1000 m from highly contaminated local sources. The PCB content in leaves of trembling aspen, along an easterly transect from the Fort Miller dump site, raged from 180 ppm at the site to 6.6 ppm at 41 m, 2.0 ppm at 92 m, 0.54 ppm at 250 m and to background levels of 0.11 to 0.14 ppm at 1000 m. Topographically, this transect is up-gradient of surface runoff and of leachate flow from the dump so that neither soil nor plants are exposed to PCB contaminated drainage. Data from current studies suggest that foliar PCB content is a reliable indicator of air concentration and that very little PCBs are translocated through the roots and stems to the leaves even when plants are growing on highly contaminated sediments. However, when using plants to assess volatile PCBs in the environment, it is important to recognize many-fold differences in foliar PCB uptake between different species (28). Some plants are readily contaminated by volatile PCBs in the atmosphere and small increases lead to PCB accumulation in forage plants that could make them unsuitable (0.2 ppm according to U.S. FDA regulations) for the dairy industry.

CONTROL TECHNIQUES AND DISCUSSION

These generalized control techniques for PCB volatilization are (29):

1. destroy the wastes, prevent the problem
2. install an impermeable cap and a gas collection system.
3. install a temporary cap of soil or organic material.

The best method is to destroy the wastes by incineration or by chemical processes. However, this is often too expensive for large volumes of wastes with low PCB concentrations. Relatively few approved facilities are available to destroy PCB wastes.

A second method of control involves use of impermeable caps and artificial liners. Gas collection systems can also be constructed under the cap to collect the PCB vapors for subsequent removal on activated carbon filters. When degradable organic material is in the waste along with PCB, biological gas generation will build up pressures and cause the PCB vapors to be flushed out. Clay caps must be checked occasionally for cracks due to shrinking, swelling, settlement or woodchucks. With artificial liners, punctures or rips due to settlement forces can be a problem. Farmer (13) noted that .01 cm of polyethylene film was equivalent to about 1.36 cm of silty soil in limiting hexachlorobenzene (HCB) waste volatilization. This would make a 30 mil PE liner equivalent to about 25 cm of soil. The HCB waste chemically affected and swelled the PE film after a period of time (13). Weeks and McLeon (30) noted that PCB wastes permeated through polyethylene quickly. The recommended viton elastomer as the best material for gloves to prevent PCB passage. Several authors have given equations for calculating vapor transmission through cap materials (31, 32, 33).

The collection pipes for a vertical type gas collection system must be under an impermeable seal or cap. The vertical wells should penetrate 3/4 of the waste depth. With the horizontal gas collection system flexible pipes should be used to avoid shearing upon settlement. Both systems work better with a pump to draw out the gases (29, 34).

Several authors have noted substantial biodegradation of the lower chlorinated PCB Aroclors under aerobic conditions in a tank with water and in a compost pile (35, 36). Others have noted adsorption of PCB vapors to organic media (12, 13). Therefore, the third technique of installing an economical temporary cap of a material high in volatile solids which is intended to diffuse the PCB vapors at a very low rate, has some merit. This was tried at the Caputo site near the Hudson River and a 99.9% reduction in air PCB concentrations was achieved. The cap consisted of 10 cm manure, 70 cm of a paper mill sludge and sand mixture and 5 cm of organic topsoil. Care must be exercised with this third type of cap, that excessive shrinkage and cracking does not occur. Mixing the organic material with top soil reduces the shrinkage and cracking potential.

Proposed criteria were developed for handling PCB dredge and dump sites along the contaminated Hudson River. For sites with less than 1 µg/g no action is presently being required. For sites with sediment with PCB concentrations in the 10 to 50 µg/g range, capping is suggested. Few new sites with greater than 50 µg/g of PCB, a secure bottom liner material is also required and incineration is recommended. Federal regulations (37) suggest ploughing of sediments and soils, containing 10-50 µg/g PCB, into the topsoil to limit volatilization and promote biodegradation, while proposed New York State regulations would allow this for only <10 µg/g PCB levels (38).

SUMMARY

A brief discussion of mass transfer theory of water-air and solid-air interfaces has been given to illustrate how PCBs can volatilize from water and sediments in the Hudson River.

Experimental data indicated that volatilization of PCBs can be an important source of air pollution under certain environmental conditions. The results of field monitoring have shown that PCB concentrations are fairly high in the ambient air and in vegetation near the PCB dump sites and certain contaminated dredge sites.

PCB volatilized at substantial rates from contaminated water and sediment. For a number of open PCB dump sites and dredge spoil sites along the Upper Hudson River, it was observed that volatilization of PCB was a worse problem than was groundwater contamination although traditional control programs have been aimed at preventing groundwater pollution. Improved methods of preventing and controlling volatilization losses are also needed and their long term costs and consequences must be considered.

ACKNOWLEDGEMENTS

Acknowledgement is given to Charles M. McFarland and staff of the General Electric Company and Richard Murdock of NYSDEC for their help in performance of the studies.

REFERENCES

1. Shen, T.T. and Tofflemire, T.J. "Air Pollution Aspects of Land Disposal of Toxic Waste", Technical Paper No. 59, NY State Department of Environmental Conservation, 50 Wolf Road, Albany, NY 12233-0001 (March 1979).

2. Tofflemire, T.J., et al. "Volatilization of PCB from Sediment and Water: Experimental and Field Data", Technical Paper No. 63, NY State Department of Environmental Conservation, 50 Wolf Road, Albany, NY 12233-0001 (1982).

3. O'Connor, D.J. "Stream and Estuarine Analysis", Workbook from Manhattan College, New York, NY (March 1967).

4. Mackay, D. and Leinonen, P.J. "Rate of Evaporation of Low-Solubility Contaminants from Water Bodies to Atmosphere." Environ. Sci. and Technology, $\underline{9}$, 1178 (Dec. 1975).

5. Cohen, Y., et al. "Laboratory Study of Liquid-Phase Controlled Volatilization Rates in Presence of Wind Waves." Environ. Sci. and Technology, $\underline{12}$, 553 (1978).

6. Southworth, G.W. "The Role of Volatilization in Removing Polycyclic Aromatic Hydrocarbons from Aquatic Environments." Bull. Environ. Contam. Toxicol., $\underline{21}$, 507 (1979).

7. Liss, P.S. and Slater, P.G. "Flux of Gases Across the Air-Sea Interface." Nature, $\underline{247}$, 181 (1974).

8. Spencer, W.F. and M.M. Cliath. "Transfer of Organic Pollutants Between the Solid-Air Interface." Fate of Pollutants in the Air and Water Environments Part I, edited by I.H. Suffet, 107-109, John Wiley & Sons (1977).

9. McFarland, C.M., Brooks, R.E. and Su, T. "Unpublished Data on PCB Volatilization." General Electric, Corp. Research & Development, Niskayuna, NY (1978).

10. Hague, R., et al. "Aqueous Solubility Adsorption and Vapor Behavior of Polychlorinated Biphenyl Aroclor 1254". Environ.Sci. & Technology, $\underline{8}$, 139 (Feb. 1979).

11. Weston Environmental Consultants. "Migration of PCBs from Landfills and Dredge Spoil Sites in the Hudson River Valley, NY. Final Rpt.", W'chester PA (Nov. 1978).

12. Griffin, R.A. and Chian, E.S.K. "Attenuation of Water Soluble Polychlorinated Biphenyls by Earth Materials." Illinois State Geological Survey, Univ. of Illinois, Urban, IL 61801 (April 1979).

13. Farmer, W.J., et al. "Problems Associated with the Land Disposal of an Organic Industrial Hazardous Waste Containing PCB", paper from Dept. of Soil Science, Univ. of Calif., Riverside, CA 92502 (July 1976) and EPA Report 600/2-80-119 (Aug. 1980).

14. McCord, A.T. and Zeigler, R.C. "Study of the Rate of Emission of Gases Leaving Industrial Solid Waste Landfills", paper available from Calspan Corp., P.O. 235, Buffalo, NY 14221.

15. Shen, T.T. "Estimating Hazardous Air Emissions from Disposal Sites." Pollution Engr. $\underline{13}$(8), 31 (Aug. 1981).

16. Tofflemire, T.J., et al. "PCB in the Upper Hudson River: Sediment Distributions, Water Interactions and Dredging." Tech. Paper No. 55, NY State Dept. of Environmental Conservation, Bureau of Water Research, Albany, NY (Uan. 1979).

17. Paris, D.F., et al. "Role of Physio-Chemical Properties of Aroclors 1016 and 1242 in Determining their Fate and Transport in Aquatic Environments." Chemosphere $\underline{4}$, 319 (1978).

18. Dodge, W. and Mercer, G. "Dam Reaeration a Case Study: The Oswego River." Report by School of Civil and Environ. Engr., Cornell University (May 1980).

19. Rathburn, R.E., et al. "Laboratory Studies of Gas Tracers for Reaeration." Proc. Amer. Soc. Civil Engr., Jour. Environ. Engr. Div., $\underline{104}$(EE2), 215 (April 1978).

20. Tsivoglou, E.C. and Neel, L.A. "Tracer Measurement of Reaeration: Predicting the Reaeration Capacity of Inland Streams." Jour.Water Poll.Control Fed. $\underline{48}$(12), 2269 (Dec. 1976).

21. Steadfast, D.A. Letter dated, November 1981 and

unpublished data. U.S. Geological Survey, Albany, NY 12201 (1981).

22. Rathburn, R.E. and Tai, D.Y. "Technique for Determining the Volatilization Coefficients of Priority Pollutants in Streams." Water Research, $\underline{15}$, 2432 (1981).

23. Dell'Aqua, B. "Polychlorinated Biphenyls in Ambient Air", part of Analytical Handbook NY State Dept. of Health, Empire State Plaza, Albany, NY (1981).

24. Murphy, T.J., et al. "Polychlorinated Biphenyls in Precipitation in the Lake Michigan Basin", EPA 600/3-/8-071, EPA Lab., Duluth, Minn. (July 1978).

25. Brinkman, M., et al. "Distribution of Polychlorinated Biphenyls in the Fort Edward, New York Water System." Environ.Mgt. $\underline{4}$(6), 511 (1980).

26. Kerr, R. and Hawley, J. Unpublished Data, Division of Resources, NY State Dept. of Environmental Conservation. 50 Wolf Rd., Albany, NY (1979).

27. Buckley, E.H. Unpublished Data Files, Boyce Thompson Institute, Cornell Univ., Ithaca, NY (1981).

28. Buckley, E.H. "Accumulation of Airborn PCBs in Foliage." Conditionally accepted for publication in Science(1982).

29. Shen, T.T. "Control Techniques for Gas Emissions from Hazardous Waste Landfills." Jour. of the Air Poll. Cont. Assn., $\underline{31}$(2), 139 (1980).

30. Macleod, K.E. "Sources of Polychlorinated Biphenyls into the Ambient Atmosphere and Indoor Air." EPA Report 600/4-78-022. Research Triangle Park, North Carolina 27711 (March 1979).

31. Thibodeaux, L.J. "Estimating the Air Emissions of Chemicals from Hazardous Waste Landfills." Jour. of Hazardous Materials, $\underline{4}$, 235 (1981).

32. Thibodeaux, L.J., et al., "Models of Mechanisms for Hazardous Chemical Emissions from Landfills." Dept. of Chemical Engr., Univ. of Arkansas, Fayetteville, Arkansas 72701 (1981).

33. Hwang, S.T. "Guidance Document for Subpart F Air

Emission Monitoring Land Disposal Toxic Air Emissions" - USEPA, Office of Solid Waste, Washington, D.C.

34. Conestoga-Rovers & Associates. "Gas Recovery and Utilization from a Municipal Waste Disposal Site." 651 Colby Dr., Waterloo Ontario, N2V 1B4 (July, 1979).

35. Bobal, R. "Fate of Polychlorinated Biphenyls During Bench Scale Composting of Solid Waste." M.S. Thesis, Rutgers Univ. (May 1981).

36. Anderson, M.L. "Degradation of PCB in Sediments of the Great Lakes." A Ph.D. Dissertation at the Univ. of Michigan (1980).

37. Environmental Protection Agency. "Criteria for Classification of Solid Waste Disposal Facilities and Practices, Part 9." Federal Register, $\underline{44}$(179) (Sept. 13, 1979).

38. NY State Dept. of Environmental Conservation. "6 NYCRR Part 360, Solid Waste Management Facilities." 50 Wolf Ro Road, Albany, NY (May 1981).

39. Dexter, R.N. "An Application of Equilibrium Adsorption Theory to the Chemical Dynamics of Organic Compounds in Marine Ecosystems." Ph.D. Dissertation, Univ. of Wash., Seattle, WA (1976).

40. Hetling, L.J., *et al*. "Summary of Hudson River PCB Study Results." Tech Paper No. 51, NY State Dept. of Environmental Conservation, 50 Wolf Rd., Albany, NY 12233-0001 (July 1978).

41. Threshold Limit Values for Chemical Substances and Physical Agents in the Workroom Environment. The ACGIH Publication, $\underline{13}$ and $\underline{26}$ (1978).

42. Occupational Exposure to PCBs. The DHEW NIOSH Publication No. $\underline{77\text{-}225}$, 3 (Sept. 1977).

CHAPTER 22

APPLICATION OF A SEDIMENT DYNAMICS MODEL FOR ESTIMATION OF VERTICAL BURIAL RATES OF PCBs IN SOUTHERN LAKE MICHIGAN

David Weininger
U.S. Environmental Protection Agency
Environmental Research Laboratory
6201 Congdon Boulevard
Duluth, Minnesota 55804

David E. Armstrong

Deborah L. Swackhamer
Water Chemistry Program
University of Wisconsin-Madison
660 North Park Street
Madison, Wisconsin 53703

IMPORTANCE OF PCBs IN LAKE MICHIGAN SEDIMENTS

PCB residues are observed in the sediments of Lake Michigan. The recovery of Lake Michigan from PCB contamination, as reflected by PCB levels in fishes, depends on the rates of reduction in PCB input from external sources and removal of PCBs present in the lake system. The loading of PCBs to the Great Lakes should be decreasing. Consequently, if the PCB burden of the lake is also declining, recovery should be occurring.

The removal of PCBs from the lake water to the bottom sediments is relatively rapid (residence time ~ 2 years) (1), but PCBs in surficial sediments are resistant to degradation (2) and are recycled into the pelagic food chain by bottom-feeding organisms and resuspension. However, the PCBs in the surficial sediments are gradually buried through the processes of sedimentation.

It seems possible that sediments could play a role as an ultimate sink for PCBs if the normal accretion/burial process is faster than the rate at which PCBs can diffuse through sediments. The mobility of PCBs in fine grained sediment is apparently very low (~5 x 10^{-10} cm^2 sec^{-1}) and is not expected to compete with the linear net sedimentation rates in the depositional zones found in any of the Great Lakes with the possible exception of Lake Superior (3).

A vertical burial model is developed here to estimate the time scale of the burial process. The model is implemented for PCBs in southern Lake Michigan. Vertical sediment dynamics are two-dimensionally interpolated from available literature data. PCB distributions over the southern basin are presented. The burial process is investigated by examining model projections of the hypothetical case where sedimentation continues as normal but with a zero PCB level.

SEDIMENT DYNAMICS

Vertical dynamics of the surficial Lake Michigan sediments has been intensively investigated by others using radionuclide profile methods (4,5) and is described briefly here. A single layer mixed zone model was calibrated with the naturally occurring Pb-210 sediment profile. For calibration, time-invariant sedimentation rates, bulk densities, and mixed zone depths were fitted to the observed profile assuming a constant Pb-210 input rate. The one layer model was found to adequately describe the observed core profiles. The model was then used to predict the shape of the Cs-137 profile given the variable levels of Cs observed in fallout. Predicted profiles matched observed zones exceedingly well. These results appear to be an adequate verification of the utility of the one layer mixed zone model for dealing with the dynamics of burial through the sediment surface.

VERTICAL BURIAL MODEL

The model used here to project vertical burial is essentially the same as used by Robbins and Edgington (4), but extended into the future. PCB is treated as a contaminant strongly bound to sediment, similar to lead and cesium. For the purposes of future projection of vertical burial, the original model can be operationally simplified while retaining the ability to use the original parameterization.

It is assumed that the surface sediment layer, of thickness S, is completely and continuously mixed. Sediment is added to this mixed zone at the surface and mixed zone depth is maintained by loss of material at the base. The mixing rate of material between the mixed zone and underlying sediment is assumed to be zero. For our purposes, there is no need to track the contamination level of material exiting the mixed zone since we are concerned solely with the uppermost sediment layer. Under these assumptions, only PCB in the mixed zone can move upwards. Similarly, we do not need to account for sediment compression. In the simulations presented here mixed zone material is assumed to have uniform and constant porosity (0.9), and solids density (2.45 g/cm^3). Originally, the vertical burial model was implemented using a finite difference approach similar to that used by Robbins and Edgington (4). For the simulations presented here, the model was reprogrammed for use on a PDP 11/70 with analytical solution. The analytical solution is presented below.

For the case where sedimenting material is contaminant-free, the decrease in contaminant surface concentration is directly related to sediment replacement in the mixed zone, and is first order in concentration, C:

$$\frac{dC}{dt} = -kC \qquad (1)$$

Separating variables, integrating over time, and solving for the integration constant given that $C_{t=\infty} = 0$ and $C_{t=0} = C_o$ yields:

$$C_t = C_o \exp(-kt) + 0 \qquad (2)$$

The time constant k is the replacement rate of mixed zone material:

$$k = \frac{R}{S} \qquad (3)$$

where R = linear sedimentation rate, cm/yr, and S = mixed zone depth, cm. Linear sedimentation rate R can be calculated directly from mass sedimentation rate:

$$R = \frac{\omega}{(1-\phi)\,\rho s} \qquad (4)$$

where ω = mass sedimentation rate, g/cm^2/yr; ϕ = porosity, cm^3/cm^3; and ρs = solids density, g/cm^3.

For the case where sedimenting material is contaminated at a level C', the differential equation is:

$$\frac{dC}{dt} = -k(C-C') \quad (5)$$

Given that $C_{t=\infty} = C'$ and $C_{t=0} = C_0$, this yields the following result by reduction:

$$C = (C_0 - C') \exp(-kt) + C' \quad (6)$$

where k is as per equation 3. Equation 6 is easily implemented given C_0, C', R, S, and a time increment. The concentration units of C' must match those of C. Data in the usual terms of μg contaminant per gram dry weight of sedimentary particulate (C_{ss}) must be corrected for mixed zone bulk density, viz:

$$C' = C_{ss} (1-\phi) s \quad (7)$$

Physical parameters used to implement the burial model were obtained from John Robbins (Table 1, ref. 6). It should be mentioned that these values represent data available in 1978; some of the values may have been revised after subsequent cores were analyzed.

The fact that the burial model is an extension of the model used to derive the parameters and S speaks strongly for the validity of their use. It is certain that the single layer model is simpler than reality. In nature, the mixing rate of sediments is a variable function of depth and net sedimentation is the difference between gross sedimentation and resuspension. But since the parameters and S are fitted to observed data (Pb-210) and demonstrably predict surface sediment dynamics (Cs-137), there is good reason to expect continued validity in predicting future dynamics. Underlying this is the further assumption that and S can be considered time-invariant; this appears to be true and is argued at length elsewhere (7).

CURRENT PCB CONCENTRATIONS IN SURFICIAL SEDIMENT

PCB analysis was conducted on sediments from 30 sites obtained during three cruises in southern Lake Michigan in the summers of 1978 and 1979. In most cases, cores were obtained with a 9 x 25 cm box corer. Sediments were obtained with a ponar sampler at a few locations. The cores and ponar samples were sectioned in the field and stored in glass jars at 4°C until analysis was conducted. The data presented here represents the upper section (0.5 or 1 cm) of the sample. The sediments were extracted by continuous steam distillation/extraction followed by cleanup by elution

TABLE 1. Physical data used to implement sediment dynamics model (from Robbins, 6). Note that this data is preliminary.

Station Location* km		Mass Sedimentation Rate (mg/cm^2/yr)	Mixed zone thickness (cm)
East	North		
106.0	203.8	59.0	NA
114.3	203.8	69.0	NA
120.0	203.8	52.0	NA
112.8	184.8	75.0	0.5
108.5	185.3	35.0	1.0
104.5	166.5	75.0	0.5
119.3	166.5	83.0	2.0
127.8	166.5	20.5	1.0
60.5	72.0	15.0	1.6
80.5	147.0	NA	0.3
93.7	147.0	57.0	1.0
122.0	148.3	14.0	2.0
135.3	148.3	95.0	3.5
81.0	128.8	13.0	2.0
95.0	128.8	73.0	2.0
108.5	129.8	74.0	1.0
122.5	129.8	76.0	2.0
136.0	130.5	102.0	1.5
95.5	108.8	9.0	3.0
109.3	109.3	11.0	3.0
113.3	115.0	30.0	2.0
102.8	120.0	25.0	1.4
123.3	110.0	31.0	3.0
128.8	107.8	33.0	2.0
136.0	110.5	26.0	3.0
38.3	91.0	23.0	0.5
55.5	91.0	35.0	1.0
82.8	92.3	9.0	1.0
96.0	92.8	17.0	2.0
110.0	93.3	53.0	2.5
112.0	86.0	64.0	3.7
126.0	80.0	46.0	0.5
98.8	73.8	20.0	2.0
108.3	73.8	35.0	2.0
90.5	53.8	18.0	1.0
100.0	54.5	29.0	1.0
97.8	43.3	22.0	2.5
123.3	93.3	74.0	2.5

* Locations in km east and north of 41°30'N by 88°0'W.

of the extract in hexane through a column containing anhydrous Na_2SO_4 (2 g) over 10% deactivated alumina (5 g) over acid-washed Cu filings (5 g) (ref. 8, for details). The extracts were analyzed by electron capture gas chromatography using a Hewlett-Packard Model 5830A or 5840A instrument equipped with 60 m SCOT column (SP 2100) or 50 m WCOT column. PCB concentrations were calculated by comparison to Monsanto Aroclor mixtures or EPA Aroclor standards using multiple linear regression analysis (9).

IMPLEMENTATION AND RESULTS

For the purposes of this work, a 33 x 39 grid was defined defined over the southern basin of Lake Michigan. The grid is oriented E-W along latitude 41°30'N with the origin being 41°30'N by 88°0'W, and each grid cell side equal to 5 km. (This scale corresponds to centimeters on NOAA Map 7 which is a 1:500000 map.) 807 of the grid cells represented lake surface, with the remainder representing land surfaces. Each of the over-water cells was parameterized for the burial model and run independently.

As noted above, the physical and chemical parameters were obtained from different core sets. Neither set covers the whole southern basin. The U. of Michigan (physical data) stations are biased towards the eastern shore and the U. of Wisconsin stations (PCB data) are concentrated near Waukegan and the basin center. Figure 1 shows the station locations, individual, and combined geographic coverages of the data sets.

Each of the 807 relevant grid cells was assigned a value for sedimentation rate, mixed zone thickness and current PCB concentration (Table 2). A modified inverse-distance-weighted averaging alogrithm was used to perform the two-dimensional interpolation (11,12). Figure 2 shows the distribution of and S in contour plot display. The grid cell values were averaged and integrated by summing. Mean sedimentation rate for the area of interest was 26.7 $mg/cm^2/yr$ which represents 5.4×10^6 metric tons per year over the area of interest. Average mixed zone thickness was found to be 1.38 cm, corresponding to a total mixed zone volume of .278 km^3. The PCB distribution was generated in a similar manner (Figure 3). PCB concentration over this area averages 61 parts per billion (wet weight). The mixed zone volume and PCB concentration data were combined and integrated cell-by-cell to yield a total mass in the mixed zone of 2760 kilograms.

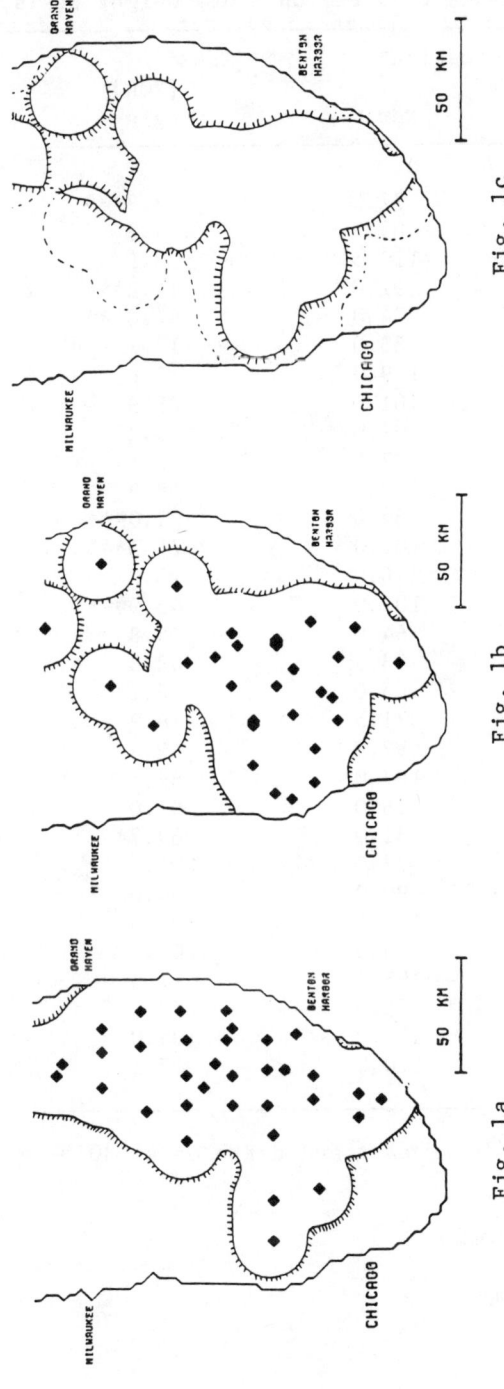

Figure 1. Data set coverage over the southern basin of Lake Michigan. 1a) Sediment dynamics model coverage (6). 1b) Surficial (PCB) sample coverage. 1c) Combined coverage.

TABLE 2. PCB concentrations in surficial Lake Michigan sediments. Concentrations reported on a dry weight basis.

Station Location* km		[PCB]
East	North	ng/g
19.0	77.5	3.3***
27.0	83.5	9.3
27.0	110.5	3.1
28.0	91.0	14.2***
32.5	74.0	67.0***
34.0	83.0	13.7
39.0	119.0	13.8
41.0	101.0	68.9
47.0	92.0	13.2
47.5	72.5	2.1
47.5	110.5	18.9
49.5	64.0	1.0***
58.5	102.0	74.5**
58.5	146.0	23.0
61.0	101.0	43.9***
61.5	64.0	25.8
62.5	83.5	48.8
70.5	65.0	6.1
73.5	71.5	16.9
75.0	92.0	69.4
76.0	167.0	52.3
81.5	136.0	91.0
85.5	131.0	69.7**
88.5	37.0	201.1
93.0	90.0	104.0
93.0	108.5	103.3
101.5	55.0	103.1
101.5	195.0	17.5**
102.5	75.0	120.0
119.0	135.0	105.0
128.0	170.0	7.1

* Locations in km east and north of 41°30'N by 88°0'W.

** 1-2 cm sample

***Ponar sample

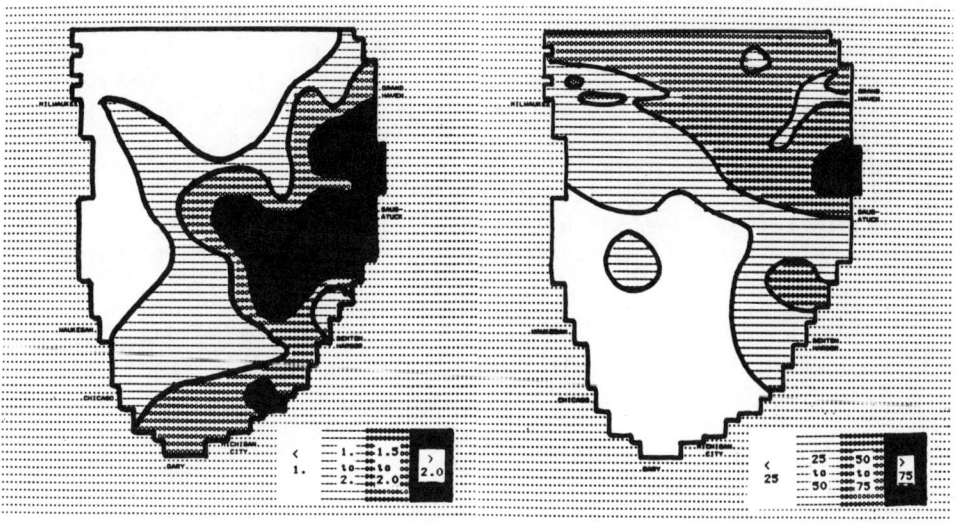

Fig. 2a. Mixed zone depth, cm.   Fig. 2b. Sedimentation rate $mg/cm^2/yr$.

Figure 2. Sediment dynamics model parameters.

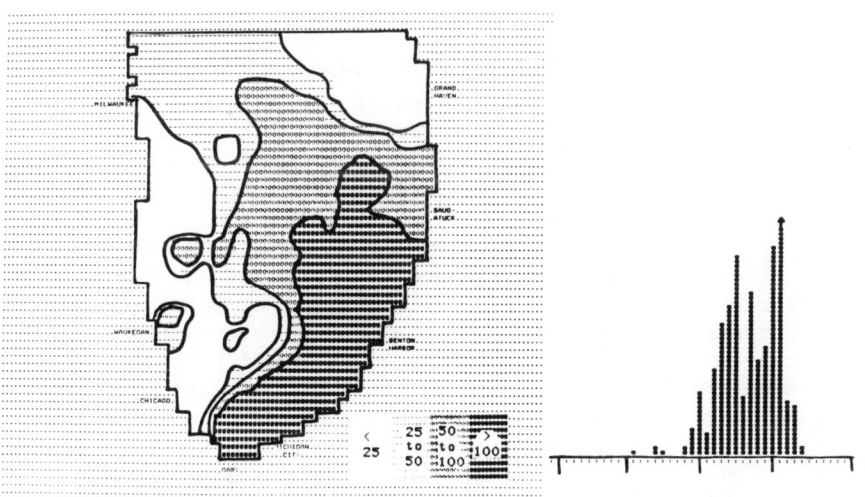

Figure 3. PCB concentration in surficial sediment, ppb.

It must be emphasized that these averages and integrations are based on sediment samples which are not evenly distributed over the lakes. The values above are extrapolated to the whole southern basin but in fact only reflect the area which was sampled. Comparison with Cahill (13) shows that these stations are biased towards depositional and transitional regions, leading to inflated estimates of basin-wide sedimentation rate. This consideration does not affect the following analysis, however, because cell-by-cell simulation will be valid for the sampled areas.

The burial rate of each cell is determined by R and S, as shown above. A simple parameter which characterizes this is the 50% response time:

$$t_{1/2} = - \frac{\ln(.5)\, S}{R} \qquad (8)$$

The values for $t_{1/2}$ were computed for each cell and are displayed in Figure 4. A histogram of the $t_{1/2}$ distribution is also included showing two modes at about 5 and 10 years, respectively. The mean $t_{1/2}$ is 7.5 years.

Simulation was set up for the best case clean-up situation. In these simulations, the PCB concentration in sedimenting particulates is set to zero. Output grids were generated and displayed for 0, 5, 10, and 20 years (Figure 5).

Comparison of the observed sediment PCB concentrations with the predicted sediment response time can provide information on PCB sources and trends in PCB concentration of the depositing sediment. If diffuse sources are of primary importance, or if suspended sediment is well mixed before permanent sedimentation, a correlation may exist between these two distributions. Four scenarios are presented: I) Deposition of sediment of constant PCB concentration has occurred sufficiently long for the PCB concentration in the surface layer to reach a maximum level. The concentration in the surface layer would be constant throughout the basin and equal to the concentration in the accumulating sediment. II) The deposition of sediment of constant PCB concentration is occurring, but the concentration in the surface layer has not reached a maximum level throughout the basin due to dilution by mixing with underlying sediment. In this case, an inverse relationship between PCB concentration and $t_{1/2}$ is expected (low PCB

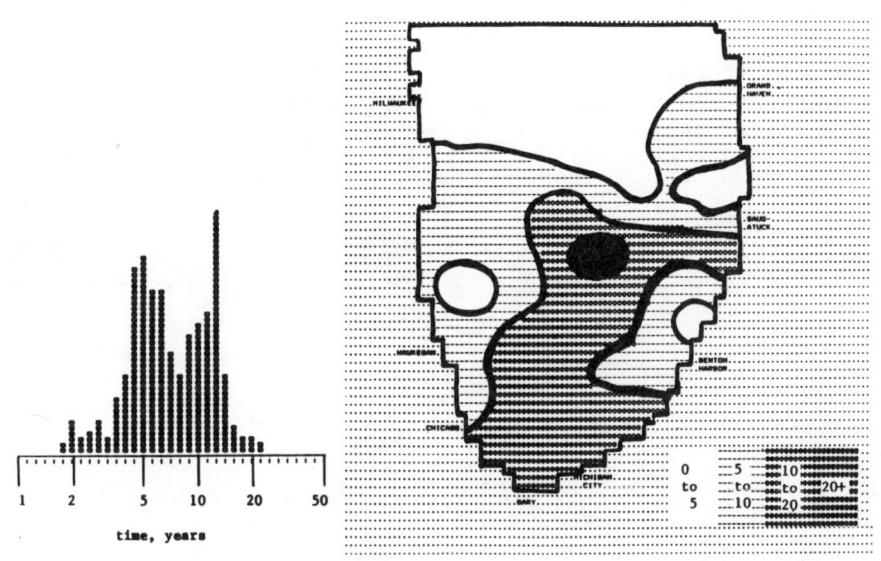

Figure 4. 50% response time for sediment in the surficial mixed zone, years.

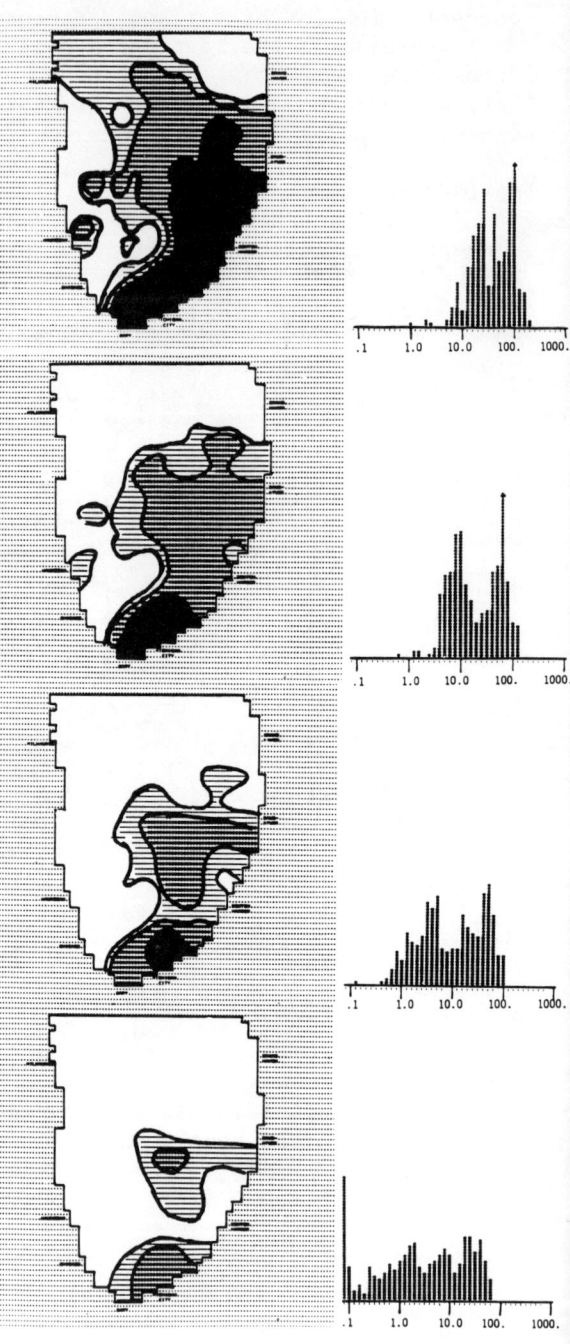

Fig. 5a. Current surficial (PCB) concentration, ppb. Contour intervals are identical to Figure 3.

Figure 5b. Best case projection; replacement of surficial sediment with uncontaminated sediment for 5 years.

Figure 5c. 10 years.

Figure 5d. 20 years.

concentration where $t_{1/2}$ is long). III) The PCB concentration has reached a maximum level as in (I), and the concentration in the surface sediment is now decreasing due to a decrease in the level of PCB contamination of the depositing sediment. In this case, a direct correlation between surface layer PCB concentration and $t_{1/2}$ is expected because concentrations would decrease most slowly in areas with the longest response time. Also, a subsurface maximum is expected. IV) The level of PCB contamination of the depositing sediment began to decline before the concentration in the surface layer had reached a maximum throughout the basin. In this case, areas where $t_{1/2}$ is long would not have reached a maximum, and the decrease in concentration would also be slow. A sub-surface maximum might occur. However, areas where $t_{1/2}$ is short would have reached higher surface concentrations in the past, and now the concentrations would be decreasing more rapidly. A sub-surface maximum would be expected. Consequently, the relationship between surface layer PCB concentration and $t_{1/2}$ would be complex for scenario IV.

The two distributions of interest here are shown in Figures 3 and 4. The observed [PCB] distribution and the computed 50% response time for the sediment mixed zone appear to be positively correlated over almost all of the sampled area. As described above (scenario III), this implies that the surface sediment PCB cocentration has reached a maximum and is now decreasing over most of the basin. This conclusion is strengthened by the fact that [PCB] is well correlated with mixed zone depth (+.57).

Higher PCB levels also tend to be found in depositional than in non-depositional zones, suggesting that much of the PCB gradient is due to lateral transport of contaminated sediment. A striking exception is the very high [PCB] region near the south-southeast shore (>100 ppb). This region is not associated with either a high or a low response time prediction. A characteristic of this area is a relatively low mixed zone depth and a thin layer of fine-grained sediment overlying sand and glacio-lauclustrine clays (low net sedimentation). The possibility of a point source is indicated here. It is not possible to differentiate between a local point source (such as Gary or Michigan City) or a remote point source because lateral sediment transport is not accounted for by this model.

SIMULATION OF FUTURE PCB BURIAL

The projected response of Lake Michigan sediments to burying surficial PCBs is shown in Figure 5. An obvious point to be made is that "hot spots" are predicted. The actual development of these areas may not occur as shown due to lateral transport. It does seem certain, however, that the sediment PCB distribution is and will remain quite heterogeneous. Note that ω and S are correlated. The areas with slowest response are those with moderate sedimentation and mixed zone thickness. Extreme mixed zone depth appears to be largely offset by high sedimentation, and doesn't lead to extreme retention. The reverse is probably not true: areas with low sedimentation and mixed zone depth are mostly "scoured" and will probably have a response time at least as fast as shown.

Projected average PCB concentrations over the basin are shown in Figure 6. Although different areas have markedly different response times, the lake can be roughly thought of as a single reactor with a 50% response time on the order of 5-10 years. The single mixed reactor model breaks down after approximately four half lives (30 years), but the explicit model probably loses validity by then due to lateral transport.

CONCLUSIONS

1. A heterogeneous distribution of PCBs was found in the southern basin of Lake Michigan. This was expected and is anticipated to remain heterogeneous. Great care is advised in sampling sediments to establish [PCB] time trends, since large variations can be expected within a few kilometers.

2. An average 50% response time of about 5-10 years is expected for the sediments of the whole basin.

3. It appears that the surficial sediments over most of the southern basin have reached their maximum PCB concentration and are now decreasing in concentration.

4. One or more significant point sources are indicated in the south reach of the lake. Further measurements could be of great value, although there is a great deal of difficulty obtaining good samples from this region.

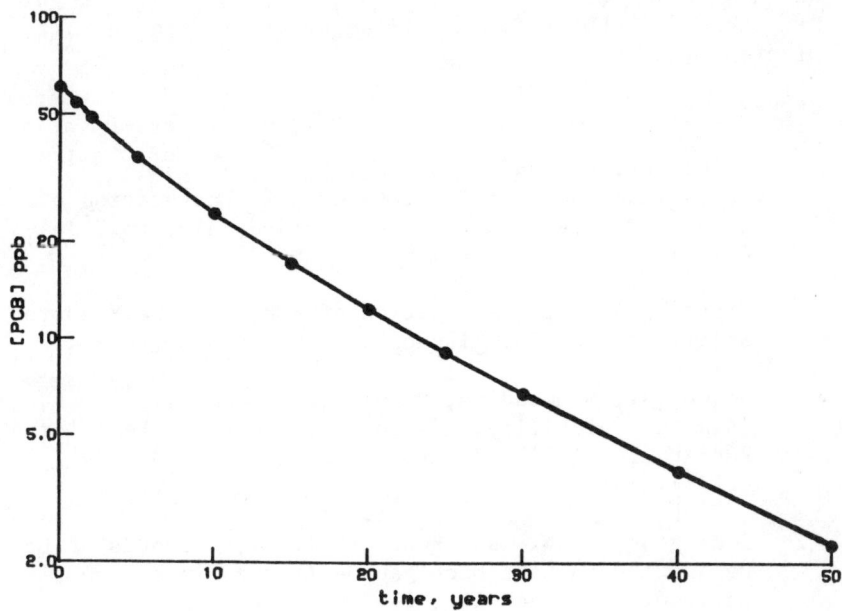

Figure 6. Best case projection mean surficial (PCB). This represents the fastest decrease which can be expected due to vertical burial.

ACKNOWLEDGMENTS

This investigation was supported in part by NOAA, Office of Sea Grant, through an institutional grant to the University of Wisconsin, and by EPA Contract No. 68-01-0502. We thank David Liebl for assistance with PCB analyses and Dr. John A. Robbins for providing physical data on Lake Michigan sediments.

REFERENCES

1. Weininger, D., and Armstrong, D. E., "Organic Contaminants in the Great Lakes", In Restoration of Lakes and Inland Waters, EPA-440/5-81-010, pp. 364-372 (1981).

2. Flotard, R. D., "Degradation of PCBs in Lake Mendota Sediments", U. of Wisconsin Ph.D. Thesis (1979).

3. Haque, R., Schmedding, D. W., and Freed, V. H., "Aqueous Solubility, Adsorbtion, and Vapor Behavior of PCB Aroclor 1254", Environ. Sci. Technol. 8: 139 (1974).

4. Edgington, D. N., and Robbins, J. A., "Records of Lead Deposition in Lake Michigan Sediments Since 1800", Environ. Sci. Technol., 10: 266 (1976).

5. Robbins, J. A., and Edgington, D. N., Geochim. Cosmochim. Acta, 39: 285 (1975).

6. Robbins, J. A. Unpublished data and original notebook copies. Note these data are preliminary. (1973).

7. Edgington, D. N., and Robbins, J. A., "Records of Lead Deposition in Lake Michigan Sediments since 1800 - reply", Environ. Sci. Technol., 13: 480 (1979).

8. Swackhamer, D. P., "The Recovery of PCBs in Sediments by Steam Distillation", M.S. Thesis, Water Chemistry Program, University of Wisconsin-Madison (1981).

9. Weininger, D., Burkhard, L., and Armstrong, D. E., "COMSTAR: Complex Mixture Statistical Reduction - Application to PCB Analysis", in preparation.

10. NOAA-Lake Survey Center, "Great Lakes Map 7 - Lake Michigan", NOAA, Detroit, MI 48226 (1971).

11. Shepard, D., "A Two-Dimension Interpolation Function for Computer Mapping of Irregularly Spaced Data", Paper No. 15, Harvard Papers in Theoretical Geography (1968).

12. Switzer, P., Mohr, C. M., and Heitman, R. E. "Project Trident Technical Report - Statistical Analyses of Ocea Terrain and Contour Plotting Procedures", Dept. Navy Bur. Ships, NObsr-81564, SS-050, 77 pp.

13. Cahill, R. A., "Geochemistry of Recent Lake Michigan Sediments", Illinois State Geological Survey, Circular 517 (1981).

INDEX

Air-water exchange - see volatilization or air deposition
air-water partitioning - see Henry's law constant
adsorbent   15, 89, 340
air concentration   34, 115, 145, 367
air deposition   56, 116, 130, 134, 141, 157
air particulates   36, 128-131
air washout   129
analysis   1, 15, 115, 141, 161, 231, 367, 385
atmosphere - see air

Benthic processes   213, 229, 245, 423
bioconcentration   62, 83, 269, 394
biodegradation   76
boiling point   60, 75
bubble processes   136

Clams   394
congeners   3, 188, 371

Desorption   89
dry deposition - see air deposition

Emissions   283, 311, 329, 391
enzymes   2, 3

Fish   83, 127, 269, 385, 396
fugacity   50, 60

Gas chromatography   2, 17, 51, 162, 184, 231, 246, 330, 369
Georgian Bay   254
Grand River   258

Henry's law constant   50, 53, 61, 74, 130, 143, 146
Hudson River   411

Input rates - see emissions
Isle Royale   116

Lake Erie   246, 250, 256
Lake Huron   215, 246, 250, 340
Lake Michigan   77, 118, 133, 134, 137, 142, 158, 214, 229,
                246, 250, 314, 423

441

Lake Ontario   246, 250, 256, 385
Lake St. Clair   250, 255
Lake Superior   115, 138, 141, 181, 240, 246, 250
lead dating   229
loadings   257, 264, 303, 306, 340

Mass spectrometry   2
melting point   74
Minneapolis   118
models   283, 313, 339, 423
monitoring   367, 385

Nepheloid layer   216
Niagara River   385, 391

Octanol-water partition coefficient   62, 73, 269

Photolysis   77
polyurethane foam   22
pore water   222
precipitation   57

Reactivity   76
resuspension   95, 314, 346
rivers   260, 388, 414

Saginaw Bay   292, 329
Saugeen Rivers   258
sediments   76, 213, 229, 245, 289, 314, 386, 392, 414
sediment deposition   229, 340, 423
solubility   61, 72
soils   412
sorbents   15
sorption coefficient   63, 82, 89, 90, 172, 222, 315, 392
surface microlayers   157
suspended sediments   245, 262, 284, 330, 392
synthesis   4

Vapor pressure   15, 17, 60, 72, 75
vegetation   416
volatilization   56, 67, 79, 134, 145, 157, 301, 340, 411, 414

Washout - see air washout
water column   76
water concentrations   167, 181, 332
wet deposition - see air deposition